Light and Light Sources

P. Flesch

Light and Light Sources

High-Intensity Discharge Lamps

With 226 Figures, 5 in Color, and 4 Tables

Springer

PD Dr.-Ing. Peter Flesch
OSRAM GmbH
Nonnendammallee 44-61
13629 Berlin
Germany
E-mail: peter@flesch.de

Cover: Kleiner Leuchtturm Borkum, Germany, North Sea
(Courtesy of Frank Toussaint, "Der Leuchtturm-Atlas", http://Leuchtturm-Atlas.de/)

Library of Congress Control Number: 2006923689

ISBN-10 3-540-32684-7 Springer Berlin Heidelberg New York
ISBN-13 978-3-540-32684-7 Springer Berlin Heidelberg New York

Springer is a part of Springer Science+Business Media.

springer.com

Printed in the Netherlands

Typesetting by SPI Publisher Services
Cover design: *design & production* GmbH, Heidelberg

Printed on acid-free paper SPIN 11534952 57/3100/SPI 5 4 3 2 1 0

Dedicated to my family

Preface

High-intensity discharge lamps (HID lamps, also high-pressure discharge lamps) are very important light sources for visible, UV, as well as IR radiation. They have captured a major share of the markets for automotive headlight lamps (D2 lamp), video projection (UHP lamp), general lighting, street/industrial lighting, commercial lighting, floodlighting, sun tanning, microscopy, endoscopy, photochemistry, lithography, etc. This renders HID lamps an interesting and seminal field of research. High-pressure discharge lamps are normally very small (typically 0.7 mm to some centimeters' electrode gap), have short time scales (typical ac frequency: 50–500 Hz or higher), and the electrical power input lies between some tens and several thousand watts. This book is concerned with the understanding of these lamps.

Beginning with the human eye and an explanation of light and color, the working principle of different light sources is explained and the light sources are compared with each other. Starting point is the incandescent lamp, then low- and high-pressure (high-intensity) discharge lamps are discussed in detail. Furthermore, a large part of this book deals with important subjects concerning HID lamps like electrode and plasma physics as well as the state of the art in HID lamp diagnostics and modeling.

The aim of this book is to give an introduction to the working principle of HID lamps and to point out challenges and problems associated with the development and operation of high-pressure discharge lamps. This book is directed at students interested in high-pressure discharge lamps as well as persons already involved in the research and development or the usage of HID lamps.

Karlsruhe, January 2006 *Peter Flesch*

Acknowledgments

Many thanks to Prof. Dr. Neiger, Prof. Dr. Dr. h.c. Thumm, and Prof. Dr. Zissis, as well as Prof. Dr. Lemmer, the Lichttechnische Institut, and the Fakultät für Elektrotechnik und Informationstechnik of the Universität Karlsruhe (TH) for all the support and the stimulating atmosphere.

Furthermore I would like to thank all my colleagues of the Lichttechnische Institut and all colleagues from other universities and industry with whom I worked in different BMBF and EU projects or who I met at conferences and workshops. Special thanks to the BMBF for supporting the projects FKZ 13N7107/0, FKZ 13N7765, and FKZ 13N8073 as well as to the EU for supporting the cost action 529 "Efficient Lighting for the 21st Century".

Last but not least, many thanks to all the people supporting the work on this book by discussions, proofreading, and providing many of the images and figures shown in this book. This book would not have been possible without this support.

Contents

Introduction

This book is concerned with high-intensity discharge (HID) lamps. High-intensity or high-pressure discharge lamps have captured a major share of the markets for automotive headlight lamps (D2 lamp), video projection (UHP lamp), general lighting, street/industrial lighting, commercial lighting, floodlighting, sun tanning, microscopy, endoscopy, photochemistry, lithography, etc. This renders HID lamps an interesting and seminal field of research. High-pressure discharge lamps are normally very small (typically 0.7 mm to some centimeters electrode gap), have short time scales (typical ac frequency: 50–500 Hz or higher), and the electrical power input lies between some tens and several thousand watts.

The aim of this book is to give an introduction to the working principle of HID lamps and to point out challenges and problems associated with the development and operation of high-pressure discharge lamps. The fundamentals of plasma and electrode physics as well as the current research on HID lamps will be treated. This book is directed at students interested in high-pressure discharge lamps as well as persons already involved in the research and development or the usage of HID lamps.

An overview of light and light sources is given in Chap. 1. This includes a short introduction to the human eye and to photometric characteristics of light sources as well as the discussion of different light sources, covering the sun, the incandescent lamp, and low-pressure and high-pressure discharge lamps. The laws of physics needed for the description and understanding of plasma and electrodes in HID lamps are the subject of Chap. 2. The plasma generates the light and must be heated to sufficiently high temperatures to achieve high efficiencies. The electrodes must supply the electric current to the plasma and must therefore have a high enough temperature, e.g., to emit electrons. On the other hand, the electrodes may cause an early end of life of the lamp, if they get too hot, or the melting of an electrode tip might change the electrode shape and thus alter the operating conditions of the lamp.

The current research on HID lamps is the focus of Chaps. 3 and 4. Chapter 3 is concerned with the experimental investigations of high-pressure discharge lamps, namely electrode and plasma diagnostics, whereas numerical simulations of high-pressure discharge lamps are discussed in Chap. 4. The starting point of the numerical simulations is the modeling of the electrodes, followed by the theoretical treatment of the plasma. However, plasma and electrodes are coupled in real lamps, and so the numerical simulation of electrodes and plasma including the interactions between them is a special focus of Chap. 4 and the research work of the author. Chapters 3 and 4 not only present the state of the art of HID lamp research, but also aim at giving a deeper insight into the working principle of high-intensity discharge lamps and the problems and challenges associated with HID lamps. Part of this work has already been published in [53–65].

The generation of light matching our demands in home, office, factory, or shop lighting, for the illumination of roads or sports stadiums, or for the usage in video projection, or for automotive headlight lamps is not the only goal of the development of HID lamps. Environmental issues are also important guidelines for the usage and improvement of high-pressure discharge lamps: Today, the generation of light by approximately 9 billion incandescent lamps, 4 billion fluorescent lamps, and 500 million high-pressure discharge lamps consumes 25% of worldwide electric energy production [111]. Therefore, there is a huge potential to save energy and thus reduce environmental pollution and save energy resources. Moreover, the replacement or reduction of mercury in high-pressure discharge lamps is also an important current issue.

The generation of light is not the only purpose of HID lamps, but it is the most visible application area of all lamps and has an obvious effect on our quality of life (cf. Figure 1). Nevertheless, high-pressure discharge lamps are also used in many other fields of application where electromagnetic radiation in the UV and IR part is important. Photolithography, disinfection, curing, or laser excitation are only some of the manifold possibilities offered by HID lamps.

Fig. 1. Earth at night. The Eastern USA, Europe, and Japan are brightly lit by their cities, while the interiors of Africa, Asia, Australia, and South America remain (for now) dark [124]

1 Light and Light Sources

High-intensity discharge (HID) lamps are very often used for lighting purposes (indoor and outdoor), automotive headlights, or video projection. To optimize HID lamps and all other light sources for these applications, one has to understand the human eye, the receptor of the light. The assessment of light sources with respect to "light quality" or efficiency as well as the working principle of video projection is strongly coupled to the characteristics of the human eye. Therefore, this chapter starts with the understanding of the human eye and the meaning of "color" (Sect. 1.1). Based on the discussion of the eye, photometric measurement units will be introduced which help to rate light sources (Sect. 1.2). Then, an overview of different light sources will be given: starting from the sun, over the incandescent lamp to discharge lamps (Sect. 1.3). The discharge lamps are divided into two classes: low-pressure and high-pressure discharge lamps, the latter being the main subject of this book.

1.1 The Human Eye

The human eye is a very complex sense organ. A cross section is shown in Fig. 1.1. The light enters the eye through the pupil (the circular hole in the iris), passes through the lens and is projected on the retina at the back of the eye. The iris is a colored ring of muscle fibres that can expand or contract over a range from 1 to 8 mm pupil diameter in response to different lighting conditions to vary the light falling on the retina. Muscles move the eyeball and allow us to focus the image at the macula lutea (yellowish central portion of the retina). Light passing through the pupil is focussed by the lens, which can change the focal length by assuming flatter or more bended shapes with the help of the ciliary muscle. In this way, the image of an object at any distance can be brought into focus on the retina at the back of the eye [72, 78].

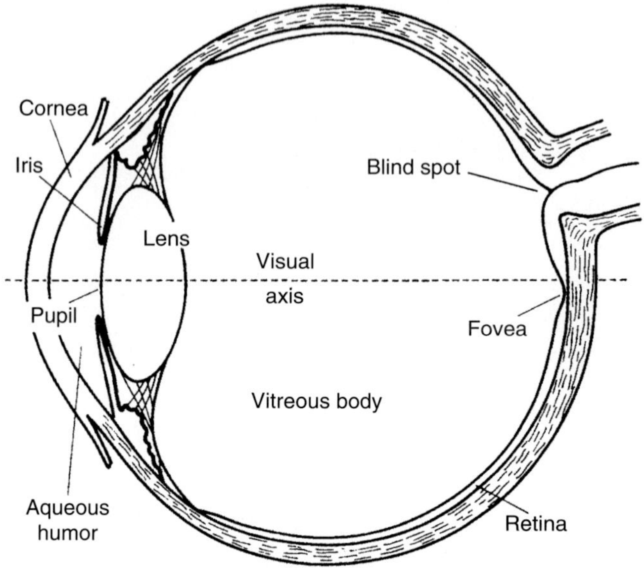

Fig. 1.1. Cross section of human eye [72]

1.1.1 Rods and Cones and More

The retina contains two types of photoreceptors: rods and cones. The rods are more numerous, about 120 million, and are more sensitive to light than the cones. Two or three photons can already be detected by a single rod [103]. Moreover, rods are responsible for the very sensitive motion detection and the peripheral vision (cf. Fig. 1.2). However, the rods are not sensitive to color. The 6–7 million cones provide the eye's color sensitivity and the highest acuteness of vision. Visual examination of small details involves focusing light from these details onto the fovea centralis, which is in the center of the macula lutea. The cones are much more concentrated in the center of the retina as shown in Fig. 1.2 (horizontal cut through the human eye). The center of this region is the fovea centralis, a 0.3-mm-diameter rod-free area with very thin, densely packed cones. The maximum concentration of cones is roughly 140,000 mm^{-2} in the fovea region, and this density decreases rapidly outside the fovea. In contrast, the rods are absent from the fovea, but a few degrees away from it their density rises up to 160,000 mm^{-2} spread over a large area of the retina. At the blind spot caused by the optic nerve no photoreceptors can be found [78].

The sensitivity as a function of wavelength of light is different for rods and cones. The rods are responsible for our dark adapted, or scotopic, vision and are more sensitive to blue-green light with peak sensitivity at a wavelength of about 500 nm (cf. Fig. 1.3). The photopic vision (light adapted eye, normal lighting conditions) is determined by the cones with a peak sensitivity at 555 nm. This photopic response curve $V(\lambda)$ (averaged over many test persons) is the basis for the assessment of the quantity of "light output," the luminous flux (cf. Sect. 1.2.1),

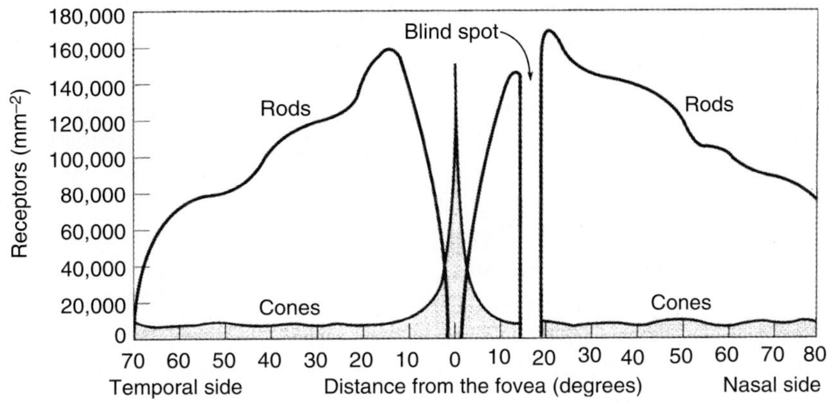

Fig. 1.2. Distribution of rods and cones across the retina [110] according to [137]

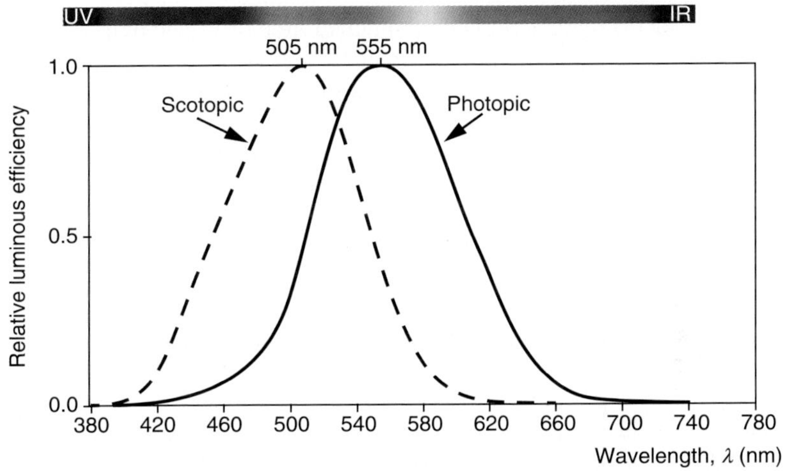

Fig. 1.3. The scotopic and the photopic vision curves of relative spectral luminous efficiency as specified by the International Commission on Illumination (CIE, normalized values)

which is important for the comparison of different light sources or assessment of the further development of a light source [127]. A third sensitivity curve is also discussed, the so-called mesopic response curve for low light levels as, e.g., street lighting. Up to now there is no well-defined mesopic response curve and the sensitivity might depend strongly on the adaptation level. The photopic response curve is valid for an illumination above 10 lux, the scotopic response curve is applicable below 0.1 lux. In-between these boundaries, the mesopic response curve is important [40].

Scotopic vision (rods) lacks completely color; a single spectral sensitivity function is color-blind and thus scotopic vision is monochromatic. Color or photopic vision is provided by the cones, of which there are three different types, each containing a different photosensitive pigment. The three pigments have maximum absorptions at about 450, 550, and 600 nm and accordingly the cones are often called blue, green, and red cones after the color of light to which the cones are most sensitive. The sensitivity curves of the cones shown in Fig. 1.4 are obtained by measuring the absorption of light of different wavelengths by the photopigment molecules and are all normalized to the same area under the curves. Thus, the combination of all three curves does not result in the photopic vision curve shown in Fig. 1.3. Another reason for this is the unequal appearance of the cones on the retina: The average percentage of the blue cones is 4%, of the green cones 32%, and of the red cones 64% [127]. Of course, all these curves (Figs. 1.2–1.4) and all the numerical values given here are average values. Even the two eyes of a single individual perceive colors slightly different, and different authors deliver different response curves or numerical values of the appearance of different cones [72, 78].

Recent publications have shown that there is a third type of photoreceptor in the human eye [23, 25, 34]. This new type of photoreceptor does not contribute to vision but is responsible for synchronizing mammalian circadian (daily) and circannual (seasonal) rhythm with environmental time using the light arriving at the eye. The sensitivity curve of this new type of retinal cells can be measured by extracting them from the retina and measuring their voltage response as a function of wavelength and intensity of the incident light [23]. Another possibility is the measurement of the effect in the human body, e.g., measuring the concentration of the hormones cortisol ("stress hormone") or melatonin ("sleep hormone") before and after exposing the human eye for 90 min to a monochromatic light stimulus.[1] Both measurements indicate a spectral biological response curve shifted to the blue compared to the visual sensitivity curve of the human eye (photopic vision).

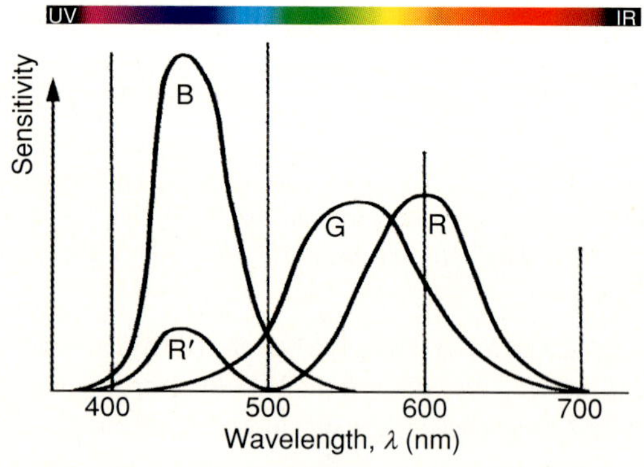

Fig. 1.4. Sensitivity curves of the three different types of cones in the human eye [72]

[1] Cortisol levels increase in the morning, preparing the body for activity. It remains at a sufficiently high level during the bright day, falling finally to a minimum at midnight. The level of the "sleep" hormone melatonin drops in the morning, reducing sleepiness. It normally rises again when it becomes dark, permitting healthy sleep.

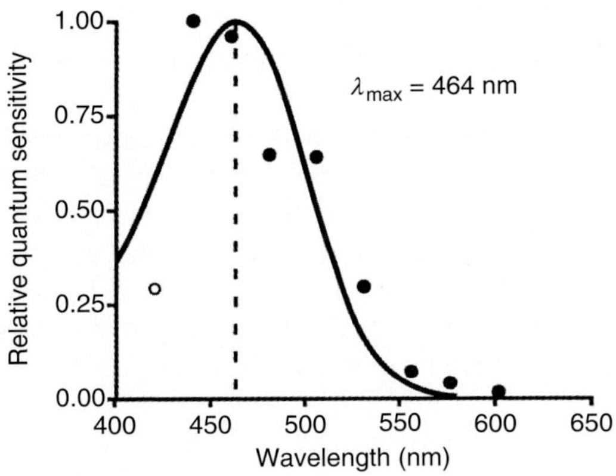

Fig. 1.5. Influence of monochromatic light of different wavelengths on the melatonin suppression for 72 healthy human subjects ([34], © 2001 by the Society for Neuroscience)

Figure 1.5 shows the influence of monochromatic light of different wavelengths on the melatonin suppression for 72 healthy human subjects. The sensitivity curve has a maximum at 464 nm [34]. But not only the wavelength and the intensity of the incident light are important, the direction seems to be very important as well. Light coming from above the line of view has a far more important "biological" meaning than light from below the line of view [76].

This new type of photoreceptor has consequences for artificial light sources. Light should not only be judged by its influence on vision caused by the reaction of rods and cones (e.g., luminous flux, color rendering, etc.) as described in Sect. 1.2. Artificial light sources must be optimized also with respect to this new type of retinal cells. This is of particular importance for applications where human beings are exposed to artificial light for longer times. The artificial light might influence the circadian and circannual rhythm in a positive or negative manner. One example is the illumination in aircrafts, especially during oversea connections. It should safeguard the pilots from getting tired and might reduce the jet lag after traveling over several time zones. The selection of appropriate light sources for long duration space flights might be quite important for the health and well-being of astronauts. Shift workers (factories, police, etc.) can be assisted by artificial light sources adjusting their circadian rhythm to their work rhythm. But also everyday life might be influenced and improved by these new findings: Artificial light sources for domestic lighting might enhance well-being and health if optimized to fit our circadian and circannual rhythm. Furthermore, street lighting with low pressure sodium lamps (monochromatic light of 589 nm, cf. Sect. 1.3.5) or with other artificial light sources with their characteristic spectral power distribution (cf. Sect. 1.3) might influence alertness either way. The discovery of this new type of photoreceptors already initiated new products, e.g., the Lumilux Skywhite fluorescent lamp described in Sect. 1.3.4.

1.1.2 Color Space

The existence of three spectral sensitivity curves provides a basis for color vision since light of each wavelength will result in a unique ratio of blue, green, and red cone responses. The cones, therefore, provide us with color vision (photopic vision) that can distinguish between many colors. The light signal is reduced by the eye to three values (tristimulus values), representing the intensity of the response of each of the cone types. Because of the overlap between the sensitivity ranges, not all combinations of stimuli are possible. For example, it is not possible to stimulate only the green cones: The other cones will be stimulated to some degree at the same time. The set of all combinations of stimuli that are possible make up the human color space.

Two different light spectra which have the same effect on the three color receptors in the human eye (same ratio of stimulation of red, green, and blue cones) will be perceived as the same color.[2] One example is the white light emitted by fluorescent lamps, which typically has a spectrum consisting of a few narrow bands, while (white) daylight has a continuous spectrum. Another example is the "production" of white light by mixing red, green, and blue on a television screen or with a video projector. All these light sources produce (at first sight) white light. The human eye cannot tell the difference between such light spectra just by looking into the light source. Nevertheless, the reflected light spectrum from objects might be very different for different "white" light sources, i.e., the effect of a light source on the color appearance of objects is important for the assessment of light sources (see Sects. 1.1.3 and 1.2.5).

Since the light signal is reduced by the eye to three values (tristimulus values), representing the intensity of the response of each of the cone types, every possible stimulation of the eye can be mapped in a three-dimensional space. This three-dimensional color space can be further reduced to a two dimensional color space by considering only colors of the same intensity[3] (cf. Fig. 1.6). The x value in Fig. 1.6 corresponds to the stimulation of the red cones, the y value to the stimulation of the green cones. Subtracting the sum of x and y from 1 corresponds to the stimulation of the blue cones.[4] Because of the overlap between the sensitivity ranges (cf. Fig. 1.4), a stimulation corresponding to $x = 1$ or $y = 1$ is not possible. The spectral colors are distributed around the edge of the "color space" as shown in Fig. 1.6 and enclose all colors perceivable by the human eye. If we start with red light of 700 nm, the red cones are stimulated most, represented by a value of approximately $x = 0.75$. Nevertheless, the green cones are stimulated as well ($y = 0.25$), only the blue cones

---

[2] This phenomenon is called metamerism (cf. Sect. 1.1.3).

[3] In the CIE system the parameter Y measures the brightness, the parameters x and y specify the chromaticity [127].

[4] On the straight line from $x = 0$, $y = 1$ to $x = 1$, $y = 0$ the sum of x and y equals 1, i.e., the stimulation of the blue cones is zero, whereas the stimulation of the blue cones would be at maximum at the point $x = 0$, $y = 0$ (cannot be reached because of the overlap of the three sensitivity curves). Therefore, the distance of a color from the straight line from $x = 0$, $y = 1$ to $x = 1$, $y = 0$ is a measurement of the stimulation of the blue cones.

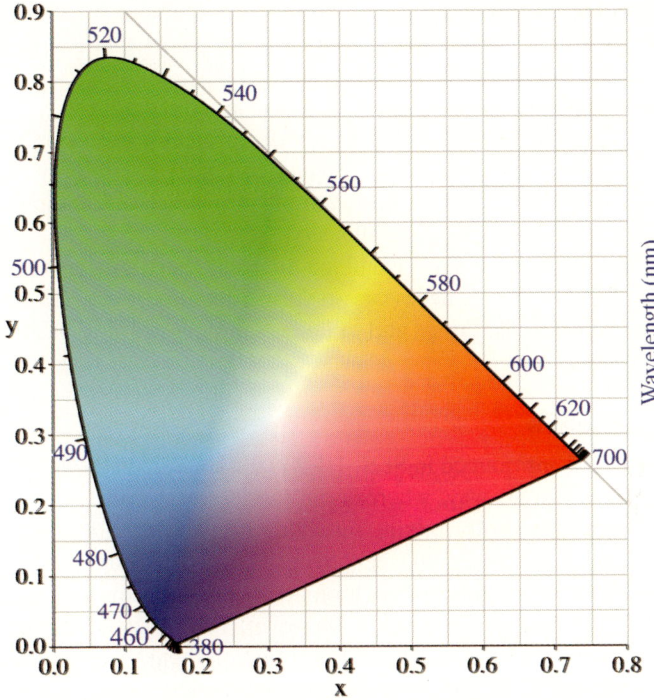

Fig. 1.6. CIE chromaticity diagram or color space [140]

are not affected. Reducing the wavelength to 550 nm, the ratio of stimulation changes continuously to $y = 0.7$ (green) and $x = 0.3$ (red). Wavelengths smaller than 550 nm stimulate the blue cones as well, i.e., the straight line from $x = 0$, $y = 1$ to $x = 1$, $y = 0$ is left for $\lambda < 550$ nm. If we reach the wavelength of 400 nm, the green cones are no longer stimulated ($y = 0$), the red cones only a little bit ($x = 0.2$) due to the small hillock between 400 and 500 nm. So the blue cones are stimulated most for 400 nm ($1 - 0.2 = 0.8$). The connecting line between 380 and 700 nm is the purple line, which cannot be represented by monochromatic light [127].

Taking two "color points" in the color space shown in Fig. 1.6, we can mix every color on the line in-between these two points by mixing the given two colors with the right proportion. Nevertheless, no mixture of colors can produce a color completely identical to a spectral color (one can get quite close for the longer wavelengths, where the chromaticity diagram has a nearly straight edge). Accordingly, the colors which can be mixed by combining a given set of three primary colors[5] (such as the blue, green, and red of a color television screen or

[5] Any set of three colors which will yield white when mixed in an appropriate combination can be considered to be primary colors.

video projector) are represented on the chromaticity diagram by a triangle joining the coordinates for these three colors. This range of colors that can be reproduced with a given color reproduction system (photography, printing, television, video projection) is called the gamut. The CIE chromaticity diagram can be used to describe the gamut. The art of developing a video projector or television set is now to reproduce a color gamut including as many colors as possible. In the case of a video projector, this means finding a projection lamp which offers three appropriate primary colors with sufficiently high intensity.

So, understanding the human eye and the different sensitivity curves for the rods and the three types of cones is necessary to define quantity (Sect. 1.2.1) and quality judgements (Sect. 1.2.5) for lamps and to understand the working principle of color reproduction systems.

1.1.3 Mixing Colors

The mixing of colors introduced in Sect. 1.1.2 is the so-called additive color mixing, involving light in a direct way, whereas the world we are normally looking at is dominated by subtractive color mixing involving substances. Additive color mixing describes how the eye interprets wavelengths of light with the help of the three types of cones and can be described in the chromaticity diagram or with a color gamut. The colors of our surrounding world, however, are created by absorbing and reflecting defined parts of the (white) light shining on them. A tomato, for example, absorbs nearly all the light with wavelengths below 600 nm, reflecting (partially) only light with wavelength above 600 nm. The resulting color is red. Cabbage, on the other hand, (partially) reflects wavelengths between 500 and 600 nm and absorbs the rest of the light. The result is the green color of cabbage. White paper reflects about 80% of the (white) light shining on it independent of the wavelength,[6] whereas black paper absorbs approximately 95% of the light shining on it independent of the wavelength [78].

The best choice of primary colors for additive color mixing are spectral colors (monochromatic light, e.g., blue (460 nm), green (530 nm), and red (660 nm)), because they lie on the border of the color space perceivable by the human eye (cf. Fig. 1.6). The resulting color gamut thus includes as many colors as possible. In contrast, the dyes for subtractive color mixing must remove a band of wavelengths. A good choice for the three primary colors for subtractive color mixing[7] are cyan (a bluish green, result of additive color mixing of green and blue), magenta (a pinkish purple, result of additive color mixing of blue and red), and yellow (result of additive color mixing of red and green). Cyan absorbs orange and red at wavelengths above 580 nm (cyan = white – red), magenta dye absorbs green and yellow light in a band of wavelengths between 490 and 580 nm (magenta = white – green), and yellow dye absorbs blue and violet light at wavelengths below about 490 nm (yellow = white –

[6] Depending on the whiteness of the paper, this value can vary.

[7] As for example in printers.

blue). Mixing now cyan and magenta, all wavelengths above 490 nm will be absorbed, resulting in blue[8] reflected light and so on.

So, light sources start with additive color mixing (spectral distribution of the light emitted by the light source), but the quality of the light source must be rated by looking at the reflected light from defined color samples, because the reflected light spectrum from these color samples might be very different for different "white" light sources. In this way, one can learn more about the light source and the light emitted from it by regarding the reflected light than by looking directly in the light source. Instead of defined color samples, one could also use colored filters for the assessment of a light source, but the method with defined color samples is easier (cf. Sect. 1.2.5).

If two colors have the same visual appearance (i.e., the same tristimulus value and therefore the same position in the CIE chromaticity diagram, cf. Fig. 1.6), but different spectral compositions, they are called metameric. They will have the same effect on the three color receptors (cones) in the human eye (same ratio of stimulation of red, green, and blue cones) and will be perceived as the same color. The light color emitted by different light sources with the same correlated color temperature (cf. Sect. 1.2.6) will usually be metameric, but they may still be distinguishable by looking at their color rendering index (cf. Sect. 1.2.5). Metameric colors will share the same location on the CIE chromaticity diagram. It might be that a pair of color samples match under one illuminant but not when viewed under another.

1.2 Photometric Characteristics of Light Sources

The assessment of light sources with respect to the quantity and quality of light output is not as easy as measuring just the spectral power distribution of different light sources. The human eye with its characteristic sensitivity curves especially of the three types of cones must be incorporated in the assessment of light sources yielding photometric characteristics of light sources like luminous flux and color rendering index. It should be kept in mind that these photometric characteristics are only important in the case of the production of visible light for human beings. These photometric characteristics do not make sense for judging, for example, discharge lamps for producing UV radiation or for the illumination of an aquarium.[9] Moreover, taking into account the additional photoreceptors responsible for the setting of our biological clock (cf. Sect. 1.1.1), the photometric characteristics do not include all important aspects of the human eye.

---

[8] cyan + magenta = white − red − green = blue.

[9] Of course, for the human beings looking at the fish, the photometric characteristics of the illuminating light source are important, but the fish themselves have different sensitivity curves for their photoreceptors and need maybe a certain light distribution for a healthy growing.

Table 1.1. Some typical light sources, their wattage, luminous flux, and luminous efficacy [54, 71, 118, 170]

light source	wattage (W)	luminous flux (lm)	luminous efficacy ($\mathrm{lm\,W^{-1}}$)
candle		5–15	
LED Luxeon white 5 W	5	150	30
compact fluorescent lamp	11	660	60
incandescent lamp	60	730	12
halogen incandescent lamp	100	1,250	13
automotive headlight lamp[a] H7	55	1,500	27
fluorescent lamp	36	2,850	79
automotive headlight lamp[b] D2	35	3,200	91
high-pressure mercury lamp	80	3,450	43
low-pressure sodium lamp	35	4,800	137
UHP 120	120	7,000	58
High-pressure sodium lamp	100	10,000	100

[a] Halogen incandescent lamp

[b] Metal halide (high-pressure) discharge lamp

1.2.1 Luminous Flux

The quantity of light or radiation of a light source can be measured with a spectrophotometer resulting in the spectral power distribution of the light source, i.e., the radiated power of a light source depending on wavelength λ (cf. Sect. 1.3). But to rate the quantity of light from a light source, it is not enough just to measure how much power is emitted between 380 (violet) and 780 nm (red). This spectral power distribution must be weighted with the photopic sensitivity curve of the human eye (daylight sensitivity or sensitivity of the light adapted eye, Fig. 1.3) to give the so-called luminous flux. So, the biological response of our eye to the incoming light must be included in the definition of the luminous flux which is measured in lumen (lm). Thus, the luminous flux is the energy per unit time ($\mathrm{d}Q/\mathrm{d}t$) radiated from a light source over visible wavelengths weighted with the sensitivity of the human eye.

The maximum luminous flux can be reached with a light source emitting only at the wavelength where maximum photopic sensitivity is given, i.e., 555 nm with $V(555\ \mathrm{nm}) = 1$. 1 W radiated power at the wavelength of 555 nm is (historically) defined as 683 lm. The luminous flux Φ_V in the general case is given by the folding or weighting of the spectral power distribution $\Phi(\lambda)$ with the normalized photopic response curve[10] $V(\lambda)$ and the constant factor $K_\mathrm{m} = 683\ \mathrm{lm\,W^{-1}}$

$$\Phi_V = K_\mathrm{m} \int \Phi(\lambda) V(\lambda)\,\mathrm{d}\lambda \tag{1.1}$$

The maximum value of 683 lm for 1 W of radiation can only be reached for monochromatic light at the wavelength of 555 nm. As soon as we take "white" light, the

[10] Maximum photopic response at 555 nm normalized to 1.

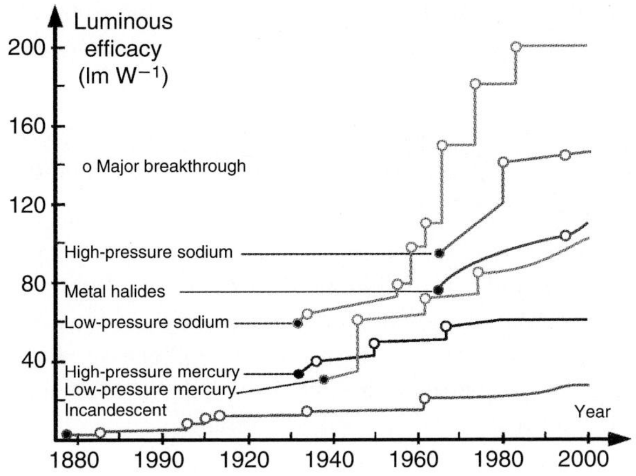

Fig. 1.7. Evolution of luminous efficacy of discharge lamps compared to the incandescent lamp [70]

luminous flux must be smaller due to the decline of $V(\lambda)$ for larger or smaller wavelength than 555 nm. Monochromatic light with wavelengths unequal to 555 nm has a lower luminous flux as well: 1 W monochromatic red light (700 nm, $V(\lambda = 700\ \text{nm}) = 0.4\%$, [104]) produces a luminous flux of 2.7 lm, 1 W monochromatic blue light (450 nm, $V(\lambda = 450\ \text{nm}) = 3.8\%$, [104]) produces a luminous flux of 26 lm. Typical light sources and their luminous flux are given in Table 1.1.

1.2.2 Luminous Efficacy

The luminous flux related to the electrical power input to a light source gives the luminous efficacy η (lm W^{-1}). The luminous efficacy may help to judge the efficiency of different lamps. Nevertheless, as the luminous flux, the luminous efficacy is only a measure related to the "quantity" of light, not to the "quality". Table 1.1 gives some examples of luminous efficacies of different light sources. A graphical representation of the evolution of the luminous efficacy of different discharge lamps compared to an incandescent lamp is given in Fig. 1.7.

1.2.3 Luminous Intensity

The luminous intensity (of a point-like light source[11]) is the luminous flux per unit solid angle and expresses the directionality of the radiated light. The luminous intensity is used to describe how bright a beam of light is. It is expressed

[11] For an extended light source, the luminous flux per unit area per unit solid angle is measured in lm m^{-2} steradian^{-1} = candela m^{-2} = nt (nit). This quantity is also called luminance.

in lumen per steradian = candela (cd). The candela is the foundation unit for the measurement of visible light. It is one of the seven foundation SI units. Its formal definition is:

The candela is the luminous intensity in a given direction of a source that emits monochromatic radiation of frequency 540×10^{12} Hz (555 nm) and that has a radiant intensity in that direction of 1/683 W per steradian.

The candela is then used to define the lumen and other quantities used in the measurement of visible light. If the luminous intensity of a light source is the same in all directions, it is called isotropic. For an isotropic light source, the relationship between the candela and lumen is 1 cd = 4π lm. Otherwise, for a flat radiating surface, known as a lambertian, the intensity falls off as the cosine of the observation angle with respect to the surface normal.

1.2.4 Illuminance

The illuminance or illumination (lux) measures the brightness of an illuminated surface. One lux is 1 lm m^{-2}. The difference between the lux and the lumen is that the lux takes the area into account over which the luminous flux is spread, i.e., it depends not only on the light source but also on the position (distance) of the illuminated surface. The illumination from the sun on an average summer day on the Earth's surface ranges from 32,000 to 100,000 lux and from 3,000 to 4,000 lux on a winter day with cloudy sky. A bright office has up to 400 lux, domestic lighting ranges from 40 to 150 lux, street lighting produces 3 to 30 lux, a full moon at the zenith on a clear night about 0.1–0.3 lux [107, 170].

1.2.5 Color Rendering Index

As shown in the preceding Sects. 1.2.1–1.2.4, the quantity or the efficacy of light can be assessed, but up to now the "color" of the light source was not accounted for. What could be meant by the "color" or the quality of a light source? Since we are often not interested in the direct light from the light source but the reflected light from the illuminated surfaces, the quality of the light source depends on what is illuminated or what is looked on. The spectral power distribution of light from an illuminated surface is the product of the reflectance of the surface and the spectral power distribution of the light which falls on the surface (cf. Sects. 1.1.2 and 1.1.3). Thus, we can rate the quality of light by comparing the color impression of a defined color sample under illumination by the light source under consideration and some sort of standard light source. The result is a "color rendering index".

The general color rendering index (CRI) measures the ability of a light source to accurately display color as compared to a standard illuminant. The CRI is a scale from 0 to 100 which rates a light source by comparing the color appearance of eight different colors viewed under the light source to the color appearance of these eight colors viewed under a standard illuminant.[12] These eight colors are defined by the

[12] The description of the exact procedure can be found at CIE publication no. 13.2-1974.

International Commission on Illumination (CIE). The higher the CRI, the better the color rendering. A CRI value of 100 means the source is identical to the standard illuminant. A CRI of more than 90 gives a very good color rendering, a CRI between 80 and 90 is good, a CRI between 60 and 80 gives a sufficient color rendering [118]. A light source for office space may have a CRI in the 60s, but in areas where color appearance is very critical, a light source with a CRI in the 90s would be more appropriate. Typical light sources and their color rendering index are given in Table 1.2.

The color rendering index is coupled to a particular reference light source, which should have the same correlated color temperature (cf. Sect. 1.2.6) as the light source under consideration. Two standard light sources for different color temperatures can show considerable differences in the color rendering. That means that two light sources with the same color rendering index CRI but different correlated color temperatures may show considerable differences in the color rendering as well. But even two light sources with the same color rendering index CRI and the same correlated color temperature can show differences in the appearance of colors. This is caused by the definition of the (general) color rendering index, which is the average of eight special color rendering indices (one for each of the eight standard colors, see earlier). One of the light sources might render color 1 most, the other color 2. The averaging does not keep track of these details. Nevertheless, for high CRI above 90 experience shows that none of the eight special color rendering indices is shifted considerably more than the others. But the lower the CRI, the more different the color rendering of two light sources with equal CRI and equal correlated color temperature.

1.2.6 Correlated Color Temperature

The correlated color temperature (K) is a simple way to characterize the color of a light source. While in reality the color of light is determined by the spectral power distribution, the color can still be summarized on a linear scale, the correlated color

Table 1.2. Some light sources and their typical color rendering index (CRI) [92, 118]

light source	general color rendering index CRI
low-pressure sodium lamp	0–18
high-pressure sodium lamp	25–82
high-pressure mercury lamp	16–58
warm white fluorescent lamp	55
cool white fluorescent lamp	65
LED Luxeon white 5 W	70
deluxe warm white fluorescent lamp	73
daylight fluorescent lamp	79
metal halide lamp (4,200 K)	85
deluxe cool white fluorescent lamp	86
incandescent lamp (100 W)	100

temperature (CCT). This value is useful, e.g., for classifying light sources for different fields of application (home, office, sports, etc.) or for specifying the right light source types in architectural lighting design. Nevertheless, light sources of the same CCT may vary widely in the quality of light emitted, i.e., in their color rendering (CRI, cf. Sect. 1.2.5). One light source may have a continuous spectrum, while the other just emits light in a few narrow bands of the spectrum, both having the same CCT. A low CCT implies warmer light (more red) while a high correlated color temperature implies a colder light (more blue). Daylight has a rather low CCT near dawn, and a higher one during the day. Nevertheless, a high CCT does not imply that the light source is actually at that temperature. Fluorescent lamps have, for example, a high CCT, but actually they are quite cool during operation compared to high-pressure discharge lamps (cf. Sect. 1.3.4).

The correlated color temperature is obtained by comparing the light source under consideration to a black body radiator (Planckian radiator) of a given temperature and characterizing the light source by the temperature of the black body radiator which has the same chromaticity, i.e., which is closest[13] to it on the CIE chromaticity diagram (cf. Fig. 1.6). That is, the CCT refers to the temperature to which one would have to heat a Planckian radiator to produce light of the same visual color. Since the (continuous) spectral power distribution of a Planckian radiator depends only on its temperature, the CCT reduces the color to a single value (cf. Sect. 2.4.3). The preferred CCT are between 2,500 and 7,500 K. Warm white light lies between 2,500 and 3,300 K, neutral white between 3,300 and 5,000 K, and daylight white between 5,000 and 7,500 K [104]. Table 1.3 gives some examples of correlated color temperatures of different light sources.

1.3 Light Sources

To understand high-pressure or high-intensity discharge (HID) lamps, it is helpful to know more about other artificial light sources. The characteristics of HID lamps can easier be classified with knowledge about other types of light sources at the back of our minds. The working principle of high-pressure discharge lamps, the photometric characteristics, or the advantages and disadvantages of HID lamps in general can be better understood and rated against the background of other light sources.

Moreover, many principles hold for several types of lamps: Comparing, for example, an incandescent lamp with a HID lamp, the black body radiation plays an important role in both cases. In incandescent lamps, the visible light is "produced"

[13] Strictly speaking, not the closest point on the Planckian locus (line in the CIE chromaticity diagram of Planckian radiators of different temperatures) is taken for reference, but iso temperature lines or Judd lines are used [118].

Table 1.3. Some light sources and their typical correlated color temperature (CCT) [92, 118, 133]

light source	correlated color temperature (CCT) (K)
sunrise/sunset	3,200
1 h before/after dusk/dawn	3,400
sunny daylight around noon	5,800
candlelight	1,500
high-pressure sodium lamp	2,000–2,500
incandescent lamp (40 W)	2,680
incandescent lamp (200 W)	3,000
LED Luxeon white 5 W	5,500
fluorescent lamp	2,700–8,000
high-pressure mercury lamp	3,300–7,600

by black body radiation. In HID lamps, the electrodes are normally very hot, producing black body radiation as well. This black body radiation contributes in fact only a little bit to the light output of the HID lamp,[14] but it can be used to measure the electrode temperature by means of pyrometry (cf. Sects. 3.1.1–3.1.7). Moreover, the tungsten–halogen cycle in halogen (incandescent) lamps and some types of high-pressure discharge lamps is similar [45]. Low-pressure discharge lamps (e.g., fluorescent lamp, low-pressure sodium lamp) and HID lamps have, for example, the electron emission mechanism at the cathode and the starting mechanism in common.

So, Sects. 1.3.1–1.3.9 will give an overview over light sources, starting with the sun (Sect. 1.3.1). The sun sets the benchmark for all artificial light sources since the human eye is adapted to its spectral power distribution. Then, the oldest artificial light source (after fire), the incandescent lamp, will be considered (Sect. 1.3.2), followed by low-pressure discharge lamps (Sects. 1.3.4 and 1.3.5) and finally arriving at high-pressure discharge lamps (Sects. 1.3.6–1.3.9).

1.3.1 Sun

The human eye is adapted to the sunlight (inside our atmosphere, i.e., at sea level), so generating light with a similar spectral power distribution as the sun will be perceived as "good light." Nevertheless, even a spectral power distribution quite dissimilar to that of the sun can produce good light with respect to color rendering, if the characteristics of the cones in the human eye are matched.

The sun is a hot body. Its core can reach 20×10^6 K, the surface temperature is about 6,000 K. The outer atmosphere of the sun (which is visible during a solar eclipse) is extremely hot again, about 10^6 K. At the center of big sunspots, the temperature can be as low as 4,300 K (dark spots on a bright background).

[14] In the first electric arc discharges (carbon arc between carbon electrodes, 1802), the incandescent of the electrodes created the light output [70, 111].

The radiation emitted by the sun is discrete, however, the emission lines are broadened by the pressure of the gases comprising the sun (cf. Sect. 2.2.3) and the numerous emission lines are so close together that the spectral power distribution in the visible can be regarded as continuous radiation. There are hundreds of absorption lines in the solar spectrum due to the absorption of specific wavelengths in the sun's atmosphere, the so-called Fraunhofer lines. The elements in the sun's atmosphere can be determined by studying these lines. Moreover, the atmosphere of the earth absorbs and scatters electromagnetic radiation[15] [143].

Despite all this, we can compare the spectrum of the sun with the radiation coming from a black body, calculated with the Planck radiation formula (cf. Sect. 2.4.3). Black body radiation of approximately 5,800 K matches the solar spectrum best, but it goes without saying that this temperature depends strongly on the position of the observer on the earth surface, the height above sea level, the position of the sun (time of day), the season, and the weather conditions (cf. Table 1.3 and Fig. 1.8).

1.3.2 Incandescent Lamp

If we want to match the solar spectrum of the sun, we could use an artificial light source consisting of a black body radiator of approximately 5,800 K. A good choice would be carbon with a melting point higher than 4,100 K. The carbon, in the form of a filament in a vacuum bulb, can be brought to very high temperatures (=incandescence). But the evaporation rate of hot carbon is very high. This would result in a fast blackening of the bulb containing the carbon filament. Alternatively, one could reduce the temperature of the carbon filament, resulting in a lower evaporation rate but a much lower efficacy as well. Around 1854, Heinrich Göbel, a German emigrant in the USA, built the first practical incandescent lamps: Carbonized bamboo filaments enclosed in evacuated bottles to prevent oxidation. They were intended to illuminate the shop window of his watch shop in New York City [70].

It turned out that tungsten is the best choice for the filament. The melting temperature of tungsten lies at about 3,680 K and the tungsten vapor pressure is about 5,000 times smaller than the carbon vapor pressure at temperatures used in incandescent lamps. Moreover, the emissivity $\varepsilon(\lambda)$ of tungsten (cf. Sect. 2.4.3, especially Fig. 2.30) is high in the visible part of the spectral power distribution (about 0.4–0.5), but low in the infrared part of the spectral power distribution (0.15–0.4). This enhances the efficacy in comparison to a theoretical black body radiator of the same temperature.[16] Nevertheless, even if radiation is the only loss mechanism, the theoretical efficacy is limited. Figure 1.9 depicts the percentage of

---

[15] Absorption causes gaps in the solar spectrum, scattering reduces the spectral power distribution in general.

[16] Carbon is practically grey, i.e., the emission coefficient $\varepsilon(\lambda)$ is nearly independent of the wavelength λ.

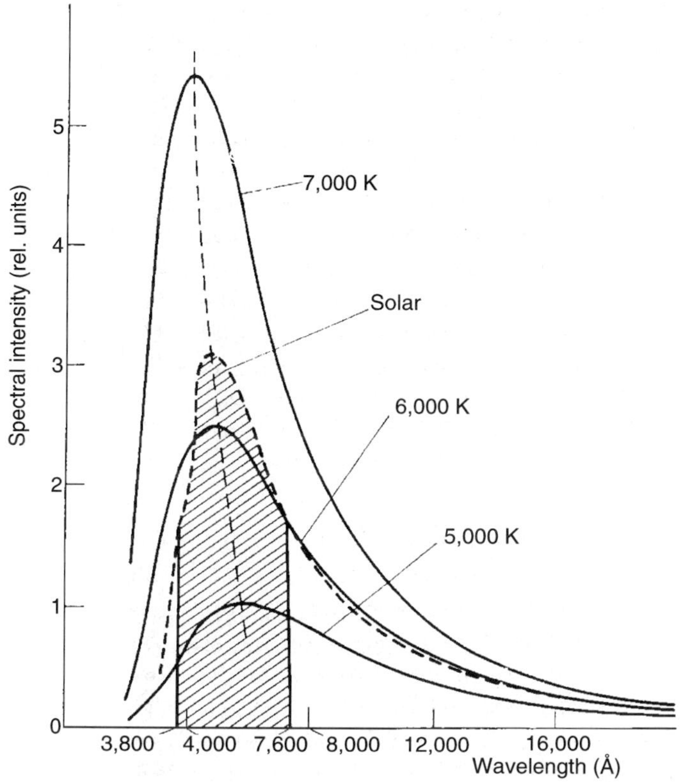

Fig. 1.8. Comparison of smoothed solar radiation curve and black body radiation for three different temperatures [143]

radiation emitted between 400 to 700 nm (visible light) for a theoretical black body radiator as a function of temperature (curve 1, right scale). The same type of curve for a tungsten filament reaches higher values, due to the enhanced emissivity in the visible part of the spectrum (curve 2). Assuming that there are no other loss mechanisms apart from radiation, the resulting efficacies are given in Fig. 1.9 by curve 3 (black body radiator) and curve 4 (tungsten). Thus, the maximum efficacy at 3,000 K is 28 lm W^{-1} for a tungsten filament [48].

The filament of an incandescent lamp is a resistor. The electrical power applied to it is converted into heat. Its temperature rises until the power loss of the filament is as high as the heat being generated in the filament. Ideally, the power loss is dominated by radiation, but thermal conduction can also be a loss mechanism. The advantage of vacuum incandescent lamps is that nearly 100% of the electrical power put in the tungsten filament is converted into radiation. Nevertheless, the temperature of the tungsten coil must be relatively cold (2,000–2,500 K say) to prevent considerable evaporation of tungsten. The evaporation of tungsten would shorten the lifetime because of narrowing of the tungsten filament. Where the filament is narrower, electrical resistance is higher (due to the smaller cross section)

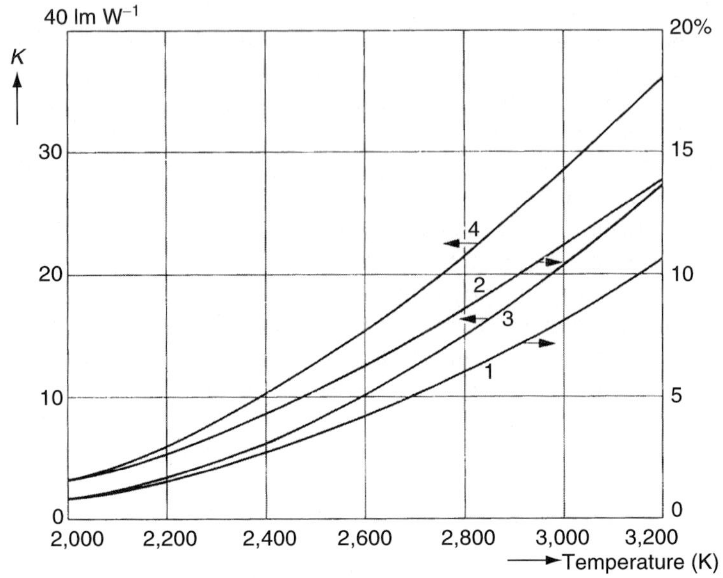

Fig. 1.9. Percentage of radiation emitted between 400 to 700 nm by a black body radiator (curve 1) and by tungsten (curve 2). The resulting efficacies are curve 3 (black body radiator) and curve 4 (tungsten). Radiation has been assumed to be the only loss mechanism for the calculation of the efficacy [48]

and less surface is available to radiate heat away. The filament heats up more, increasing the rate of evaporation even further at that point until finally the filament either melts or becomes weak and breaks. Moreover, the evaporated tungsten blackens the walls of the bulb, reducing the light output.

Filling the bulb with an inert, noble, or rare gas, such as argon, xenon, or krypton, slows down evaporation of the filament. Tungsten atoms evaporated from the tungsten filament will be bounced back to the filament by the gas atoms. The heavier the noble gas atoms, the better this effect works, giving preference to xenon (131.3 u) or krypton (83.8 u). Xenon and krypton also have a smaller thermal conductivity than argon, i.e., the energy loss by heat conduction is lower for xenon and krypton compared to argon, but argon is much cheaper. The filament can be operated at a higher temperature with a fill gas compared to operation in a vacuum.[17] This results in more efficient radiation of visible light (cf. Fig. 1.9).

[17] Some incandescent light sources are still operated with a vacuum inside. The reason is that in low-wattage incandescent lamps with thin filaments and lower currents the thermal heat conduction is too high, resulting in low efficacies despite higher filament temperatures.

In spite of the gas filling of the lamp, there is still evaporated tungsten blackening the bulb, a problem which is also known in HID lamps.[18] This reduces light output over the lifetime of the lamp and prevents, for example, scaling down of incandescent lamps (narrow tubes blacken faster, because the same amount of tungsten condenses on a smaller area). A solution used in incandescent lamps as well as in HID lamps (e.g., UHP lamp, cf. [45] and Sect. 1.3.6) is the so-called tungsten–halogen cycle.

Traces of a halogen, e.g., iodine or bromine, are added to the gas filling of the lamp. As tungsten evaporates from the tungsten filament, it usually condenses on the inner surface of the bulb. But the halogen, being chemically reactive, combines with this tungsten deposited on the wall and forms tungsten halide, which evaporates more easily, i.e., tungsten halide is more volatile than pure tungsten. When the tungsten halide gets near the filament, the high temperature of the filament causes the halide to dissociate. Thus, the tungsten concentration near the hot filament is higher than without a tungsten–halogen cycle, increasing the back diffusion of tungsten.

Nevertheless, this process extends the lifetime of the filament only somewhat, because problems with uneven filament evaporation and uneven deposition of tungsten onto the filament by the tungsten–halogen cycle do occur. This limits the ability of the tungsten–halogen cycle to prolong the life of the bulb. However, the tungsten–halogen cycle keeps the inner surface of the bulb clean. This lets halogen bulbs stay close to full brightness during lifetime, i.e., the efficacy is constant over lifetime. Moreover, smaller envelopes around the filament can be chosen, which is not possible without a tungsten–halogen cycle (due to the increased blackening of the smaller inner wall area without a tungsten–halogen cycle). The resulting smaller volume of the lamps permits the use of expensive rare gases like xenon or krypton with a higher pressure. This reduces even more the evaporation of tungsten, resulting in longer lifetimes or higher filament temperatures with higher efficacy (see earlier). In order for the tungsten–halogen cycle to work, the bulb surface must be very hot, generally over 500 K. The halogen may not adequately vaporize or fail to adequately react with condensed tungsten if the bulb is too cool. This means that the bulb must be small (which is desired anyway, see earlier) and made of either quartz glass, vycor, or very hard glass [48].

One failure mode of halogen incandescent lamps is filament notching or necking. Since the ends of the filament are somewhat cooler where the filament is attached to the lead wires, the halogen attacks the filament at these points, forming tungsten halogen which evaporates. This effect, which makes the coldest tungsten filament parts thinner, can also be observed in HID lamps using a tungsten–halogen cycle (cold end attack, cf. [75]).

The incandescent light bulb is still widely used in domestic applications and many portable lighting devices are equipped with incandescent lamps (for instance electric torches, car interior lighting, etc.). A typical spectral power distribution of

[18] HID lamps are normally operated with tungsten electrodes.

an incandescent lamp compared to the $V(\lambda)$ curve of the light adapted human eye is shown in Fig. 1.10. Halogen incandescent lamps have become more common in automotive headlights (replaced more and more by HID lamps, cf. Sect. 1.3.8) and domestic situations, particularly where light is to be concentrated on a particular point.

1.3.3 Gas Discharge Lamps

The electromagnetic radiation emitted by an incandescent lamp is produced by a hot solid, namely a tungsten filament. The physical process behind this way of generating light is the black body radiation (cf. Sect. 2.4.3). The fluorescent lamp, the low-pressure sodium lamp, the high-pressure mercury, sodium, or metal halide lamp are in contrast gas discharge lamps, i.e., the radiation originates from a gas (plasma) instead of a solid state. The gas is located in general in a discharge tube with two electrodes. During operation, the filling consists of an inert gas and often a metal vapor. The electrons emitted by one of the electrodes (the cathode) gain energy in the electric field between the electrodes. This energy is transformed in some way or other in excitation, ionization, or dissociation of gas atoms or molecules and heat. The excitation energy obtained by an atom can be transmitted after a while as electromagnetic radiation (radiative de-excitation[19]). The wavelength λ of this radiation is coupled to the excitation energy E via the relation $\lambda = ch/E$, where c is the velocity of light and h is Planck's constant. In the

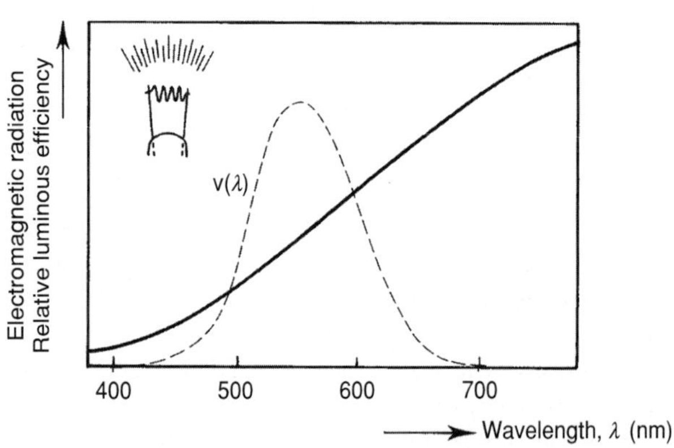

Fig. 1.10. Typical spectral power distribution of tungsten filament (incandescent lamp [118]) superimposed by the sensitivity curve $V(\lambda)$ of the light adapted human eye

[19] There are other possibilities of de-excitation like the so-called quenching collisions, where the excitation energy of an atom is transferred to an electron as kinetic energy. These quenching collisions and inelastic collisions of atoms outnumber the radiative de-excitations in HID lamps.

same way the recombination of an ion and an electron can lead to continuous electromagnetic radiation with a lower wavelength limit corresponding to the ionization energy of the atom. Thus, in contrast to the continuous spectral power distribution of a black body radiator (solid state), the electromagnetic radiation originating from a gas discharge consists of a discrete spectral power distribution[20] (corresponding to the possible (discrete) excitation states of the gas atoms or molecules), i.e., the spectral power distribution consists of a number of more or less separated spectral lines which may be broadened by certain mechanisms (cf. Sect. 2.2.3).

Starting of Discharge Lamps

Many gas discharge lamps have a filling consisting of an inert, noble, or rare gas plus a metal vapor. In the first place, the inert gas is needed to start the lamp, because at room or ambient temperature the metal is not vaporized, i.e., the vapor pressure of the metal is too low. In the UHP lamp for example, the starting gas is argon with a pressure of about 10^4 Pa = 0.1 bar. To heat up the discharge tube and therefore to get the solid or liquid metal vaporized, a noble gas discharge is started. The so-called breakdown of the (initially insulating) gas is necessary to create a plasma. By applying a high starting voltage (ignition voltage), up to some tens of kV, single electrons in the discharge or from the preheated cathode are accelerated and thus create new electron–ion pairs by ionizing gas atoms during inelastic collisions.[21] If the accelerating voltage is high enough, this process produces sufficient electrons resulting in a self-sustained discharge.

The starting voltage depends strongly on gas composition, gas pressure, and electrode gap. The lower this voltage, the easier the discharge can be started. The best choice is a so-called penning mixture: Taking neon or argon alone as the starting or auxiliary gas, the starting voltage is considerably higher than for a mixture of both of them in a ratio of 99:1 (neon/argon). In this mixture, the electrons might excite neon to a metastable state (16.6 eV), whereupon neon passes this energy to ground-state argon atoms, which become ionized (15.7 eV). This produces an additional electron, which can be accelerated in the electric field, yielding more electrons: The discharge has been started.

Besides the gas composition, the gas pressure is an important factor. In general, the starting voltage decreases for increasing pressure, reaches a minimum value and rises again with further increasing pressure. This can be explained with the mean

[20] There are also physical processes within the discharge producing continuous spectral power distributions which may be important for certain types of lamps. This will be discussed in the following and in Chap. 2.

[21] Elastic collisions preserve the kinetic energy of the participating partners (exchange of kinetic energy), inelastic collisions convert the kinetic energy or parts of it in excitation energy of the atoms or molecules or the energy is used to ionize or dissociate atoms or molecules, etc.

free path of the electrons in the gas (cf. Sect. 2.1.6). If the pressure of the gas is too low, the mean free path is too large, i.e., the electrons might not even hit one gas atom at their way through the discharge or their energy gain between two collision is much higher than the ionization energy, reducing the efficiency of ion and electron production. Increasing the pressure leads to a smaller mean free path, increasing the efficiency of the ionization process. However, if the mean free path of the electrons gets too small, the electrons might not gain enough energy between two collisions, decreasing the efficiency of electron and ion production again (Paschen's law).

The electrons gain energy in the electric field between the two electrodes and either produce new electrons (inelastic collisions) or heat up the inert gas by elastic collisions. As a result, the temperature in the discharge rises, so that the metal filling starts to evaporate. The coldest spot at the inside wall of the discharge tube determines the vapor pressure of the filling, since there (at the so-called cold spot) the gaseous metals will condense. Depending on the desired pressure of the filling, this cold spot temperature might be as low as, e.g., 310 K (low-pressure discharge) or higher than 1,100 K for certain high-pressure discharge lamps (e.g., UHP lamp, cf. Sects. 1.3.6 and 4.3).

Starting a discharge lamp requires free electrons which are accelerated in the electric field produced by the applied starting voltage. To provide as many electrons as possible, the electrodes (usually made of tungsten with small additives) can be preheated before starting the lamp. The electron emission current due to thermal electron emission from the cathode is thus increased (cf. Sect. 2.4.4). The heating of the electrodes can be switched off after starting or the electrodes can be heated continuously during operation. A second possibility to increase the number of free electrons is the usage of so-called emitters at the electrode surface (e.g., barium oxide or thorium oxide, cf. Sect. 2.4.4). They reduce the work function of the electrodes and enhance therefore the electron emission, i.e., they enhance the number of emitted electrons.

Auxiliary Gas

In low-pressure discharge lamps the inert, auxiliary, or buffer gas has a second duty during lamp operation. Without an inert, noble, or rare gas, the density of metal atoms in the discharge is so low that electrons have a mean free path length comparable to the order of the tube radius.[22] Thus, a lot of energy gained by the electrons in the electric field between the two electrodes is "wasted" and transformed to heat when the electrons hit the discharge wall. To prevent this, the pressure of the inert gas might be much higher than the pressure of the metal (e.g., 200–1,000 Pa), leading to frequent elastic collisions between electrons and

[22] Typical pressures of metal vapor in low-pressure discharge lamps are below 1 Pa during operation, the mean free path of electrons in a mercury vapor at room temperature is about 50 mm.

inert gas. This reduces the mean free path length considerably to the order of about 0.1 mm [48]. The electrons then follow a zigzag path through the discharge which enhances the probability of collisions with metal atoms. The energy loss by the elastic collisions with the rare gas atoms does not prevent the electrons from gaining enough energy for the excitation of the metal atoms. Since excitation and ionization potential of the inert gas is in general much higher than the excitation potential of the metal, the electrons do not excite or ionize excessively the inert gas during operation. Nevertheless, the rare gas pressure should not be too high to prevent too much energy loss by elastic collisions.

In the D2 automotive headlight lamp the starting gas (7 bar xenon in the cold lamp, 50 bar in the operating lamp) has a further function: It is responsible for the so-called "Xenon-Instant-Light" at the cold start of the lamp, to provide the required minimum output of 800 lm one second after switch on (cf. Sect. 1.3.8). In general, the run-up phase of a discharge lamp, during which the metal is vaporized, can take a couple of seconds up to a couple of minutes, depending on the desired vapor pressure of the metals and the geometry and extent of the discharge. This warm-up time can often be observed when street lights (high-pressure sodium or high-pressure mercury) are turned on and luminous flux and light color change over time [92, 118].

Negative Voltage–Current Characteristic

In a metallic conductor, the current is transported by a given number of electrons. Doubling the current means doubling the voltage and with this the velocity of the electrons. The result is Ohm's law ($U = RI$). In a gas discharge lamp, however, the number of electrons carrying the current is not fixed. The higher the lamp current, the higher the number of free electrons. In fact, the number of free electrons generally increases more than linearly with the lamp current, resulting in a negative voltage–current characteristic. The consequence of this is the following: A discharge connected directly to the mains voltage V_0 will carry the lamp current I_0. If the lamp current is increased by a small amount ΔI, a smaller lamp voltage $V_0 - \Delta V$ is needed. But since the mains voltage V_0 is still applied to the discharge, the voltage drop over the discharge is ΔV higher than necessary, the electron velocity will increase and with this, the lamp current I will even further increase. This leads to an unstable situation until the fuse blows.

To prevent this unstable situation, an impedance can be included in the electrical circuit, usually an inductive ballast in the ac case or a series resistance in the dc case, which limits the current. Instead of these conventional ballasts, modern discharge lamps are often operated with so-called electronic ballasts. An electronic ballast usually uses high frequency switching controlled by active components (transistor, thyristor, etc.) and is responsible for both the starting and the stable and efficient steady-state operation of the lamp. The advantage of high frequencies (typically between 18 and 50 kHz, cf. [118]) compared to the normal mains supply frequency of 50 or 60 Hz is the higher efficiency of the discharge: At 50 or 60 Hz most electrons and ions will be lost during current reversal by recombination of electrons and ions. The lamp must consequently reignited after each half cycle.

A solution of this problem is the operation of the discharge lamp with rectangular shaped lamp currents or high frequency operation with sinusoidal lamp current.

Low-Pressure Discharge Lamps

The electrons emitted by one of the electrodes (the cathode) gain energy in the electric field between the electrodes. This energy is transformed in some way or other in excitation, ionization, or dissociation of gas atoms or molecules and heat. This transformation of energy is different in low-pressure and high-pressure discharge lamps.

In low-pressure discharge lamps, the mean free path of electrons is so large (cf. Sect. 2.1.6), that the electrons are able to gain enough energy in the applied electric field between two collisions to excite the metal atoms in the discharge (mercury or sodium). The most probable excitations are excitations from the ground state (which is most probable for the metal atoms in the discharge) to the first excited states. The excited atoms emit resonance lines or resonance radiation.[23] If the pressure is too low, the mean free path is too large, i.e., the electrons might not even hit one gas atom at their way through the discharge or their energy gain between two collision is much higher than the excitation energy, reducing the efficiency of the excitation process (cf. "Auxiliary Gas" in this section). However, if the pressure is too high and the mean free path of the electrons gets too small, the electrons might not gain enough energy between two collisions for the excitation of the metal atoms, decreasing the efficiency again. Moreover, the resonance radiation is reabsorbed with a higher probability by a metal atom[24] if the vapor pressure and thus the density of metal atoms gets higher. The optimum pressure of metal vapor is below 1 Pa in low-pressure discharge lamps (mercury or sodium).

Typical low-pressure discharge lamps contain either mercury (low-pressure mercury lamps or fluorescent lamps, cf. Sect. 1.3.4) or sodium (low-pressure sodium lamps, cf. Sect. 1.3.5). The advantage of mercury is the sufficiently high vapor pressure at relatively low temperatures, the advantage of sodium is the wavelength of the resonance lines (589.0 and 589.6 nm), which are very close to the maximum of the sensitivity curve of the human eye. The typical spectral power distribution of low-pressure discharge lamps is a spectrum with very few, sharp spectral lines due to the resonance radiation. In the case of low-pressure mercury discharge lamps UV radiation is produced and transformed into visible light with the help of fluorescent powder. In the case of low-pressure sodium discharge lamps, the wavelength of the resonance radiation is already very close to the maximum of the sensitivity curve of the human eye.

---

[23] Lines produced by transitions involving the ground state of an atom are referred to as resonance lines.

[24] Since resonance radiation originates from an excited atom de-excitating to the ground state, an atom in the ground state can be easily excited by this radiation (corresponding to the energy difference between ground state and excited state).

The low pressure combined with low current densities leads to large discharge dimensions to ensure sufficient electrical power input. Typical electric field strengths in a low-pressure discharge are around 1–2 V cm^{-1}. Thus, increasing the wattage of a lamp correlates with increasing the size of the lamp, i.e., the discharge tube dimension is the limiting factor for this type of discharge.

A typical characteristic of low-pressure discharge lamps is a considerably higher electron temperature compared to the heavy particle temperature[25] (temperature of atoms and ions). This is caused by the more effective energy gain of electrons compared to heavy particles in the discharge. The electric field accelerates ions and electrons, but the ions lose a substantial part of this kinetic energy at each elastic collision with another heavy particle due to the equal mass of the collision partners. The electrons lose only a very small percentage of their kinetic energy during elastic collisions with heavy particles due to the tremendous mass difference between electrons and atoms/ions. Moreover, the electrons quite probably excite atoms during inelastic collisions. If these excited atoms emitted the energy as radiation, the kinetic energy of the atom does not increase, i.e., the heavy particle temperature stays low. Typical heavy particle or gas temperatures in fluorescent lamps (low-pressure mercury lamps) are between 300 and 700 K, the electron temperature normally exceeds 10,000 K. In low-pressure sodium lamps, the gas temperature is about 500 K, electron temperature about 10,000 K [48, 80, 111, 118].

High-Pressure Discharge Lamps

Increasing the pressure from below 10^{-5} bar (1 Pa) to pressures around 1 bar (10^5 Pa) and up to 200 bar (20 MPa) changes the energy transformation process compared to low-pressure lamps. The mean free path of electrons decreases with increasing pressure, increasing as well the number of elastic collisions. Although the electron passes only a small percentage of its kinetic energy to heavy particles during elastic collisions, the huge number of collisions ensures a considerable energy transfer from electrons to heavy particles. This results in an increasing heavy particle or gas temperature with a simultaneously decreasing electron temperature (cf. Fig. 1.11). In high-pressure discharge lamps, the electrons and the heavy particles have approximately temperatures between 1,000 and 11,000 K. At these temperatures, the thermal excitation of atoms is sufficiently high, resulting in radiation from transitions from excited states to the ground state and from excited states to other excited states (cf. Sect. 2.2.3). The temperature profile normally shows a hot columnar temperature area between the electrodes (plasma column) with distinct temperature maxima near the electrode tips (hot plasma spots) as shown in Fig. 1.35.

The resulting spectral power distribution of high-pressure discharge lamps consists therefore not only of resonance lines but also of spectral lines due to

---

[25] The temperature concept in discharge lamps correlates strongly with the mean kinetic energy of the particles (cf. Sect. 2.1.4).

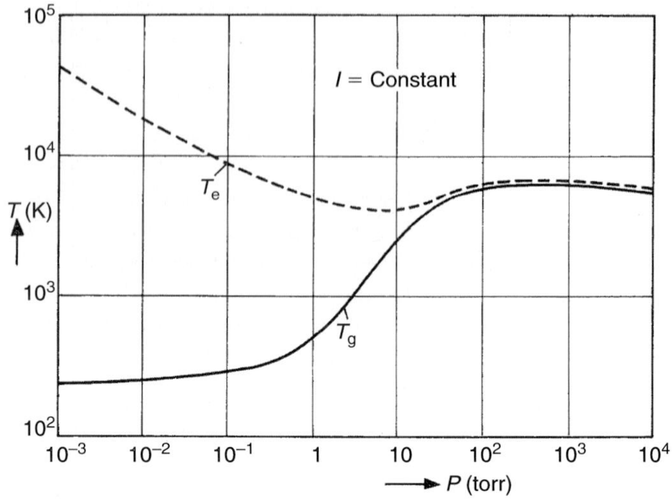

Fig. 1.11. Schematic of electron (T_e) and heavy particle (T_g) temperature in a mercury discharge as a function of pressure p (from 10^{-3} torr = 0.13 Pa to 10^4 torr = 1.3×10^6 Pa = 13 bar) at constant current I [48]

transitions between excited states. In fact, the resonance lines are even missing in high-pressure discharge lamps of sufficiently high pressure, because they are most probably reabsorbed in the outer (and colder) part of the plasma in the discharge tube. This is caused by the high density of metal atoms in high-pressure discharge lamps, which is even higher in the colder plasma regions. Since these atoms are normally in the ground state, the absorption of resonance radiation is very probable. Moreover, the spectral lines in a high-pressure discharge lamp are no longer sharp spectral lines with a line width according to the natural line width. Due to the high pressure, the line width is broadened by different mechanisms including pressure broadening, collision broadening, and Doppler broadening (cf. Sect. 2.2.3).

Typical high-pressure discharge lamps contain either mercury (high-pressure mercury lamp, Sect. 1.3.6, or metal halide lamp, Sect. 1.3.8) or sodium (high-pressure sodium lamp, cf. Sect. 1.3.7). The advantages of mercury are the high vapor pressure, i.e., mercury is very volatile, and the high resistance of a mercury plasma due to the large cross section for collisions between electrons and heavy particles. The latter makes it possible to operate the lamps at low currents and high voltages, which is desirable with respect to low strain and long lifetime of the electrodes and which enables the production of very small light sources, i.e., high power input into small volumes (high-intensity discharge lamps, point like light sources). The advantage of sodium is, as in the low-pressure discharge lamps, the wavelength of the resonance lines, which lie very close to the maximum of the sensitivity curve of the human eye. In contrast to the sharp resonance lines in the low-pressure discharge lamp at 589.0 and 589.6 nm, the high sodium pressure results in a considerable widening and an accompanying self-reversal of the sodium resonance lines (cf. Sect. 1.3.7).

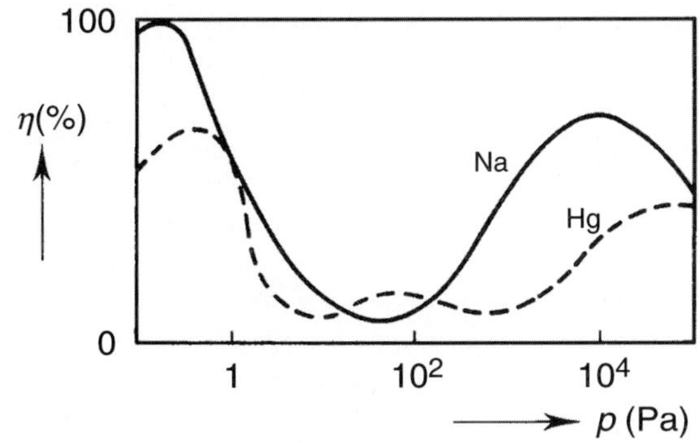

Fig. 1.12. Relative luminous efficacies of sodium and mercury discharge lamps depending on vapor pressure of sodium and mercury, respectively. In the case of the mercury discharge, the use of a fluorescent powder to convert UV radiation into visible light is assumed [118]

The high pressure leads to large electric field strengths up to 600 V cm^{-1}. Thus, high electrical power input into small discharge volumes is possible, e.g., 120 W for the UHP 120 super high-pressure mercury discharge lamp with an electrode gap of only 1 mm. This is an important feature for light sources for, e.g., projection applications or automotive headlight lamps (high-intensity discharge lamps, point like light sources).

The change of discharge efficiency with respect to luminous efficacy as a function of vapor pressure is shown in Fig. 1.12. Mercury vapor pressure around 0.8 Pa leads to efficient discharge lamps (low-pressure mercury discharge or fluorescent lamps), where the UV radiation of the discharge is transformed into visible light with the help of a fluorescent powder. Increasing the mercury pressure above 10^5 Pa (1 bar) enables efficient discharge lamps as well (high-pressure mercury discharge lamps). In the case of sodium, the optimum vapor pressure for a low-pressure sodium lamp lies around 0.4 Pa, the optimum vapor pressure for a high-pressure sodium lamp lies around 10^4 Pa.

1.3.4 Fluorescent Lamp

The fluorescent lamp is a low-pressure mercury discharge lamp (Fig. 1.13). The two resonance lines of mercury (185.0 and 253.7 nm) in the UV part of the electromagnetic spectrum are converted into visible light with the help of fluorescent material[26] (cf. Fig. 1.14). The fluorescent material is coated on the inside of the discharge tube and should have an absorption maximum at 185.0 and 253.7 nm. The emission band of the fluorescent powder should be a wide one in the

---

[26] There are two types of luminescence: the fluorescence (light emission as long as material is irradiated) and phosphorescence (light emission some time after irradiation). The fluorescent materials used in fluorescent lamps have both properties (fluorescence and phosphorescence), which is favorable to "bridge" the darker periods in the case of 50 or 60 Hz sinus operation.

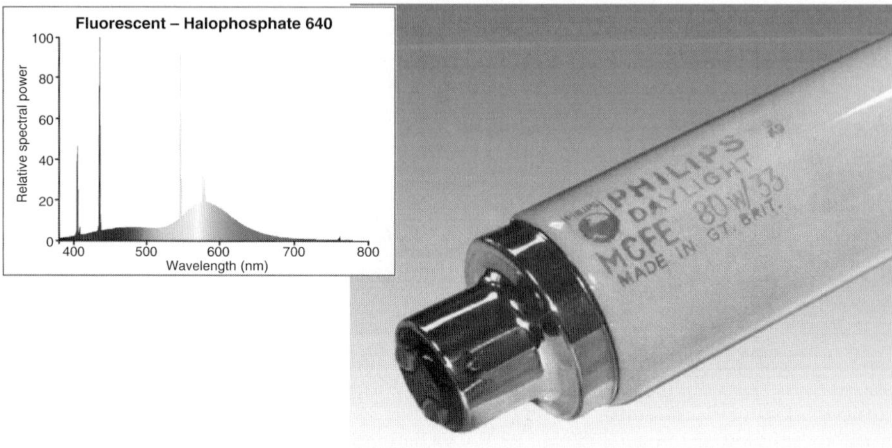

Fig. 1.13. Fluorescent lamp (FL): 80 W, $\eta = 61$ lm W^{-1}, CCT 4,080 K, CRI 62, 1,500 mm overall length ([92], the spectral power distribution is from a comparable fluorescent lamp)

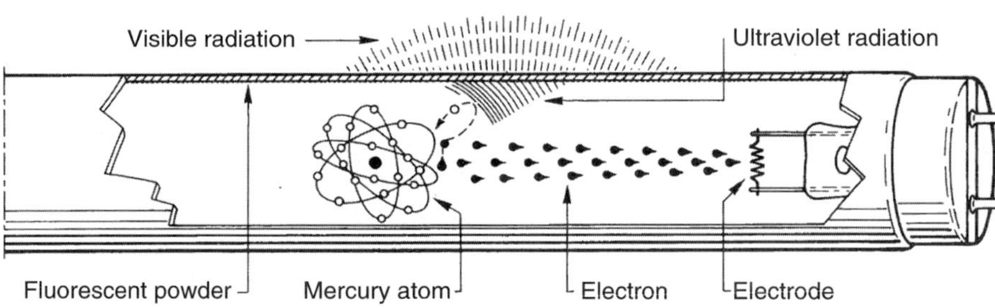

Fig. 1.14. Working principle of fluorescent lamp [118]. The mercury atoms are excited by accelerated electrons emitting UV radiation. The UV radiation is converted into visible light by a fluorescent powder

visible part of the electromagnetic spectrum if the lamp is used for lighting purpose. With up to 70% conversion of electrical power input into UV radiation, the generation of UV radiation is very efficient. Nevertheless, the transformation of a UV photon of 185.0 nm (6.7 eV) or 253.7 nm (4.9 eV) into a photon in the visible part of the electromagnetic spectrum (1.5–3 eV) is a great loss mechanism, resulting in a total conversion efficiency of electrical power to visible light of 20–30%.

The spectral power distribution of a fluorescent lamp can be adjusted to the specific demand of the consumer by the proper choice of the fluorescent powder. There is a wide range of available fluorescent materials, which may be used in combination or in layers to provide the desired color temperature, color rendering index, or spectral power distribution. Thus, the disadvantage of the use of fluorescent powder (loss of energy which ends up as heat in the fluorescent material) is also an advantage due to the flexible adjustment of the spectral power

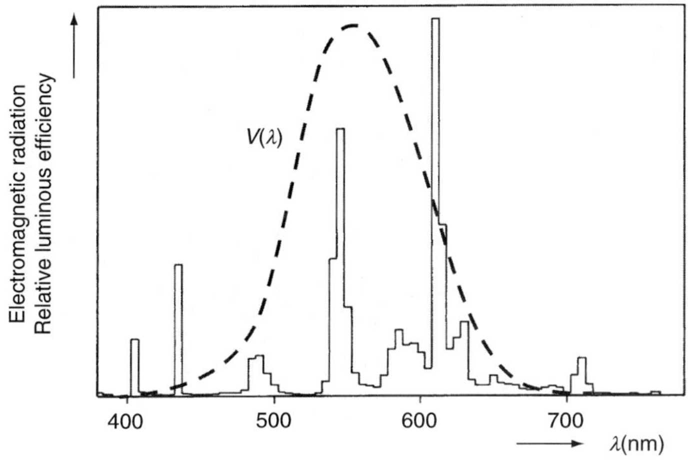

Fig. 1.15. Typical spectral power distribution of a fluorescent lamp with so-called three-band phosphors [118] superimposed by the sensitivity curve $V(\lambda)$ of the light adapted human eye

distribution with the help of the fluorescent powder. A typical spectral power distribution of a fluorescent lamp with a so-called three-band phosphor is shown in Fig. 1.15. This distribution approximates the color rendering of incandescent light sources. Without a fluorescent powder, the low-pressure discharge lamp is a very efficient UV source,[27] emitting 85% of the radiation at 253.7 nm and 15% at a wavelength of 185.0 nm.

A new product in the field of fluorescent lamps is the Lumilux Skywhite fluorescent lamp from Osram (cf. Fig. 1.16). The radiation in the blue part of the visible spectrum (410–460 nm) has been increased to match the sensitivity curve of the new discovered third type of photoreceptor described in Sect. 1.1.1. This demonstrates the advantage of fluorescent lamps: The spectral power distribution can be adjusted by the proper choice of fluorescent material.

The optimum vapor pressure of low-pressure mercury discharge lamps is about 0.8 Pa (cf. Fig. 1.12). This pressure is determined by the coldest temperature at the inner wall of the discharge tube, the so-called cold spot temperature. The mercury will condense[28] at this coldest spot and the pressure in the discharge corresponds to the vapor pressure of mercury at the cold spot temperature (cf. Fig. 1.17). At approximately 40°C, the optimum vapor pressure is reached. The discharge in a fluorescent lamp is, under normal operating conditions, able to generate this temperature. A change in the ambient temperature might change the cold spot

[27] UV sources are used, e.g., for industrial applications like lithography, disinfection, sterilization, curing, or surface cleaning [152].

[28] During operation, only around 50 μg of the approximately 8 mg mercury in the discharge is in the vapor phase [111].

Fig. 1.16. Fluorescent lamp (FL): 58 W, $\eta = 84$ lm W^{-1}, CCT 8,000 K, CRI >80, 1,500 mm overall length [133]

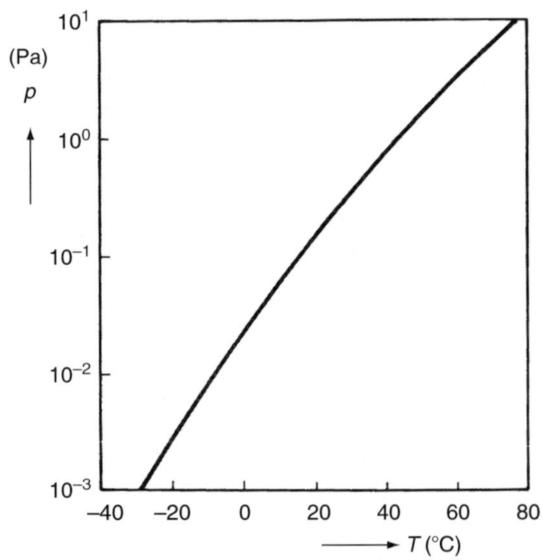

Fig. 1.17. Saturated mercury vapor pressure as a function of cold spot temperature [118]

temperature and thus the vapor pressure. Since efficiency is strongly coupled to the vapor pressure (cf. Fig. 1.12), the lumen output is very temperature dependent.

Using fluorescent lamps as outdoor lighting in the winter, the fluorescent lamp should be operated in a closed luminaire to reduce heat transfer from the fluorescent lamp to the environment with low ambient temperatures. Operating a fluorescent lamp in a hot environment, the vapor pressure of mercury might rise above the optimum vapor pressure of 0.8 Pa. The usage of a mercury amalgam instead of pure mercury is a possibility to reduce the mercury pressure, because the mercury vapor pressure above an amalgam is lower than the vapor pressure of pure mercury. Moreover, there are amalgams with much less temperature dependent mercury vapor pressures in the temperature region interesting for fluorescent lamp operation [118].

Besides the mercury, the auxiliary gas is important for starting and operating a fluorescent lamp (cf. "Auxiliary Gas" in Sect. 1.3.3). The optimum working pressure of the auxiliary gas is between 200 and 300 Pa [118]. The higher the atomic weight of the auxiliary gas atoms, the smaller the energy transfer during elastic collisions between electrons and atoms. This would make xenon to the perfect auxiliary gas. But since the first excitation energy of xenon is relatively low (8.43 eV, 147 nm), the excitation of xenon is relatively frequent but does not contribute to the desired radiation. Moreover, xenon is a very expensive gas. A better choice is krypton with a higher excitation energy of 10.0 eV (123.6 nm). Krypton is cheaper but also quite expensive. However, by reducing the arc tube diameter from 38 to 26 mm, less of the very expensive fluorescent powder is needed, compensating the additional costs of a dose of krypton.

Fluorescent lamps are typically long, thin cylinders, e.g., 1,200 mm long and 38 mm in diameter. Short fluorescent lamps would be inefficient because the power loss in the near-electrode region would be too high compared to total power consumption (electrode fall voltages, cf. Sects. 3.2.2, 4.1, and 4.4). Compact fluorescent lamps are made by folding the discharge tube (cf. Fig. 1.18). Another way of reducing or even abandon the electrode losses is to use electrodeless fluorescent lamps, where the power is coupled into the discharge by an electric or magnetic field. In the case of inductively coupled fluorescent lamps, an ac current produces an alternating magnetic field. This alternating magnetic field creates an electric field in which the electrons are accelerated and gain energy. The advantage is not only the avoidance of the electrode losses, which are replaced by losses during the generation of the alternating magnetic field for example in ferrite cores. But since the electrodes in a discharge lamp are often the limiting factor for lamp life, the absence of electrodes might increase the lamp life considerably [77]. This is of particular importance for lamp operation in areas difficult to access, where the replacement of a lamp is very costly (tunnel illumination, halls with high ceiling, etc.). The fluorescent lamp has replaced many applications of the incandescent lamp with its superior life and energy efficiency. A large application area of all fluorescent, compact fluorescent, and electrodeless fluorescent lamps is indoor lighting in offices, factories and stores, and domestic lighting. Fluorescent lamps are also used for outdoor lighting, especially motorway, street, tunnel, or car-park lighting. In addition to general lighting, low-pressure mercury lamps are used for example for background lighting of LCD displays or in reprography. Without a fluorescent powder, low-pressure mercury lamps are used as UV radiators (e.g., photochemistry, ozone production).

1.3.5 Low-Pressure Sodium Lamp

The low-pressure sodium discharge lamp (LPS or SOX, SO/H, or SOI) is up to now the most efficient artificial light source (Fig. 1.19). The radiation is produced in the same way as in the fluorescent lamp, i.e., electrons are accelerated in an electric field and excite the sodium atoms. The advantage of sodium is that the resonance lines (589.0 and 589.6 nm (sodium D-lines), corresponding to 2.1 eV) are very

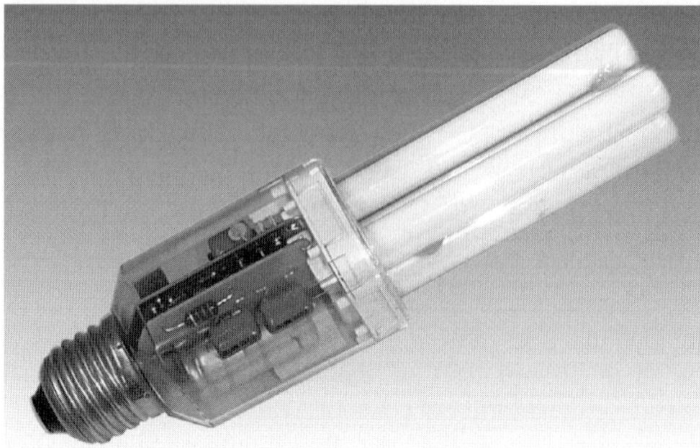

Fig. 1.18. Compact fluorescent lamp (CFL): 7 W, $\eta = 57$ lm W^{-1}, CCT 2,700 K, CRI 82, 161 mm overall length [92]

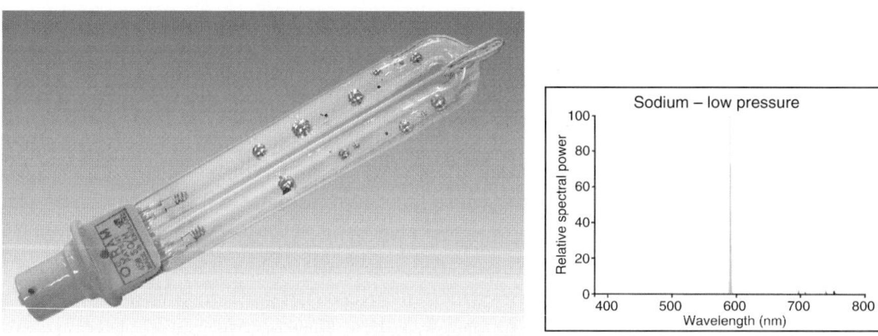

Fig. 1.19. Low-pressure sodium lamp, 45 W, $\eta = 49$ lm W^{-1}, CCT 1,700 K, 248 mm overall length [92]

close to the maximum of the sensitivity curve of the human eye at 555 nm (photopic vision, cf. Fig. 1.20). With up to 40% conversion of electrical power input into visible radiation, the luminous efficacy can be up to 200 lm W^{-1} [111, 118].

Nevertheless, since the radiation power is concentrated to one single line (monochromatic light, cf. Fig. 1.20), colors can not be distinguished under the light of a low-pressure sodium lamp. However, contrasts between moving and stationary objects are perceived faster under this monochromatic light compared to illumination with other (white) light sources.[29] Moreover, the light of the low-pressure sodium

[29] There are more recent investigation indicating that there is no difference between "white" and "yellow" light or that "white" light is better with respect to response time [41].

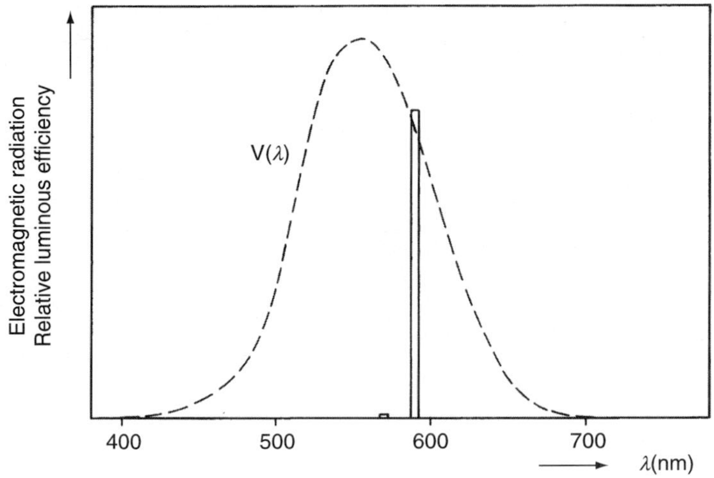

Fig. 1.20. Typical spectral power distribution of a low-pressure sodium lamp superimposed by the sensitivity curve $V(\lambda)$ of the light adapted human eye [118]

discharge lamp pierces mist with minimal scattering, thus improving the visibility in bad weather conditions when used for street lighting [118].

The optimum vapor pressure of sodium in low-pressure sodium lamps is about 0.4 Pa (cf. Fig. 1.12). As the sodium in low-pressure sodium lamps is typically overdosed (like the mercury in fluorescent lamps), the pressure is determined by the cold spot temperature. A typical low-pressure sodium lamp contains 100,000–1,000,000 times more sodium than required for operation. The sodium will accumulate at the coldest areas of the discharge tube during lamp life, thus decreasing the density of sodium in the discharge near hot areas of the discharge tube. There, the auxiliary gas (typically neon) will take over the discharge, decreasing the efficacy and changing the spectral power distribution. To ensure that the sodium is evenly distributed within the discharge tube, a number of small dimples are evenly distributed over the discharge tube. These dimples are naturally colder, helping to maintain an even sodium distribution. Especially after switching off the lamp, the sodium preferably settles in the small dimples during cooling down. This ensures not only the even distribution of sodium but also the formation of sodium mirrors on the discharge walls is prevented. These sodium mirrors would otherwise reduce or alter the light output [92, 118].

Sodium is not as volatile as mercury, i.e., at the same cold spot temperature the mercury vapor pressure is much higher than the sodium vapor pressure (compare Figs. 1.17 and 1.21). The optimum vapor pressure of sodium in low-pressure sodium lamps of about 0.4 Pa is attained at 260°C cold spot temperature, i.e., the temperature of the discharge tube is much higher in a low-pressure sodium lamp compared to a fluorescent lamp, which entails higher thermal losses in sodium low-pressure lamps. To reduce heat losses to the surrounding, the discharge tube is normally mounted in an outer bulb. Additionally, this measure reduces the effect of

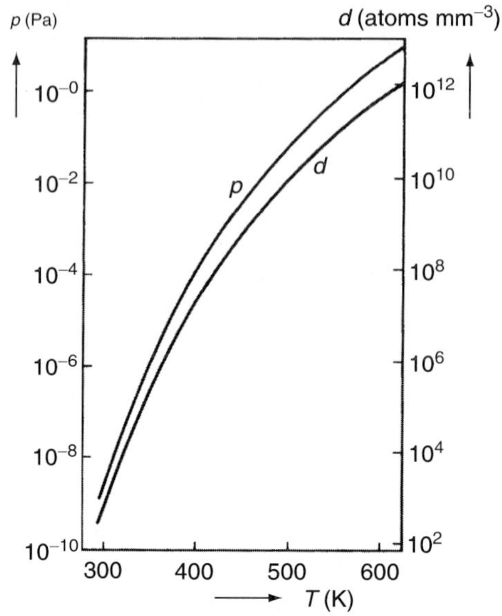

Fig. 1.21. Saturated sodium vapor pressure (p) and sodium atom density (d) as a function of cold spot temperature [118]

the ambient temperature on the operating pressure of the discharge and therefore on the efficacy of the lamp. This is important, since a temperature change of only 6 K results in a 20% change in vapor pressure, resulting in lower efficacies (cf. Figs. 1.12 and 1.21). As auxiliary gas, which is needed for starting the discharge and to reduce the mean free path of the electrons, a penning mixture of 99% neon and 1% argon with a total pressure of around 1 kPa is used (cf. "Staring of Discharge Lamps" and "Auxiliary Gas" in Sect. 1.3.3).

Since ionization of sodium in low-pressure sodium lamps (and mercury in fluorescent lamps) is more likely than ionization of the buffer or auxiliary gas, a "demixing" of the active species (sodium in LPS or mercury in fluorescent lamps) from the buffer gas takes place. This occurs when ions of the active species leave the center of the discharge faster than they can be replaced by diffusion of neutral atoms or molecules through the buffer gas. This phenomenon is called cataphoresis. Axial cataphoresis occurs in discharge lamps operated with direct current, causing the accumulation of the active species near the cathode. This can be avoided by operation with alternating current. Radial cataphoresis leads to depletion of the active species at the center of the discharge.

At the high operating temperature of the lamp, sodium is chemically very active. Ordinary glasses are rapidly stained brown and block light. Borate glass is quite resistant to sodium vapor but difficult to process and sensitive to moisture. A solution to these problems is the use of a double-layer glass tube. Inside the discharge tube, a very thin layer of borate glass is used, the main part of the discharge tube consists of an ordinary soda-lime glass tube.

The low-pressure sodium lamp has its main application area in the field of outdoor lighting, especially motorway, street, tunnel, or car-park lighting and illumination of buildings. The main advantage of low-pressure sodium lamps is that they are the most efficient light sources available. This means that they have the highest lumen output for each watt of electrical power compared to any other type of lamp.

1.3.6 High-Pressure Mercury Lamp

In a low-pressure mercury lamp with heavy particle temperatures between 300 and 700 K and electron temperatures above 10,000 K, the resonance lines of mercury at 185.0 and 253.7 nm play a central role. Increasing the pressure, the heavy particle temperature approaches the electron temperature, both typically between 4,000 and 11,000 K, depending on lamp current, pressure, and position within the plasma[30] (cf. Fig. 1.11 [99, 111]). At these temperatures, the thermal excitation of atoms is sufficiently high, resulting in radiation from transitions from excited states to the ground state and from excited states to other excited states. Especially transitions from excited states to other excited states produce strong lines in the visible part of the electromagnetic spectrum at 405, 436, 546, 577, and 579 nm (cf. Figs. 1.22 and 2.9 and Sect. 2.2.3). Figure 1.23 shows the spectral power distributions of a mercury discharge as a function of mercury vapor pressure. The mercury discharges are very efficient at low pressures, emitting the resonance lines at 185.0 and 253.7 nm, and at high pressures, emitting spectral lines in the visible part of the electromagnetic

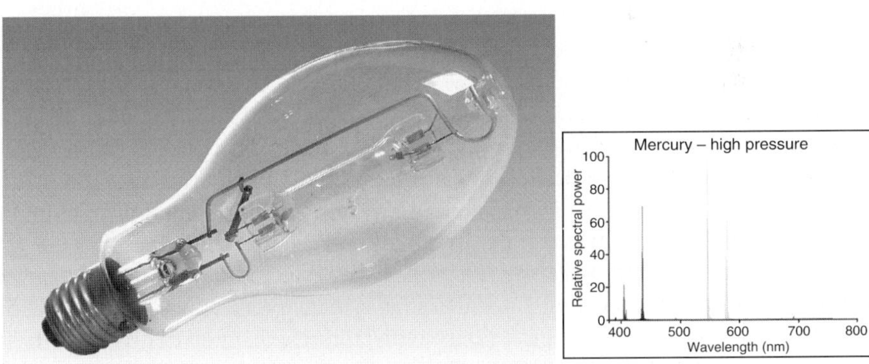

Fig. 1.22. High-pressure mercury lamp: 125 W, $\eta = 42$ lm W^{-1}, CCT 6,000 K, CRI 16, 170 mm overall length [92]

[30] The plasma temperature in the vicinity of the electrodes is in general higher than the temperature in the so-called plasma column between the electrodes (cf. Fig. 1.35).

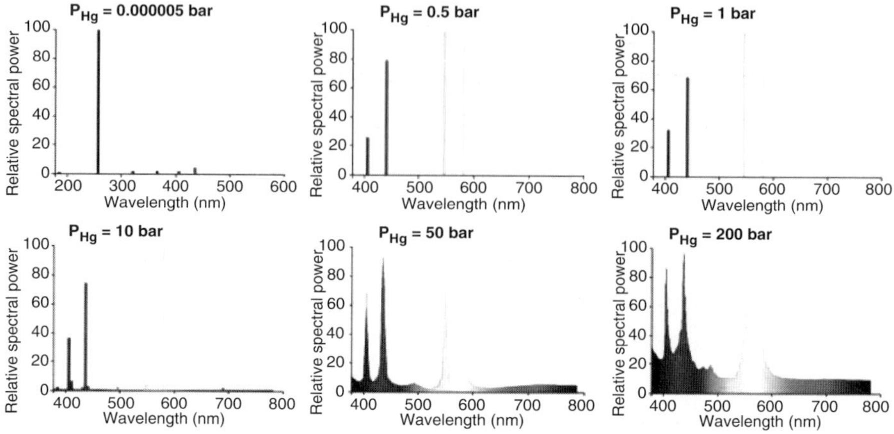

Fig. 1.23. Effect of mercury vapor pressure on the spectral power distribution of low-pressure and high-pressure mercury discharge lamps [92]

spectrum (cf. Fig. 1.12). Increasing pressure results in broader spectral lines (cf. Sect. 2.2.3). At very high pressures, bremsstrahlung and Hg_2 molecular radiation increase the luminous flux and the color rendering, which is of particular importance in the red part of the emitted spectrum, where mercury atomic radiation lacks spectral lines (cf. Sect. 2.2.3). In general, increasing pressure increases luminous efficacy and improves color rendering of high-pressure mercury discharge lamps (compare Figs. 1.22 and 1.26).

Another way of improving the radiation of a high-pressure mercury discharge with respect to color rendering and luminous flux is the use of a fluorescent powder as in low-pressure mercury discharge lamps (fluorescent lamps). Depending on the pressure of the discharge, a considerable portion of the radiation is emitted in the ultraviolet part of the electromagnetic spectrum, which can be transformed using a fluorescent powder coated on the inside walls of the outer bulb. This fluorescent powder should have optimum absorption of the UV light but should not absorb visible light up to 579 nm. The emission of the fluorescent powder should be in the red part of the visible radiation to compensate for the lack of mercury spectral lines there. A typical spectral power distribution of such a mercury high-pressure lamp with fluorescent powder is shown in Fig. 1.24.

In contrast to the low-pressure discharge lamps described earlier, the entire dose of liquid mercury in high-pressure discharge lamps is evaporated during operation, thus the mercury vapor is described as unsaturated. This means that the number of mercury atoms remains constant even if the cold spot temperature changes (as long as the cold spot temperature stays above the saturation temperature T_s, which depends on the lamp volume and the amount of mercury filled into the lamp). Therefore, high-pressure mercury lamps are much less temperature dependent than most other discharge lamps (cf. Fig. 1.25). The higher the pressure, the higher the

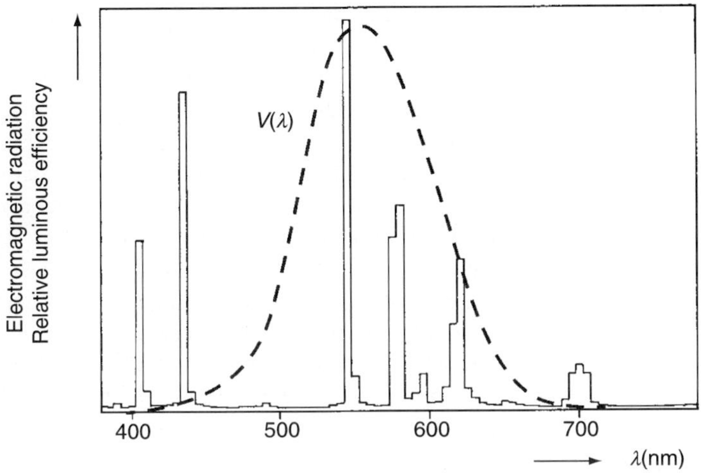

Fig. 1.24. Typical spectral power distribution of a high-pressure mercury lamp with atomic lines at 405, 436, 546, 577, and 579 nm and a contribution due to fluorescent powder for light emission above 600 nm [118] superimposed by the sensitivity curve $V(\lambda)$ of the light adapted human eye

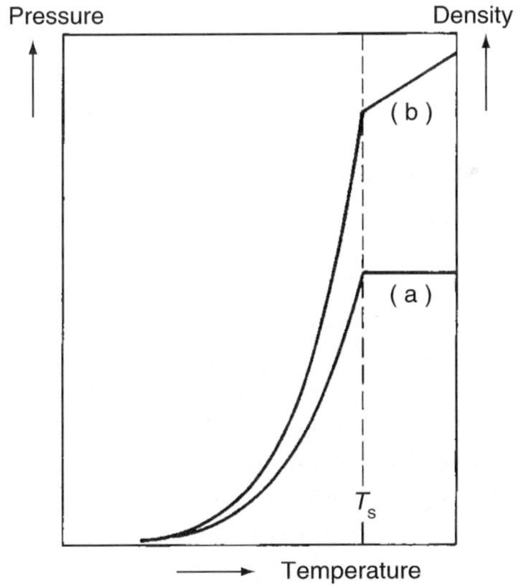

Fig. 1.25. Density (a) and pressure (b) of mercury in a high-pressure mercury discharge as a function of cold spot temperature. For temperatures above the saturation temperature T_S, the density of mercury in the discharge volume remains constant [118]

luminous efficacy of high-pressure mercury discharge lamps (cf. Figs. 1.12 and 1.23).

One example of a very or super high-pressure mercury lamp is the UHP lamp from Philips (UHP = ultra-high performance or ultra-high pressure [44–46, 52, 88, 142]).

With a pressure of around 20 MPa (=200 bar[31]) and a luminous efficacy of around 60 lm W^{-1} at electrical power input in the range from 100 to more than 200 W, this lamp is ideal for the use in video and data projection systems. The small electrode gap of 1–1.3 mm allows the usage of small optical components like small LCD displays or small digital micromirror device chips (DMDs), thus decreasing the overall size and weight of projection systems and simultaneously increasing the lumen output. The wall temperature (cold spot temperature) must be above 1,100 K to achieve complete evaporation of the mercury and thus to obtain a mercury pressure of about 200 bar. At the same time, the temperature of the walls must stay below 1,400 K to avoid quartz recrystallization and softening of the quartz glass. The wall temperature of the discharge is not to be confused with the temperature of the plasma, which is around 7,000–8,000 K between the electrodes. To reach lifetimes of up to 12,000 burning hours, small amounts of oxygen and halogen are added to the UHP lamp, establishing a tungsten–halogen cycle which prevents tungsten (evaporated from the tungsten electrodes) to blacken the walls by condensing on the inside of the quartz vessel[32] (cf. Sect. 1.3.2, [44–46, 75]). The high electrode tip temperatures above the melting point of tungsten are not only responsible for the high evaporation rate of tungsten, but the electrodes even change their shape noticeable during operation. By using a special shape of the lamp current[33], this effect can be controlled, thus preventing unstable arc operation due to arc jumps on the electrode tip (e.g., [119]). Images of the electrodes can be found in Figs. 1.26, 3.19, and 3.20.

The advantage of the very high pressure in UHP lamps is not only the high luminous efficacy but also the high resistance of the mercury plasma. This results in high burning voltages and therefore high electrical power input into a very small volume (e.g., 65 V for an electrode gap of 1 mm resulting in an electrical power input of 120 W). From 100 W electrical power input, 65 W is transformed in the plasma into radiation (25 W in the visible, 34 W in the infrared part, and 6 W UV radiation) and 11 W is conducted through heat conduction and convection from the plasma to the quartz walls. The remaining 24 W of the 100 W electrical power input is necessary for the heating of the electrodes. 12 W therefrom is radiated from the electrode surfaces, the remaining 12 W is conducted into the quartz through the electrode ends (e.g., [52, 88]). Of equal importance is the high contribution of Hg_2 molecular radiation and electron-atom bremsstrahlung due to the very high pressure, increasing the light output in the red part of the emitted spectrum, which is needed for a well-balanced color mixing (cf. Sect. 1.1.3, [45, 46, 105]). The resonance lines of mercury at 185.0 and 253.7 nm are totally absorbed within

[31] Corresponding to a mercury dose of approximately 10 mg.

[32] For this purpose, a small amount of bromine is added to the discharge (<1 μg according to [44]).

[33] Normal lamp current (e.g., rectangular wave, 2.0 A, 90 Hz) supplemented by short current pulses at the end of each half-cycle [119].

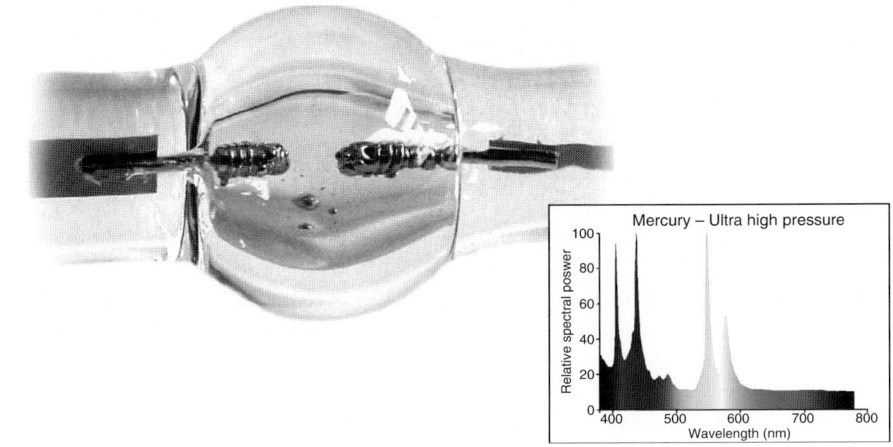

Fig. 1.26. High-pressure mercury lamp [8]: UHP lamp, 120 W, $\eta = 58$ lm W^{-1}, CCT 7,600 K, CRI 57, 1.0 mm arc gap (spectral power distribution from [92])

the plasma. More details concerning the UHP lamp can be found in [44–46, 52, 88, 92, 105, 119] and in Sects. 3.1.7, 3.1.8, and 4.3.

Due to the dominance of atomic lines in the blue and green part of the visible light spectrum and the lack of red atomic lines, the color rendering of mercury high-pressure discharge lamps is in general not very good and the correlated color temperature (CCT) is in general very high (cf. Figs. 1.22 and 1.23, and Tables 1.2 and 1.3). Thus, besides the UHP lamp for video and data projection systems, exterior lighting (motorway, street, tunnel, or car-park lighting and illumination of buildings) is the main application area for high-pressure mercury discharge lamps. In addition to general lighting and projection, high-pressure mercury lamps are used for example for microscopy, sun tanning lamps, photochemistry, or black light bulbs.

1.3.7 High-Pressure Sodium Lamp

In both, the low-pressure (LPS or SOX, SOI, or SO/H) and high-pressure sodium discharge lamp (HPS or SON), the resonance lines at 589.0 and 589.6 nm play a central role (sodium D-lines). The increasing pressure considerably broadens these resonance lines and may lead to the so-called self-reversal of the resonance lines (cf. Fig. 1.29 and Sect. 2.2.3), resulting in sufficient color rendering properties compared to the low-pressure sodium lamp. However, the luminous efficacy of high-pressure sodium lamps is lower than that of low-pressure sodium lamps. Increasing the pressure in high-pressure sodium lamps improves color rendering and increases the correlated color temperature[34] by spreading the emitted radiation more over the blue and red parts of the visible spectrum (cf. Fig. 1.29).

[34] Correlated color temperature in the range between 2,000 and 3,000 K [118].

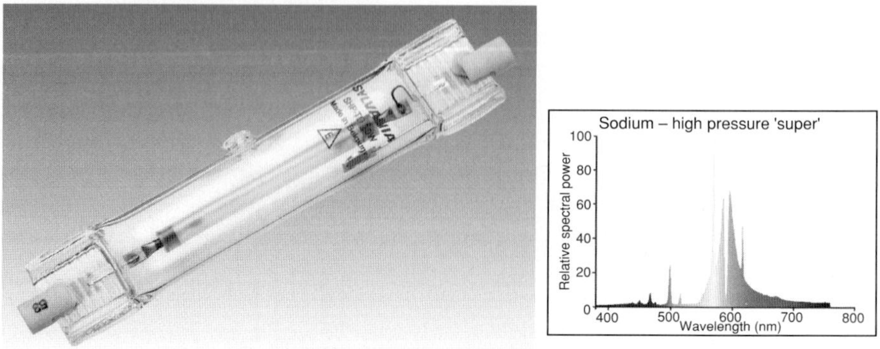

Fig. 1.27. High-pressure sodium lamp: 150 W, $\eta = 100$ lm W^{-1}, CCT 2,000 K, CRI 25, 132 mm overall length [92]

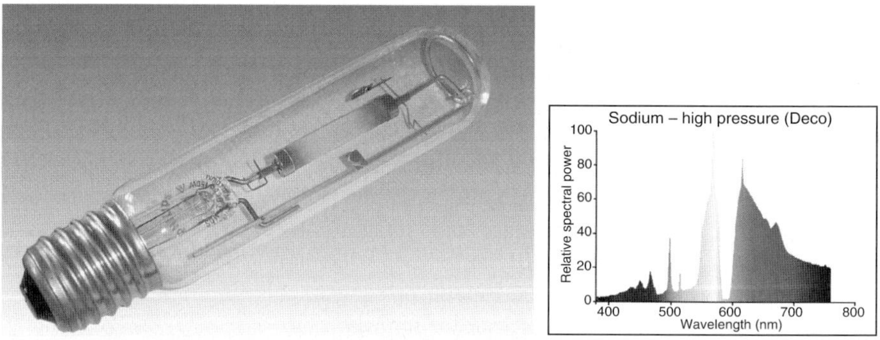

Fig. 1.28. High-pressure sodium lamp: 150 W, $\eta = 48$ lm W^{-1}, CCT 2,500 K, CRI > 80, 211 mm overall length ([92], the spectral power distribution is from a comparable high pressure sodium lamp)

Unfortunately, the luminous efficacy decreases with increasing sodium pressure due to the lower sensitivity of the human eye for blue and red light compared to yellow light. Hence, the sodium pressure has to be optimized for good color rendering *or* high luminous efficacy. This a major difference to high-pressure mercury lamps, where increasing pressure increases luminous efficacy *and* improves color rendering (cf. Sect. 1.3.6). Figure 1.27 shows a high-pressure sodium lamp optimized for high luminous efficacy, emitting the characteristic golden light of high-pressure sodium lamps ($\eta = 100$ lm W^{-1}, CCT 2,000 K, CRI 25). In Fig. 1.28 a "white" high-pressure sodium lamp is shown, optimized for good color rendering and therefore with lower luminous efficacy ($\eta = 48$ lm W^{-1}, CCT 2,500 K, CRI > 80).

The high pressure of the sodium increases the probability of absorption of resonance radiation in the outer and colder region of the plasma, resulting in the so

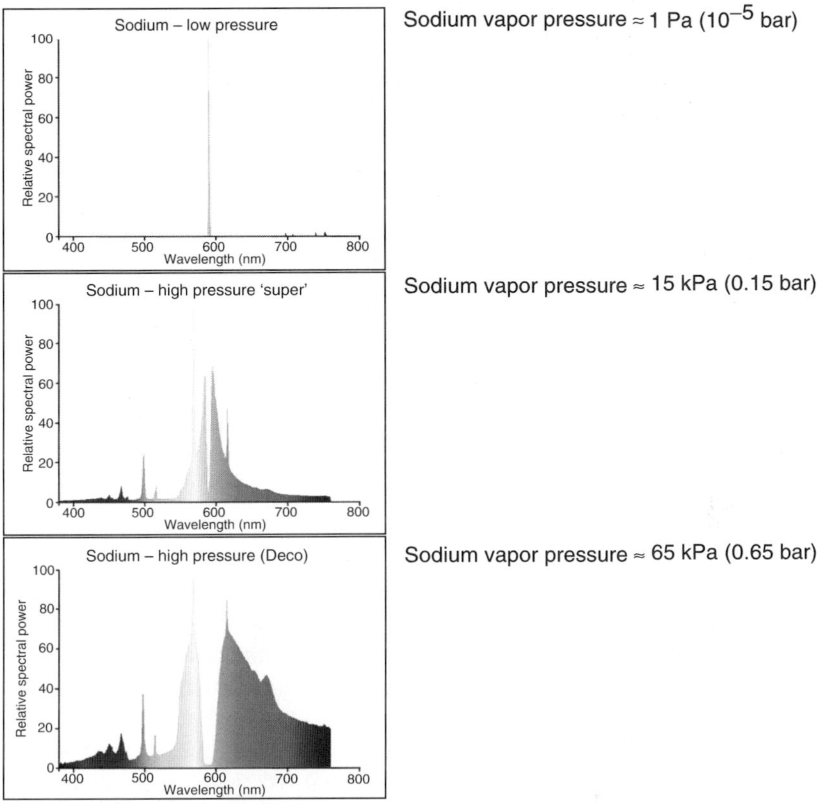

Fig. 1.29. Effect of sodium vapor pressure on the spectral power distribution of low-pressure and high-pressure sodium discharge lamps [92]

called self-reversal of the sodium resonance lines. Since the self-reversal occurs in the wavelength range where the human eye is most sensitive, this effect also reduces the luminous efficacy with increasing pressure (cf. Fig. 1.29 [80, 117, 118]).

The optimum vapor pressure of high-pressure sodium discharge lamps is around 10^4 Pa, which is low compared to high-pressure mercury lamps (cf. Fig. 1.12). The consequence is a low voltage gradient, i.e., small electrical field strengths within the discharge. Thus, a high electrical power input requires large discharge tube dimensions. To increase the voltage gradient and by that the power density, a so-called buffer gas can be added. Since mercury is very volatile and has a large cross section for elastic electron scattering, mercury is the first choice. The radiation is still dominated by sodium, since the resonance lines of sodium need less excitation energy than the mercury resonance lines. A typical spectral power distribution of such a high-pressure sodium lamp with mercury as the buffer gas is shown in Fig. 1.30. Adding 100 kPa (1 bar) of mercury to 15 kPa of sodium doubles the voltage gradient from 7.5 Vcm^{-1} to 15 Vcm^{-1}. The mixture of sodium and mercury

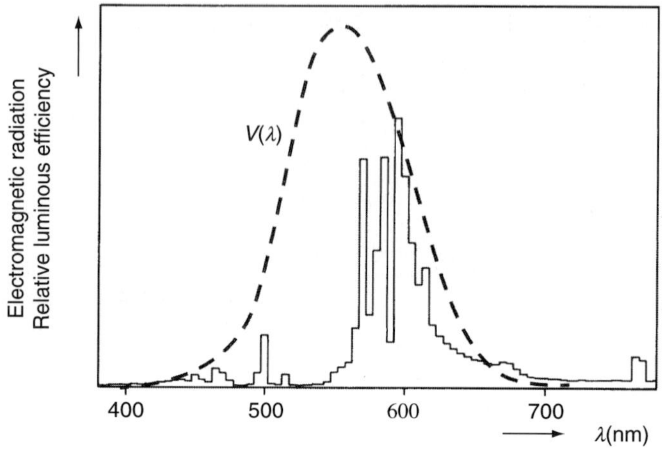

Fig. 1.30. Typical spectral power distribution of a high-pressure sodium lamp with mercury as a buffer gas [118] superimposed by the sensitivity curve $V(\lambda)$ of the light adapted human eye

is typically filled in the lamp in the form of a sodium amalgam. Since high-pressure sodium discharge lamps are typically operated with a saturated vapor pressure, the correct adjustment of the cold spot temperature is important for efficient lamp operation. A second possibility for a buffer gas is xenon, which is used as a starting gas as well [118].

To obtain the high sodium vapor pressures, wall temperatures of around 1,500 K must be obtained[35]. These high wall temperatures and the chemically very active and aggressive sodium are a problem for discharge tube materials (usually quartz) used for other low and high-pressure discharge tubes. The solution to this problem is translucent polycrystalline sintered alumina (aluminium oxide Al_2O_3), a ceramic material withstanding corrosion by hot molten and gaseous sodium and able to operate at the high temperatures necessary. Typical plasma temperatures in high-pressure sodium discharge lamps (depending on lamp current and pressure) are between 3,300 and 4,500 K for the plasma between the electrodes in the so-called plasma column [80, 99].

The high-pressure sodium lamp has its main application area (as the low-pressure sodium lamp) in the field of outdoor lighting, especially motorway, street, tunnel, or car-park lighting and illumination of buildings. Other fields of application are lighting of indoor sport fields or indoor lighting of factory buildings. The main advantage of high-pressure sodium lamps compared to low-pressure sodium lamps is the higher color rendering index combined with a sufficiently high luminous efficacy (cf. Tables 1.1 and 1.2).

---

[35] Sodium is less volatile than mercury, i.e., even for relatively small pressures of around 10^4 Pa, high cold spot temperatures are necessary.

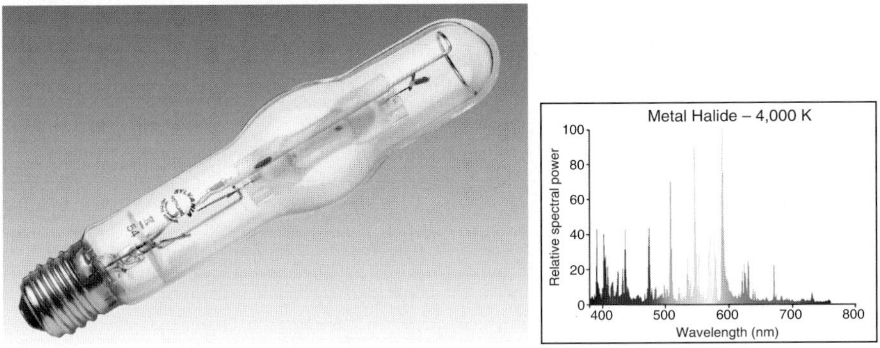

Fig. 1.31. Metal halide lamp: 420 W, $\eta = 76$ lm W^{-1}, CCT 5,200 K, CRI >90, 285 mm overall length [92]

1.3.8 Metal Halide Lamp

The high-pressure mercury lamp is an efficient high-pressure discharge lamp with the main disadvantage of unevenly distributed spectral lines. This leads to medium luminous efficiencies and poor color rendering with high correlated color temperatures (cf. Tables 1.1–1.3). The transformation of UV radiation from the resonance lines of mercury may improve this situation, but this transformation always goes hand in hand with energy losses. A solution to this problem would be the addition of metals to the discharge, because they have in general a lot of atomic lines in the visible part of the electromagnetic spectrum (Fig. 1.31). Unfortunately, most metals have low vapor pressures and attack the quartz glass at the high temperature necessary for high pressure operation. A solution to this problem is the usage of metal halide salts, because the halogen compounds have higher vapor pressures than the metals themselves (cf. Fig. 1.32). A blend of sodium and scandium iodides is widely used in the USA and Japan, while mixtures of indium (blue), thallium (green), and sodium iodides are popular in Europe. Mixtures including rare-earth iodides (e.g., dysprosium, holmium, thulium) are also used [111, 118].

The additional metal atoms in the discharge have several advantages [118]. The visible lines emitted by the metals are often resonance lines. Since resonance lines require the least energy to become excited, the luminous efficacy increases. Moreover, resonance lines are usually very strong lines, thus vapor pressure does not need to be too high. Furthermore, the color rendering index can be increased and adjusted by choosing appropriate combinations of different metal halides comparable to the fluorescent powder in fluorescent lamps. A typical spectral power distribution of such a (high-pressure) metal halide lamp (MH) is shown in Fig. 1.33. Another advantage is the high voltage gradient due to the high-pressure mercury (buffer) gas. This allows high electrical power densities and small lamp currents. Unfortunately, most common additives to metal halide lamps act as getters for either oxygen or halogen and, therefore, do not allow the operation of a halogen cycle to prevent the lamp walls from blackening as in the case of the UHP lamp (cf. Sect. 1.3.6 and [45]).

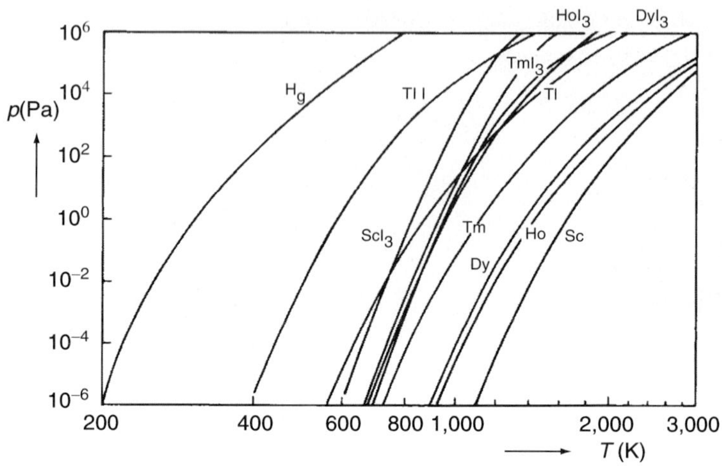

Fig. 1.32. Vapor pressure of mercury compared to vapor pressure of some metals and metal halides (here iodides) vs cold spot temperature. The vapor pressure of the metal iodides is higher than that of the corresponding metals, but lower than that of mercury [118]

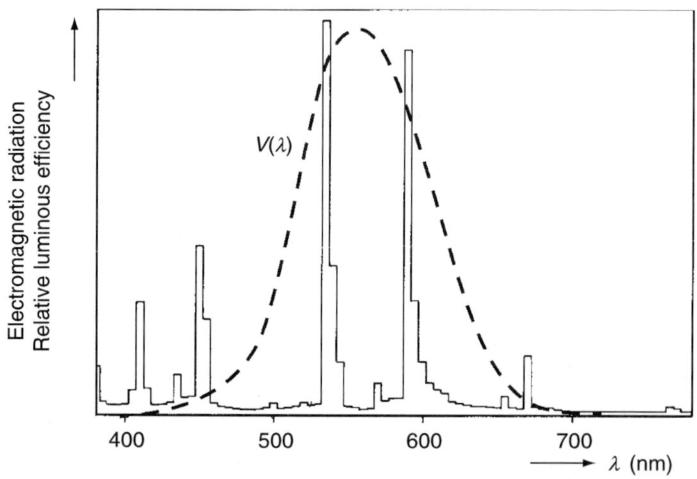

Fig. 1.33. Typical spectral power distribution of a (high-pressure) metal halide lamp [118] superimposed by the sensitivity curve $V(\lambda)$ of the light adapted human eye

A prominent example of metal halide lamps is the new automotive headlight lamp: The D2 or Xenon lamp (cf. Fig. 1.34). The first generation of D2 lamps was introduced in 1991/1992.[36] This novel automotive headlight lamp brought a

---

[36] The D2 lamp was developed within the framework of the VeDiLiS (vehicle discharge lighting system) project of the EU. A first prototype of the D2 lamp was introduced 1990 [92].

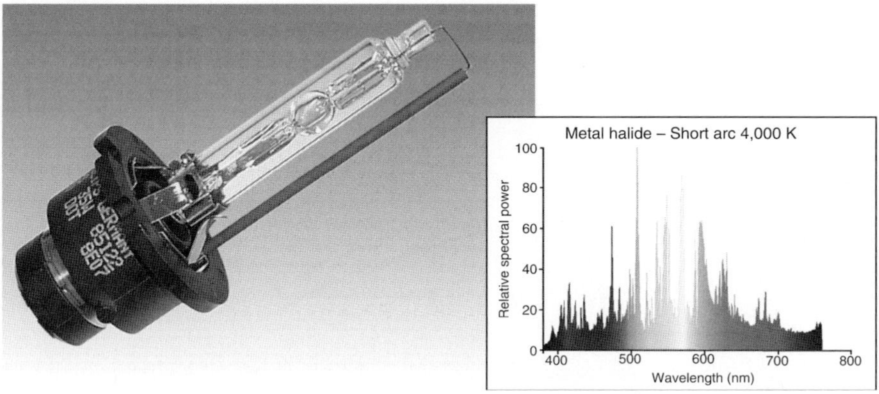

Fig. 1.34. Metal halide lamp: D2 automotive headlight lamp, 35 W, $\eta = 91$ lm W^{-1}, CCT 4,100 K, CRI 65, 4 mm arc gap ([92], the spectral power distribution is from a comparable short arc metal halide lamp for general lighting)

quantum leap with respect to light intensity and lumen output (about 3,200 lm) compared to halogen (incandescent) lights (about 1,500 lm) with power consumption of only 35 W instead of about 50 W in the case of halogen lights (H7). This increase in light intensity can be used e.g., to offer different light distributions depending on the traffic situation like town light, cross country light, and motorway light (adaptive headlamp technologies [67]).

The filling of the D2 lamp consists of xenon (several bar in the cold lamp, about 50 bar operating pressure), mercury (approximately 20 bar operating pressure), NaI (operating pressure about 0.03 bar), and ScI_3 (about 0.05 bar operating pressure). Xenon is responsible for the so-called "Xenon-Instant-Light" at the cold start of the lamp, to provide the required minimum output of 800 lm one second after switch on. After a couple of seconds the mercury and the metal halides are vaporized. Mercury increases the impedance of the discharge, making it possible to operate the lamp at low currents (0.4 A, 400 Hz rectangular-wave current) and high voltages (85 V), which is desirable with respect to low strain and long lifetime of the electrodes. Moreover, the high voltage gradient enables high electrical power input into the small electrode gap of 4 mm required for the operation in automotive headlight systems. The excitation energies of sodium and scandium lie mainly in the visible part of the electromagnetic spectrum, rendering light output largely in the visible [59].

Typical temperatures in the D2 lamp are electrode tip temperatures of about 2,500 K, plasma temperatures of about 5,000–5,500 K (so-called plasma column) up to 7,500 K in the very hot plasma immediately in front of the electrode tips (hot plasma spots). A typical temperature distribution in plasma and electrodes calculated for a D2 lamp is shown in Fig. 1.35. More details concerning the D2 automotive headlight lamp can be found in [59, 67, 71, 85, 86, 92, 100] and Sect. 4.4.6.

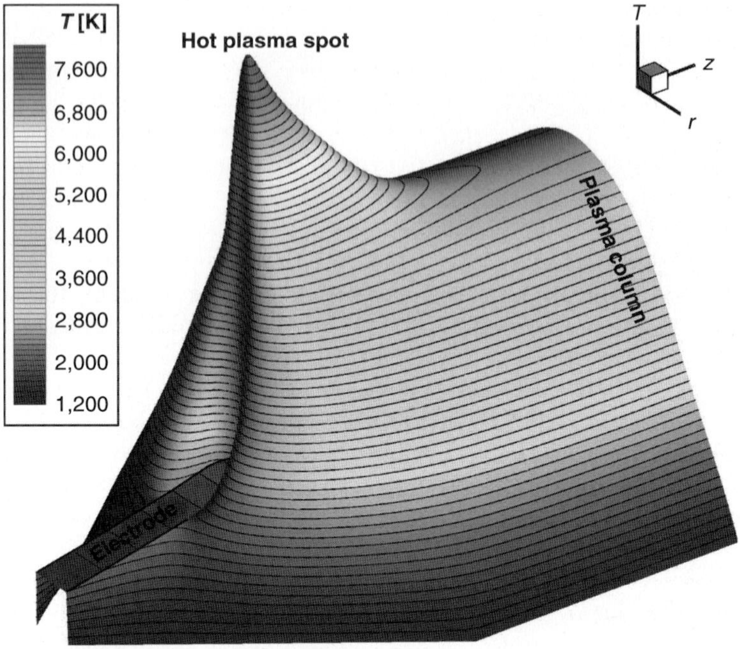

Fig. 1.35. Calculated temperature distribution (plasma and electrodes, rotational symmetry) near electrode tip during anode phase with hot plasma spot and plasma column in a D2 automotive headlight lamp (Xe/Hg/NaI/ScI_3, $p = 7$ MPa, $I = 0.4$ A, 400 Hz rectangular-wave lamp current, $\varphi_{emission} = 2.6$ eV, $\varphi_{cool/heat} = 4.5$ eV [59], © 2004 IEEE)

The high temperature and the high thermal heat load onto the lamp walls is a general problem using high-pressure discharge lamps. The lamp material not only has to withstand the high temperatures and the high pressure without quartz recrystallization, softening of the quartz glass, or bursting, but also has to be gas tight. A special problem in this context is the low coefficient of expansion of quartz. The metallic electrodes necessary for the feed through of the electrical current have much higher coefficients of expansion. Due to the large temperature difference between the operating and the switched off lamp, this makes it difficult to seal the current leads gas tight. This problem is solved by the use of thin molybdenum foils, having a thickness of about 20–40 μm (good visible in Figs. 1.26, 3.19, and 1.27 (outer bulb), also visible in Figs. 1.22, 1.31, and 1.34, cf., e.g., [70]).

A new trend for metal halide lamps is the usage of polycrystalline alumina (PCA) for the burner instead of quartz (ceramic metal halide lamps (CMH)). This ceramic material, which brought the major breakthrough for high-pressure sodium lamps (cf. Sect. 1.3.7), allows for higher wall temperatures as compared to quartz (over 1,500 K compared to 1,300 K for quartz). This implies the opportunity to operate lamps with higher metal halide vapor pressures, resulting in improved efficacies and better color rendering. Moreover, polycrystalline alumina has a high chemical resistance. The disadvantage of PCA is that it is not transparent, but only translucent. Translucent materials allow light to pass through them only diffusely, thus cannot be clearly seen through. Transparent materials (e.g., quartz) can be seen through, i.e., they allow images to pass undisturbed. The reason for this is the microstructure of PCA, which consist of many randomly distributed grains. As light passes through the PCA it is scattered at the many grain boundaries along the optical path due to different effective refractive indices (birefringence). Other factors responsible for light scattering are surface roughness and residual pores. New techniques allow the production of transparent ceramic burners either by using single crystal PCA (no grain boundaries) or submicron-grained structures (grain size less than 1 μm). These new materials open the field of focused-beam applications (such as projection lamps) to ceramic metal halide lamps [81, 97, 98, 111, 166].

Mercury is a very important element for discharge lamps. Low-pressure mercury or fluorescent lamps, high-pressure mercury, metal halide, and even high-pressure sodium lamps contain (usually) mercury as the active (i.e., emitting) gas component or as a buffer gas. In recent years an increasing tendency to reducing or replacing the environmental critical mercury (mercury is toxic) in discharge lamps could be observed.

Improvements in the production process of fluorescent lamps, where mercury is overdosed, were able to reduce the quantity of mercury considerably. Based on a 1999 NEMA survey (National Electrical Manufacturers Association, USA), the average 120 cm fluorescent lamp contains about 11.6 mg mercury. This quantity has been steadily reduced, for example by introducing mercury with the help of a closed capsule containing mercury instead of filling the mercury directly in the still hot lamp during the lamp manufacturing process. The latter method made it necessary to use more mercury than needed and some of the mercury was driven out of the lamp during the further production process entering the environment. In contrast, the mercury capsule was opened as the last step in the production process, avoiding the problems described earlier. The average 120 cm fluorescent lamp contains today over 75% less mercury than the same lamp would have contained in 1985 [118, 125, 164].

The replacement or reduction of mercury in high-pressure discharge lamps is also an important up-to-date issue. The D2 automotive headlight lamp, e.g., contains approximately 0.5 mg mercury. New mercury free alternative products are currently under investigation [27, 28, 81, 129, 158]. In 2001, US lamp manufacturers[37] used 9 tons of mercury in all lamps produced, a 67% reduction from the 27 tons manufacturers used in 1990 [126].

Nevertheless, replacing or reducing the amount of mercury without providing sufficient luminous efficacy would be counterproductive. The additional power consumption must be produced in power stations, many of them using coal or other fossil fuel containing mercury. This mercury would enter the environment with the combustion products. Replacing fluorescent lamps with incandescent lamps, the

[37] NEMA lamp manufacturers.

twentyfold amount of mercury would enter the environment compared to the amount of mercury contained in the fluorescent lamps, if the additional power consumption of the less efficient incandescent lamps would be provided by burning fossil fuels [118]. Moreover, the use of energy-efficient light sources reduces the amount of greenhouse gases, toxic pollutants, and gases that form smog and acid rain.

Due to the flexible adjustment of the spectral power distribution, the correlated color temperature, and the color rendering index by choosing the appropriate mixture of metal halides, the application areas of metal halide lamps are very diversified: automotive headlight lamps, commercial lighting, general lighting (indoor and outdoor), video and data projection, studio lighting, etc. [71, 92]. An application area with large future potential is the roadway lighting with metal halide lamps. For Europe, energy savings of more than 20 TWh might be expected, corresponding to the energy production of several large power stations and a reduction of CO_2 emission by more than 10 million tons per annum [139].

1.3.9 Other Discharge Lamps

The low-pressure and high-pressure discharge lamps discussed in the earlier sections are the most common discharge lamps used today. Nevertheless, there are many more discharge lamps used for special applications. One example is the high-pressure krypton lamp used for optical pumping of neodymium doped crystal laser. This laser has a strong absorption around 810 nm, perfectly matched by the emission lines of krypton at 810, 811, and 819 nm. As a light source however, the krypton discharge is rather inefficient, achieving only some 30 lm W^{-1} [92, 95].

Other applications of discharge lamps and especially high-pressure discharge lamps are: super high-pressure xenon lamps for microscopy and movie projection (e.g., [134]), xenon high-pressure lamps for light houses (see book cover) and as searchlight lamps, cadmium or zinc lamps for spectroscopy, sulfur electrodeless lamp for general lighting, low and high-pressure molecular discharge lamps (e.g., sulfur S_2, selenium Se_2, tellurium Te_2, etc.), and many more lamps for special application areas [71, 89, 92, 108].

2

Plasma and Electrode Physics

Chapter 1 introduced high-intensity discharge (HID) lamps, explained the basic principles, and pointed out the differences compared to low-pressure discharge lamps. A deeper insight into the gas or plasma properties and the characteristics of the electrodes will be given in this chapter. This insight helps to understand HID lamps in general and the topics addressed briefly in Chap. 1, e.g., the spectral power distribution of HID lamps or the electrical conductivity of different plasma compositions. Moreover, the plasma and electrode properties discussed within this chapter are the foundations for the understanding of the following two chapters, dealing with the experimental and theoretical investigations of HID lamps (Chaps. 3 and 4). They are necessary for carrying out numerical simulations and to analyze and interpret experimental results. Section 2.1 deals with general gas laws, which are not restricted to plasma physics, Sects. 2.2 and 2.3 are concerned with plasma physics, i.e., with the properties of the plasma in the inside of HID lamps. Finally, Sect. 2.4 deals with the properties of the electrodes in HID lamps.

2.1 Gas Laws

The gas laws discussed in this section are general laws which are not restricted to plasma physics. They provide the basic principles and form the starting point for further investigations of high-pressure plasmas.

2.1.1 Boltzmann Distribution

The Boltzmann or Maxwell–Boltzmann distribution is a probability distribution with many applications in physics and chemistry. It forms the basis of the kinetic theory of gases, which explains many fundamental gas properties. It is used to calculate electron or ion densities, the density of atoms in excited states can be calculated with the Boltzmann distribution, the Planck formula for black body radiation or the barometric formula can be derived with it, etc.

The Boltzmann distribution is the classical probability distribution function for distribution of an amount of energy between identical but distinguishable particles in a closed system. It is assumed that there is no restriction on the number of particles which can occupy a given state[1] and that a thermal equilibrium exists, i.e., the particles interact with each other by elastic collisions and exchange energy without energy loss from or energy input into the system. In such a system, the probability of a particle to have the energy E_i is given by

$$\frac{N_\mathrm{i}}{N} = \frac{\mathrm{e}^{-\frac{E_i}{kT}}}{\sum_\mathrm{j} \mathrm{e}^{-\frac{E_j}{kT}}} \tag{2.1}$$

where N_i is the number of particles at equilibrium temperature T having energy level E_i, N is the total number of particles in the system, and $k = 1.38066 \times 10^{-23}$ J K^{-1} is the Boltzmann constant. The sum in the denominator of the fraction is a normalization constant (partition function), providing that the sum over all N_i equals the total number of particles. Equation (2.1) provides a means for calculating the fraction of particles (N_i/N) that have energy E_i at a given temperature T. With increasing energy E_i it is less likely that any particle will attain that energy, so more particles will be found with lower energies (cf. Fig. 2.1). Increasing temperature increases the probability of particles with higher energies. The energy might, for example, be the excitation energy of an atom (giving the density of atoms in excited states), potential energy of gas atoms (resulting in the barometric formula), or kinetic energy (giving the velocity distribution of gas atoms).

2.1.2 Population Density of Excited Atoms

To calculate the (thermal equilibrium) density of atoms in an excited state of energy E_i at a given temperature T with the Boltzmann distribution (2.1), we need to know the possible discrete energy levels E_i (cf. "Line Radiation" in Sect. 2.2.3) and their statistical weights[2] g_i. Then, we are able to calculate the density n_i of atoms being in the excited state with energy E_i by

$$n_i = n_0 \frac{g_i}{Z} \mathrm{e}^{-\frac{E_i}{kT}}, \tag{2.2}$$

where n_0 is the total particle density of atoms and Z the so-called partition function, i.e., the normalization constant described earlier. Z is the sum over all possible $g_i \exp(-E_i/(kT))$ and is therefore a function of temperature T. Z ensures that the sum

---

[1] Examples: (1) In a plasma, the number of atoms in a certain excited state is not restricted. (2) The number of particles in a gas/plasma with a certain velocity is not restricted. Restricted are, for example, the number of electrons in metals having a certain velocity [3].

[2] More than one quantum state may have the same energy E_i. The number of possible ways to obtain the excitation energy E_i is the statistical weight g_i, also known as the degeneracy of available states for a given energy level E_i.

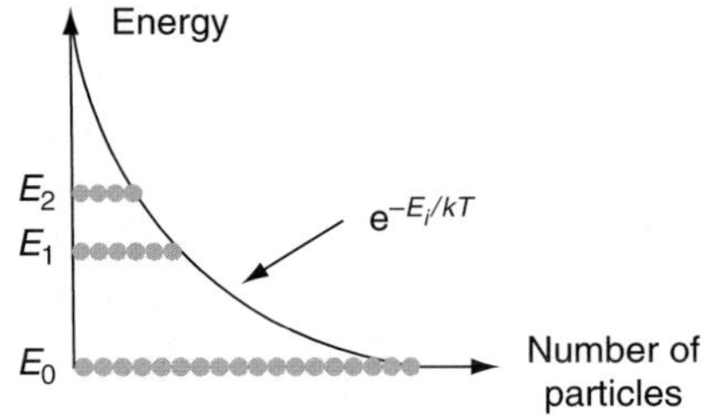

Fig. 2.1. Illustration of the Boltzmann probability distribution

over all possible n_i equals the total particle density of atoms n_0. We will need this population density of energy levels to calculate the line radiation from a high-pressure plasma (cf. "Line Radiation" in Sect. 2.2.3).

2.1.3 Maxwell Velocity Distribution

The velocity is also related to energy, i.e., (2.1) can be used to derive a relationship between temperature and velocity of atoms or molecules in a gas or plasma. Assuming an ideal gas (cf. Sect. 2.1.5), the energy E_i is given by the kinetic energy $0.5mv^2$, i.e., we have a continuous spectrum of energies instead of discrete energy states as in an exited atom (cf. Sect. 2.1.2). Thus, we switch from discrete population densities n_i to a continuous distribution function $f(v)$ where $f(v)\,\mathrm{d}v$ is the probability for a particle to have a velocity in the range between v and $v + \mathrm{d}v$. The statistical weight g of the velocity in the range between v and $v + \mathrm{d}v$ is given by $4\pi v^2\mathrm{d}v$, i.e., the "volume" of the spherical shell in the velocity space with radius v and thickness $\mathrm{d}v$. This means that we assume that the velocity space can be uniformly occupied by the velocity vectors. If the kinetic energy v_2 is $2v_1$, then the statistical weight or the number of possible ways to obtain the corresponding kinetic energy $E_2 = 0.5m(v_2)^2$ is four times as high as the statistical weight for the kinetic energy $E_1 = 0.5m(v_1)^2$, as the statistical weight goes with v^2. This means that we have four times more possibilities to find velocity vectors of norm or magnitude v_2 than possibilities to find velocity vectors of norm or magnitude v_1. An illustration of this statistical weight is given in Fig. 2.2.

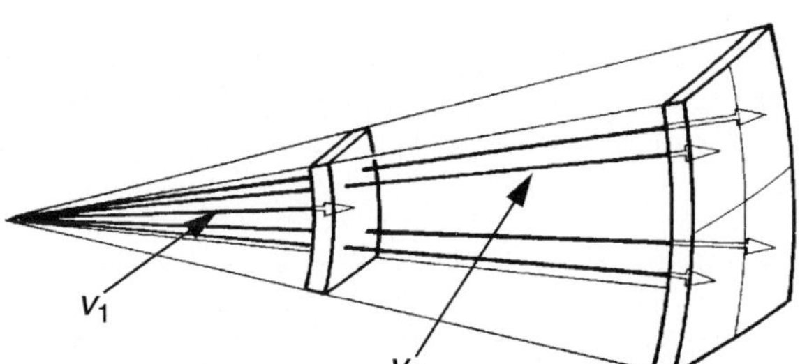

Fig. 2.2. Illustration of the increase of the statistical weight of the kinetic energy with v^2 [72]

The resulting probability to find a particle with a velocity in the range between v and $v + \mathrm{d}v$ is thus

$$f(v)\,\mathrm{d}v = C\,4\pi v^2\,\mathrm{e}^{-\frac{mv^2}{2kT}}\,\mathrm{d}v. \tag{2.3}$$

The constant C corresponds to the partition function and has to ensure that the sum (integral) over the probability of all possible velocities equals one:

$$\int_0^\infty f(v)\,\mathrm{d}v = 1. \tag{2.4}$$

After calculating C with (2.4) we get the so-called Maxwell velocity distribution, i.e., the probability to find a particle with a velocity in the range between v and $v + \mathrm{d}v$:

$$f(v)\,\mathrm{d}v = \sqrt{\frac{2}{\pi}}\left(\frac{m}{kT}\right)^{\frac{3}{2}}\mathrm{e}^{-\frac{mv^2}{2kT}}\,\mathrm{d}v. \tag{2.5}$$

The probability of finding particles with certain finite velocities is thus greater than zero. Nevertheless, the average velocity of all particles is zero, since the directions of the velocities are distributed randomly.

An illustration of the resulting particle velocities in air (N_2) and xenon is given in Figs. 2.3 and 2.4, respectively. Comparing the probability functions for the velocity distribution of these two gases at $T = 1{,}000$ K for the velocity of 1,000 m s^{-1} one reads 10^{-3} s m^{-1} for N_2 and almost zero for Xe. The reason for this difference is the different mass[3] of N_2 ($m = 28.0$ u) and Xe ($m = 131.3$ u) and that the Maxwell velocity distribution is derived from an energy distribution. Since both gases have the same energy distribution at the same temperature T, the resulting velocity distribution must be different because of the relation $E_{\mathrm{kin}} = 0.5mv^2$. The higher the mass, the smaller the corresponding velocity for the same kinetic energy. This difference is very pronounced comparing the velocity distribution of electrons and atoms. For example, the mass of a xenon atom is approximately 240,000 times higher than the mass of an electron. This is important in high-pressure plasmas, where electrons, atoms, and ions are supposed to have the same energy distribution (cf. Sect. 2.1.8). Despite this equal energy distribution, the resulting velocity distributions are quite different comparing electrons and heavy particles (i.e., atoms and ions).

Using (2.5) we can calculate the most probable (norm or magnitude of) velocity $\hat{v}$ (maximum of Maxwell velocity distribution $f(v)$), the average or mean (norm or magnitude of) velocity $\overline{v}$ (by $\int_0^\infty v f(v)\mathrm{d}v$), and the root mean square velocity $\sqrt{v^2}$ (by $\int_0^\infty v^2 f(v)\mathrm{d}v$):

$$\hat{v} = \sqrt{\frac{2kT}{m}}, \tag{2.6}$$

[3] The mass is given in atomic mass units $\mathrm{u} = 1.66057 \times 10^{-27}$ kg ($= 1/N_{\mathrm{A}}$g, with $N_{\mathrm{A}} = 6.022045 \times 10^{23}$).

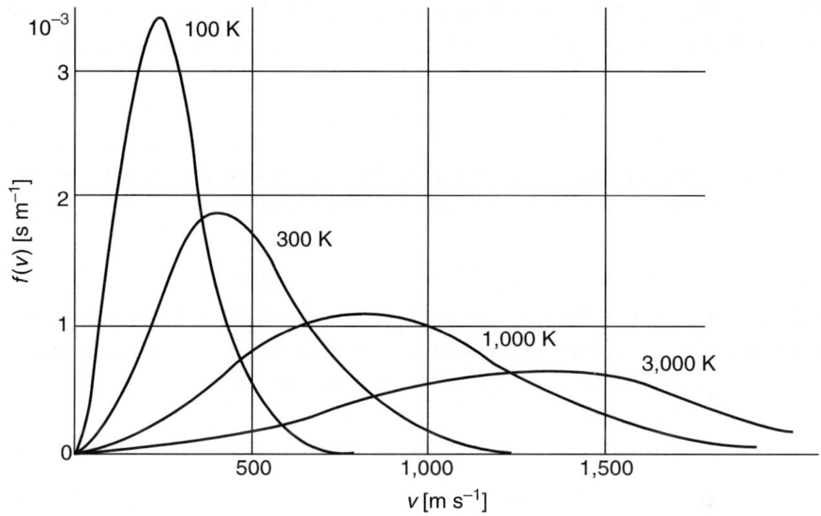

Fig. 2.3. Maxwell probability distribution for the velocity of N_2 molecules (air) for four different temperatures [72]

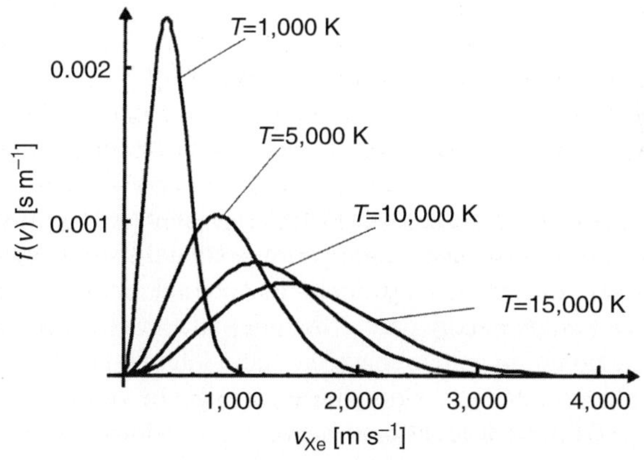

Fig. 2.4. Maxwell probability distribution for the velocity of Xe atoms for four different temperatures [171]

$$\overline{v} = \sqrt{\frac{8kT}{\pi m}} = 1.128\,\hat{v}, \tag{2.7}$$

$$\sqrt{\overline{v^2}} = \sqrt{\frac{3kT}{m}} = 1.225\,\hat{v}. \tag{2.8}$$

As described above, the mass has an important effect on the velocities. Moreover, the higher the temperature, the higher the velocities and thus the kinetic energy (cf. Sect. 2.1.4).

Converting the Maxwell velocity distribution given in (2.5) to an energy distribution, we have to transform the infinitesimal small dv into an appropriate dE. Since dv/dE = $(2E/m)^{-0.5}$, we get the following energy distribution function:

$$f(E)\mathrm{d}E = 2\sqrt{\frac{1}{\pi}}(kT)^{-3/2}\sqrt{E}\,\mathrm{e}^{-E/kT}\mathrm{d}E. \quad (2.9)$$

Here, $f(E)$dE is the probability to find a particle with the energy in the range between E and E + dE. This distribution function depends only on the temperature but not on the mass of the particles under consideration. This is consistent with the considerations given earlier.

The Maxwell velocity distribution or the resulting mean or root mean square velocities will be used in the calculations of many plasma properties, such as mean free path, pressure, or plasma transport coefficients.

2.1.4 Temperature

Temperature can be defined as a measure of the average translational kinetic energy associated with the disordered microscopic motion of atoms and molecules. The temperature defined from kinetic theory is called the kinetic temperature. For each degree of freedom, the average energy is given by 0.5kT, i.e., the translational kinetic energy with three degrees of freedom (e.g., x, y, z) is given by

$$E_{\mathrm{kin}} = \frac{1}{2}m\overline{v^2} = \frac{3}{2}kT. \quad (2.10)$$

This is a very general definition of temperature. It can be obtained, for example, by using (2.8) to replace the $\overline{v^2}$ in (2.10), i.e., the Maxwell velocity distribution and the definition of the temperature given earlier are consistent. If the temperature of two objects is the same, then their average translational kinetic energy is the same. Temperature is not directly proportional to internal energy, since temperature measures only the kinetic energy part of the internal energy, so two objects with the same temperature do not in general have the same internal energy. If a molecule has more degrees of freedom, e.g., the possibility to rotate or vibrate, the internal energy increases by 0.5kT for each additional degree of freedom. An even more general definition of the temperature is the following:

Temperature is a measure of the tendency of an object to spontaneously give up energy to its surroundings. When two objects are in thermal contact, the one that tends to spontaneously lose energy is at the higher temperature.

Temperatures are measured in Kelvin [K], sometimes Celsius [°C] is used. The Kelvin scale is called absolute temperature and the Kelvin is the SI unit for temperature. The temperature in Kelvin is related to the temperature in Celsius by the following simple relation:

$$T[\mathrm{K}] = T[^\circ\mathrm{C}] + 273.15. \tag{2.11}$$

2.1.5 Pressure or Ideal Gas Law

In an ideal gas all collisions between atoms or molecules are perfectly elastic and no intermolecular attractive forces are present. One can visualize the ideal gas as a collection of perfectly hard spheres which collide but which do not interact with each other otherwise. In such a gas, the entire internal energy is in the form of kinetic energy and any change in internal energy is accompanied by a change in temperature.

The pressure of a gas can be defined as the momentum transfer of the gas particles per unit time to an area A. Assuming (to simplify matters) that all particles have a fixed constant velocity v in arbitrary direction x, $-x$, y, $-y$, z, or $-z$, one-sixth of all particles in a virtual box with surface area A and side length $v\,\mathrm{d}t$ will reach the surface area A (cf. Fig. 2.5). If n is the particle density, the total number of particles in the box is $nAv\mathrm{d}t$, i.e., $nAv\mathrm{d}t/6$ particles will hit the wall. There, they bounce off with opposite direction of velocity, i.e., they change momentum by $2mv$. This change in momentum $\mathrm{d}I$ per unit time is transferred to the wall, i.e., the momentum transfer per unit time and area A is given by

$$p = \frac{\mathrm{d}I}{A\mathrm{d}t} = \frac{(1/6)n(Av\,\mathrm{d}t)(2mv)}{A\mathrm{d}t} = \frac{1}{3}nmv^2. \tag{2.12}$$

In a gas, the velocity of the particles can be described by the Maxwell velocity distribution, resulting in an average kinetic energy of $1.5kT = 0.5mv^2$ (cf. (2.10)). Combining (2.10) and (2.12), we get the so-called ideal gas law:

$$p = nkT. \tag{2.13}$$

Knowing for example pressure and temperature, we can calculate the particle density. If a plasma consists of different components (e.g., Hg, Hg^+, and e^- in a high-pressure mercury plasma or even more components in a metal halide lamp, see e.g., Figs. 2.6–2.8), the total pressure is the sum over the partial pressures p_i of the individual plasma components with particle densities n_i (Daltons law):

$$p = \sum_i p_i = \sum_i n_i kT. \tag{2.14}$$

The SI unit of pressure is Pa = kg $\mathrm{m}^{-1}\,\mathrm{s}^{-2}$. Other units of the pressure regularly used in the context of discharge lamps are bar (=10^5 Pa) and Torr (=133.322 Pa). Within the pressure and temperature range of the plasma in most high-pressure discharge lamps, real gas effects can be neglected and the pressure can be described by the ideal gas law [33, 168].

The ideal gas law will be used in the following to replace particle densities with pressure or to replace pressure by particle densities.

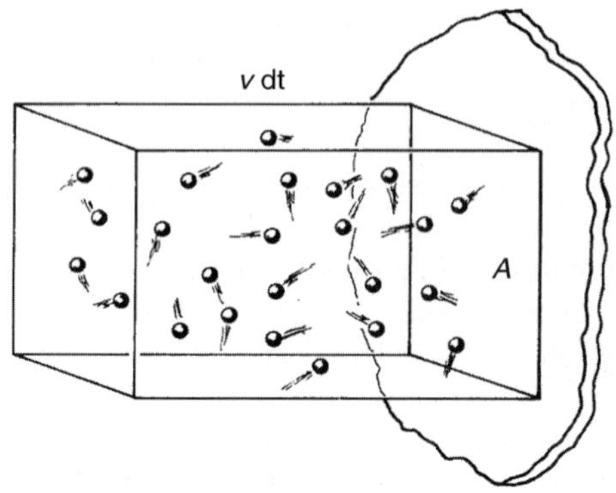

Fig. 2.5. Illustration of the pressure of an ideal gas [72]

2.1.6 Collision Frequency and Mean Free Path

The collision frequency and the mean free path or average distance between collisions for a gas atom or molecule may be estimated in the following way: Consider a single particle (atom, molecule, ion, or electron) moving fast through a gas consisting of atoms or molecules with small velocity compared to the fast particle. If the fast particle has radius r_1 and the gas particles have the radius r_2, the fast particle will hit a gas particle if the distance between the two particles perpendicular to the line of motion is less than $r_1 + r_2$. Thus, the fast particle can be treated as having the radius $r = r_1 + r_2$ while the gas particles are treated as point masses. The fast particle has the effective collision area or cross section with respect to the gas particles of $Q = \pi(r_1 + r_2)^2$. This cross section is often referred to by Q or σ. During the time Δt, the particle with the effective collision area of Q would sweep out a volume of $\pi r^2 \overline{v} \Delta t$ and the number of collisions during the time Δt can be estimated from the number N of gas molecules being in that volume: $N = nQ\overline{v}\Delta t$, where n is the particle density of the gas. If N is equal to one, exact one collision has been taken place (statistically). We can thus calculate the time τ between two collisions by setting $N = 1$ and get:

$$\tau = \frac{1}{nQ\overline{v}}. \tag{2.15}$$

The collision frequency f is thus

$$f = \frac{1}{\tau} = nQ\overline{v}. \tag{2.16}$$

After the time τ between two collisions, the fast particle has covered the distance ℓ, the mean free path:

$$\ell = \overline{v}\tau = \frac{1}{nQ}. \tag{2.17}$$

This illustrative derivation of the mean free path suffers from the assumption that the target gas particles are at rest. If the particle, which was the fast particle in the earlier derivation, and the particles of the gas have the same velocity distribution, the relative velocity between them must be taken into account. This yields an additional factor $\sqrt{2}$:

$$\ell = \bar{v}\tau = \frac{1}{\sqrt{2}nQ}. \tag{2.18}$$

Therefore, the higher the particle density, the smaller the mean free path.

2.1.7 Thermal Equilibrium

A gas or plasma is in (total) thermal equilibrium (or total thermodynamic equilibrium) if all particles constantly exchange energy, for example, by elastic and inelastic collisions, emission and absorption of radiation, etc. This constant energy transfer between all particles settles a uniform energy distribution among all particles. This results, for example, in an equal temperature of all constituents in the gas or plasma, especially the electron and heavy particle (atoms, ions, molecules) temperatures are equal. Thermal equilibrium requires that no energy is put into or taken out of the system, i.e., the gas or plasma is isolated. This results in a uniform temperature, i.e., there are no temperature gradients in a gas or plasma in thermal equilibrium. For a gas or plasma in thermal equilibrium, the Boltzmann distribution describes:

1. The velocity of all plasma or gas constituents (Maxwell probability distribution for the velocity, cf. Sect. 2.1.3)
2. The population density of excited states of all atoms or molecules in the gas or plasma (cf. Sect. 2.1.2)
3. The particle densities of all constituents which might be generated or annihilated by dissociation or formation of molecules (law of mass action, chemical equilibrium)
4. The electron and ion density (e.g., Saha equation cf. Sect. 2.2.2)
5. The radiation within the plasma (Planck radiation formula, cf. Sect. 2.4.3)

In thermal equilibrium, the thermal state of the plasma or gas can be calculated solely by knowing the temperature.[4] Needless to say that gas discharge lamps in general and especially high-pressure or HID lamps are not in (total) thermal equilibrium. Discharge lamps are used to generate radiation, i.e., energy is taken out of the system in terms of radiation energy and put (replaced) in the system by electrical power. In low-pressure discharge lamps (mercury or sodium), typical heavy particle or gas temperatures are between 300 and 700 K, the electron temperature is around or exceeds 10,000 K, i.e., the energy is far away from being uniformly distributed among all constituents (cf. Sect. 1.3.3). In HID lamps, the electron and heavy particle temperatures are approximately equal as shown in

[4] Ionization and dissociation energies, masses of all particles, etc. must be known as well.

Fig. 1.11. But there is a considerable temperature gradient from the plasma column or the hot plasma spots near the electrode tips to the inner walls of the quartz or ceramic envelope. Moreover, the radiation from gas discharge lamps cannot be described by the Planck radiation formula (continuous spectral power distribution of the radiation, cf. Sect. 2.4.3), but shows spectral lines in the spectral power distribution [48, 80, 138].

2.1.8 Local Thermal Equilibrium

In HID lamps, thermal equilibrium is not fulfilled, but with the notion of a *local* thermal equilibrium, *local* temperature equilibrium, or *local* thermodynamic equilibrium (LTE) the plasma in HID lamps can be described very effectively. In local thermal equilibrium, thermal equilibrium is fulfilled in infinitesimal small volumes only, in which the temperature is assumed to be constant. This allows for temperature gradients within the plasma as present in HID lamps. Moreover, the 5th point in Sect. 2.1.7 concerning radiation being described by the Planck radiation formula is replaced by calculating the plasma radiation and presuming that this radiation will escape from the infinitesimal small volumes (optically thin radiation, see also Sect. 2.2.3). The four other points of the thermal equilibrium (Points 1 to 4 in Sect. 2.1.7) are fulfilled in infinitesimal small volumes in local thermal equilibrium.

LTE can be assumed in HID lamps with pressure of 10^4–10^5 Pa (0.1–1 bar) upwards (depending on gas, current, etc.). In these high-pressure plasmas numerous collisions between electrons and atoms, ions, and molecules (heavy particles) provide sufficient energy transfer between electrons and heavy particles, leading to a uniform energy distribution among all constituents within an infinitesimal small volume. As a result, the electron and heavy particle temperatures are approximately the same ($T_e = T_h$). All plasma properties within these small volumes can be described using the Boltzmann distribution, i.e., velocities of plasma constituents, population densities of excited states of plasma particles, and particle densities including ions and electrons can be calculated using the local plasma temperature $T = T_e = T_h$. Moreover, radiation emission density and transport properties like electrical and thermal conductivity of the plasma can be calculated using particle densities, velocities, and population densities of excited states calculated with the Boltzmann distribution (cf. Sects. 2.2 and 2.3). The assumption of a collision-dominated plasma includes the preponderance of collisions compared to radiation processes, i.e., the radiation escaping from the plasma does not disturb the energy distribution among the plasma constituents and the population densities of excited states of plasma particles.

The requirements for local thermal equilibrium can be expressed in terms of equations, two of them will be given here to illustrate the assumptions given earlier. An important factor is that the temperature in an infinitesimal small volume can be assumed to be constant. But how small is "infinitesimal small"? Since collisions between electrons and heavy particles determine the energy transfer, the mean free path of electrons in the high-pressure plasma are a measure for "infinitesimal small", i.e., the temperature change along the mean free path should be small. In other words; the local temperature of the starting point of the electron should not differ from the temperature of the point where the next collision will take place:

$$\ell \operatorname{grad} T \ll T. \tag{2.19}$$

Typical values for the mean free path ℓ of electrons in, e.g., a 4 MPa (40 bar) mercury plasma are $\ell \approx 10$ nm, typical temperature gradients are given by temperature changes of 5,000 K over a distance of 1 mm, i.e., grad $T \approx 5\times10^6$ Km^{-1}. The temperature difference for an electron moving parallel to the temperature gradient would then be 0.05 K, i.e., (2.19) is fulfilled.

But not only the temperature difference between two collisions of an electron must be small, the energy gain of an electron in the electric field between two collisions must also be small. The energy gain can be described by $E_{kin} \approx \ell eE$ if the electron moves in the direction of the electric field. This energy must be small compared to the kinetic energy due to the thermal energy of the electron (cf. (2.10)), i.e.,

$$\ell eE \ll \frac{3}{2} kT. \tag{2.20}$$

Typical electric fields in high-pressure mercury lamps are between 1,000 and 60,000 V m^{-1}, i.e., with a mean free path of $\ell \approx 10$ nm or smaller, the energy gain of an electron in the electric field ranges between 10^{-5} to 6×10^{-4} eV. Compared to typical plasma temperatures of 4,000–7,000 K (corresponding to 0.52–0.90 eV), this is a small energy gain between two collisions.

Electrons gain energy more effectively compared to heavy particles in the discharge. The electric field accelerates ions and electrons, but the ions lose a substantial part of the kinetic energy at each elastic collision with another heavy particle due to the equal mass of the collision partners. The electrons lose only a very small percentage of their kinetic energy during elastic collisions with heavy particles due to the enormous mass difference between electrons and heavy particle. Thus, the electron temperature is always higher than the gas or heavy particle temperature. Only this temperature difference makes it possible to transfer energy by collisions from one group of particles (electrons) to another group of particles (heavy particles). Due to a very small number of collisions or a large mean free path of electrons, this temperature difference is very large in low-pressure discharge lamps ($T_h \approx 300$–700 K, $T_e \geq 10{,}000$ K, see earlier), i.e., local thermal equilibrium cannot be assumed in low-pressure discharge lamps. But in high-pressure discharge lamps, this temperature difference reduces to 1–100 K at discharge temperatures between 4,000 and 7,000 K. For plasmas with moderate temperature difference between electrons and heavy particles (i.e., $T_e - T_h \approx 200$–2,000 K), a so-called partial LTE (pLTE) concept can be applied. In pLTE, electrons and heavy particles are treated as different fluids (two-fluid model) with different temperatures $T_e \neq T_h$. The energy transfer between these two fluids can be treated as some kind of friction term in the otherwise decoupled energy or power balances of electrons and heavy particles [48, 53, 80, 138, 168].

2.2 Plasma Properties

In this section and in Sect. 2.3, plasma properties and transport coefficients of the plasma like particle densities, radiation emission densities, electrical conductivity, and thermal conductivity are calculated and discussed for different high-pressure discharges. These plasma properties and transport coefficients are helpful to understand HID lamps and necessary to carry out numerical simulations or to analyze and interpret experimental results. LTE is assumed in all the following sections.

2.2.1 Evaporation and Partial Pressure

In a rare gas high-pressure discharge lamp like the xenon or krypton high-pressure lamp no "filling material" is condensed at the inner lamp wall during operation. The same is true for certain high-pressure mercury lamps like the UHP lamp (cf. Sect. 1.3.6). If all the mercury is in the discharge and no condensed mercury is left at the inner tube wall, the mercury vapor is described as unsaturated. This means that the number of mercury atoms remains constant even if the cold spot temperature changes (as long as the cold spot temperature stays above the so-called saturation temperature T_s, which depends on the lamp volume and the amount of mercury filled into the lamp). Therefore, high-pressure mercury lamps with unsaturated mercury vapor are much less temperature dependent than most other discharge lamps.

Nevertheless, in the majority of high-pressure discharge lamps the vapor pressure of the filling materials is saturated, i.e., the filling material is evaporated only partly and a pool of liquid or solid filling material is left at the cold spot of the lamp. The vapor pressures above the melt of the different plasma constituents thus depend on the temperature of the cold spot. To get the plasma composition during operation, the cold spot temperature and the vapor pressures must be known or calculated. With the resulting partial pressures of the plasma constituents, other plasma properties and transport coefficients can be calculated (cf. Sect. 2.1.5). Of course, the partial pressures of the different plasma constituents may change considerably during ignition and the following run-up (heating up of the lamp and the walls by the plasma) until a stable situation is reached (cf. Figs. 1.25 and 1.32 and [75]).

2.2.2 Saha Equation

With the Saha or Eggert–Saha equation, electron and ion densities in plasmas can be calculated. The Saha equation is based on the Boltzmann distribution, taking into account that the ionized particles have energies composed of kinetic energy and ionization energy E_i. The resulting equation for the electron, ion, and atom densities (n_e, n_i, n_0) is given by

$$\frac{n_e n_i}{n_0} = \frac{2g_i}{g_0} \frac{(2\pi m_e kT)^{3/2}}{h^3} \exp\left(-\frac{E_i}{kT}\right). \tag{2.21}$$

g_i and g_0 are the statistical weights for ion and atom (typically $g_i/g_0 = 2$ since ions have mostly doublet structures, [141]), the 2 in front of the factor g_i originates from the statistical weight of electrons (two possible electron spins). The masses of ions and atoms are practically equal and reduce from the fraction. The ionization energy E_i measures the energy difference between ionized particles and neutral particles (atoms).

Apart from small areas near the electrodes (space charge region), the plasma is quasineutral, i.e., the density of electrons equals the density of all ions (weighted with the valency of the ions) at every point in the plasma. This can be translated to the equation $n_e = n_i$. Moreover, we can use the ideal gas law to replace the atom density by the pressure p in the case of small ionization degrees less than 1% (weakly ionized plasma). As a result we get

$$n_e^2 = p \frac{2g_i}{g_0} \frac{(2\pi m_e)^{3/2} (kT)^{1/2}}{h^3} \exp\left(-\frac{E_i}{kT}\right). \tag{2.22}$$

From this equation we can deduce that the electron (and ion) density is proportional to the square root of the pressure p and proportional to $\exp(-E_i/kT)$. Since the atom density is proportional to the pressure p, the ionization degree $n_e/n_0 = n_i/n_0$ decreases with increasing pressure proportional to one over the square root of pressure p. Increasing ionization energy E_i also leads to a lowering of the ionization degree. Ionization energies E_i of typical plasma constituents are 5.1 eV for sodium, 10.44 eV for mercury, and 12.13 eV for xenon.

The particle densities in a high-pressure mercury plasma of 4 and 8 MPa and a xenon plasma of 1 and 4 MPa vs. plasma temperature are shown in Figs. 2.6 and 2.7, respectively (logarithmic scale in Figs. 2.6–2.8). The ionization degree $n_e/n_0 = n_i/n_0$ is well below 1% for typical plasma temperatures between 5,000 and 8,000 K for these types of plasmas, i.e., the assumption of a weakly ionized plasma used for (2.22) is fulfilled. The neutral particle densities are proportional to the pressure, the ion and electron densities increase only with the square root of the pressure. Due to the higher ionization energy of xenon, the electron and ion densities are lower compared to a mercury plasma of the same pressure and plasma temperature.

Taking more complex plasma composition as in the D2 automotive headlight lamp (cf. Sect. 1.3.8) into account, the calculation of particle densities becomes more complex [75, 160, 171]. The particle densities can be calculated using chemical equilibrium calculations or by a law of mass action approach: Finding the LTE plasma composition in chemical equilibrium means that the second law of thermodynamics has to be obeyed. The entropy of the system consisting of many possible combinations of different particles (i.e., Na, I, NaI, Na_2I_2, etc. in a D2 lamp) must be maximized, i.e., one has to find the combination of given particles minimizing the free energy F (prescribed volume) or the Gibbs energy G (prescribed pressure). This has to be done taking further constraints into account. One of the constraints is the conservation of the total number of so-called basis

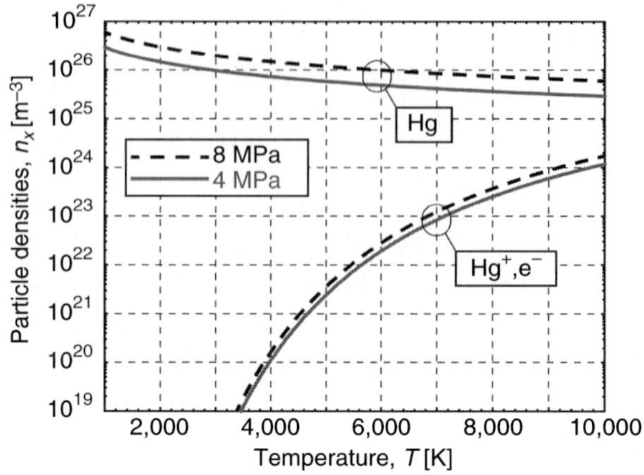

Fig. 2.6. Particle densities of Hg, Hg^+, and e^- ($n_i = n_e$) calculated using the Saha equation for a 4 and 8 MPa mercury plasma vs. plasma temperature [171]

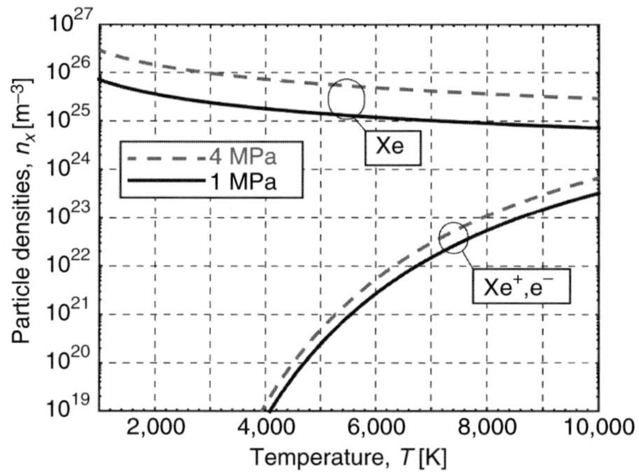

Fig. 2.7. Particle densities of Xe, Xe^+, and e^- ($n_i = n_e$) calculated using the Saha equation for a 1 and 4 MPa xenon plasma vs. plasma temperature [171]

particles, e.g., Hg, Xe, Na, Sc, I, and e^- in the case of the D2 lamp, as combinations of all these particles do not change the total numbers of these basis particles.

An example of the complex situation in such a D2 automotive headlight lamp is given in Fig. 2.8. According to the high partial pressures of 5 and 2 MPa (50 and 20 bar), the xenon and mercury atoms dominate the plasma. The overall ionization rate (electrons or ions compared to neutral particles) is well below 1% even at the highest temperatures in the hot plasma spots immediately in front of the electrodes (<8,000 K, cf. Fig. 1.35). The molecules ScI_3, Sc_2I_6, NaI, and Na_2I_2 are present at temperatures up to 3,000 or 4,000 K, then they dissociate and the Sc, Na, and I atoms outstrip the molecules (3,000–6,000 K). For even higher temperatures

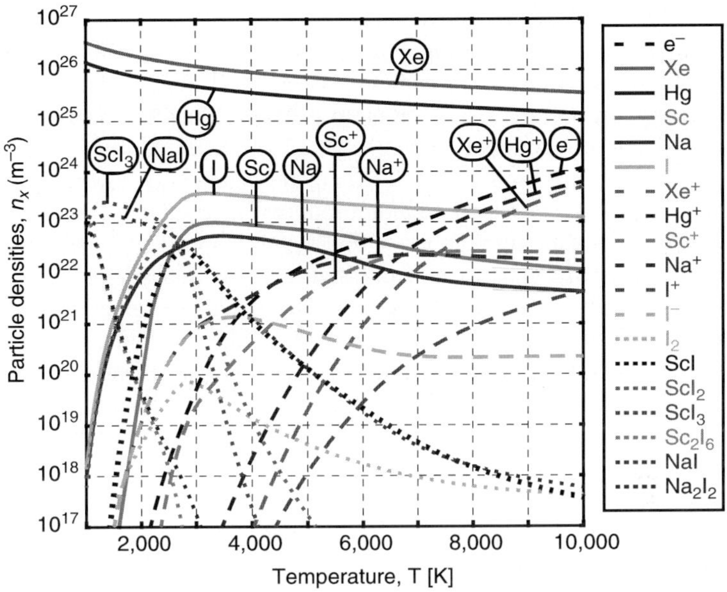

Fig. 2.8. Particle densities n_x of a D2 automotive headlight lamp plasma (Xe, Hg, NaI, ScI_3 with partial pressures of 50, 20, 0.03, and 0.05 bar, respectively,) vs. plasma temperature [171]

(6,000 K), the Sc^+ (ionization energy 6.54 eV) and Na^+ ions (ionization energy 5.12 eV) get ahead of the neutrals Sc and Na. The electron density, which is important for the electrical conductivity (cf. Sect. 2.3.2), is the sum over all ion densities. Within the temperature range up to 6,000 K, the electron density is determined by the ionization of Sc and Na, for higher temperatures, the ionization of Xe and Hg atoms becomes important (ionization energies of 12.13 and 10.44 eV, respectively).

2.2.3 Plasma Radiation

The goal of high-pressure discharge lamps is the efficient generation of light, so understanding and optimizing HID lamps requires the knowledge of the radiation processes in the plasma. The mechanisms of light generation in HID lamps include line emission, molecular radiation, and photon emission due to recombination and bremsstrahlung. The most important mechanism, the line radiation, will be discussed in the following in detail. Besides these light emitting processes, reabsorption of especially resonance radiation and line broadening mechanisms have to be taken into account. Taking all light emitting and absorbing mechanisms at the end of this section together, the net radiation emission coefficient u can be calculated, which is important for the modeling of the energy or power balance of the plasma. The net radiation emission coefficient determines the radiation energy per unit time and volume as a function of plasma temperature in local thermal equilibrium.

Line Radiation

The most important light emitting process in HID lamps is line radiation. The reason for the occurrence of line radiation or spectral lines is the discrete energy levels electrons can occupy in a given atom (or ion). Since electrons are fermions, they cannot exist in identical energy states, which is described by the Pauli exclusion principle. Thus, only one electron can occupy a given energy level in an atom (characterized by principal quantum number n, angular momentum quantum number l, magnetic quantum number m, and electron spin s, [82]. Electrons can be excited from the ground state to an excite state e.g., by inelastic collisions between (free) electrons and atoms. If the electron falls back to the ground state or to any other excited state of the atom, radiation might be emitted.[5] Nevertheless, there are other possibilities of de-excitation, e.g., the excited atom might transfer the excitation energy to a (free) electron in the form of kinetic energy without emitting radiation (quenching collisions). These quenching collisions outnumber the radiative de-excitations in HID lamps, thus the radiation processes do not disturb local thermal equilibrium. Despite this, the line radiation is a strong light emitting mechanism due to the enormous total number of different types of collisions, excitation, and de-excitation.

Energy level diagrams of mercury and sodium atoms are given in Figs. 2.9 and 2.10, respectively, important radiative transitions between energy levels and their corresponding photon wavelength are marked with circles. The important resonance lines (transition between excited states and ground state) are 185.0 and 253.7 nm (UV radiation) in the case of mercury, and 589.0 and 589.6 nm in the case of sodium (yellow low-pressure sodium line, sodium D-lines). They are important for all discharge lamps containing mercury or sodium. In the case of high-pressure mercury discharges, certain transitions between excited states are also important for the emission spectrum: The corresponding photon wavelengths are 404.7, 435.8, 546.1, 577.0, and 579.0 nm (cf. Sects. 1.3.3–1.3.9). If both sodium and mercury are in a discharge as in many sodium high-pressure (Sect. 1.3.7) and metal halide lamps (Sect. 1.3.8), the radiation is dominated by sodium, since the resonance lines of sodium need less excitation energy (2.1 eV) than the mercury resonance lines (4.9 eV). The plasma temperature in such lamps is smaller compared to pure mercury lamps, because the plasma emits more radiation power at the same temperature if the excitation energies of the plasma constituents are lower.

Small differences in the energy levels lead to long wavelengths of the emitted photons in the case of a radiative transition, large energy differences result in short wavelengths. Rare gases, which would be perfect as lamp filling as they have no problems with evaporation, starting, or temperature dependence, have energy levels with the first excited state at two thirds or more of the ionization energy (cf.

[5] Not all transitions are allowed, i.e., some of the transitions are forbidden. Selection rules state that electrons might not change all their characterizing quantum numbers in an arbitrary way, e.g., the angular momentum quantum number l must change by +1 or −1 because the emitted photon has spin 1 [82, 103].

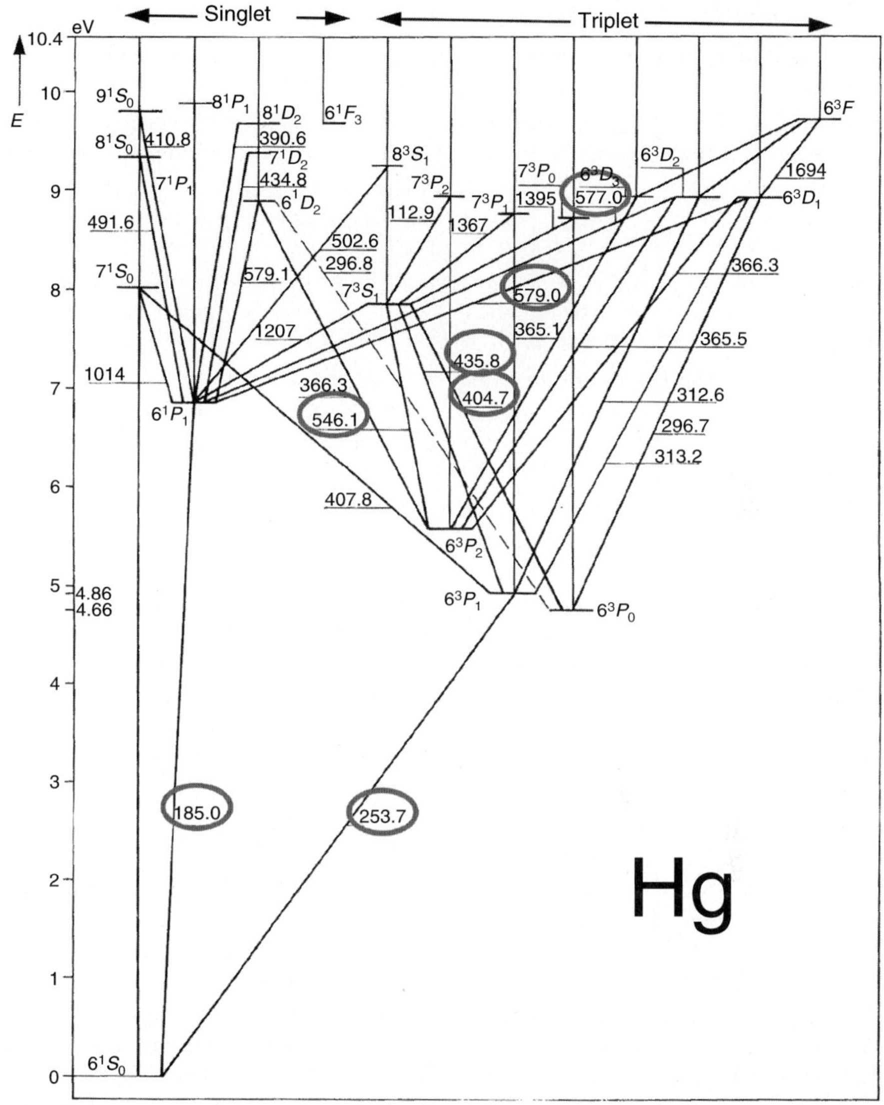

Fig. 2.9. Energy level diagram of mercury atom. The excitation energy in eV is given at the left axis (ionization energy: 10.44 eV), the numbers near the lines connecting the different energy levels are the corresponding photon wavelengths in nm for a radiative transition (excitation or de-excitation) [48]

Fig. 2.11). The resulting resonance radiation has a very high energy and cannot be converted into visible radiation without major losses. The radiation originating from transitions between excited states however, has often wavelengths in the infrared part of the electromagnetic spectrum, i.e., they are not in the visible spectrum. Thus, rare gas discharges are not very efficient light sources.

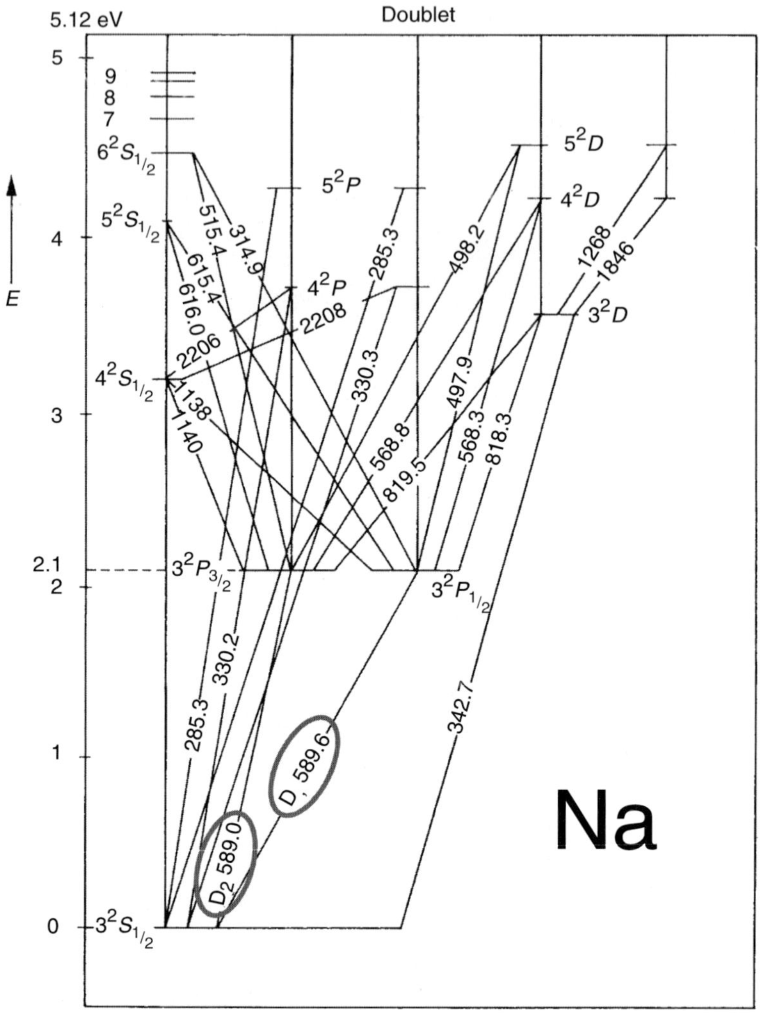

Fig. 2.10. Energy level diagram of sodium atom. The excitation energy in eV is given at the left axis (ionization energy: 5.12 eV), the numbers near the lines connecting the different energy levels are the corresponding photon wavelengths in nm for a radiative transition (excitation or de-excitation) [48]

The radiation produced by transitions of an electron from energy level n to energy level m of an atom can be calculated ($E_n > E_m$). The energy of a single photon is given by the energy difference between the excited states, i.e., $E_{nm} = E_n - E_m = h\nu_{nm} = hc/\lambda_{nm}$, where λ_{nm} is the wavelength of the emitted photon for a transition from energy level n to energy level m. The transition probability A_{nm} for such a transition must be known, the population density of excited state n can be calculated with the Boltzmann distribution if we assume LTE (cf. Sects. 2.1.1, 2.1.2, and 2.1.8). The population density of energy level m does not play a role. The

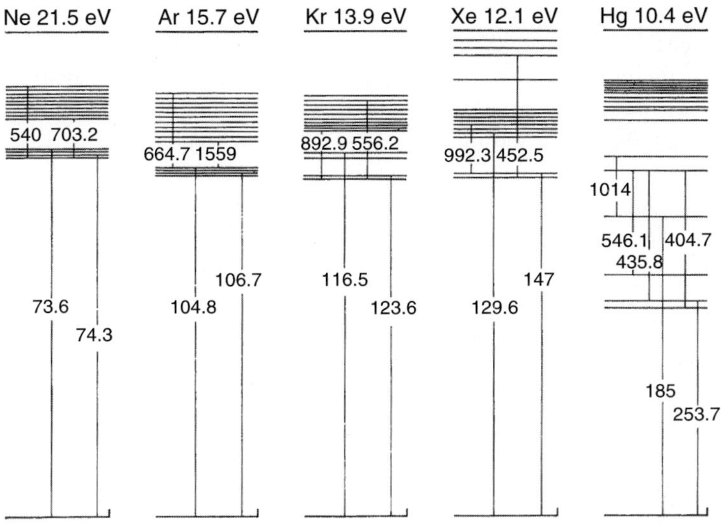

Fig. 2.11. Energy level diagram of rare gas atoms compared to mercury atom. The ionization energies are drawn at the same height, the numbers at the lines connecting different energy levels are the corresponding photon wavelengths in nm for a radiative transition (excitation or de-excitation) [48]

resulting radiation emission coefficient e_{nm} (radiation energy per unit time and volume) for transitions from energy level n to energy level m is thus given by

$$e_{nm} = \frac{hc}{\lambda_{nm}} A_{nm} \, n \frac{g_n}{Z} \mathrm{e}^{-E_n/kT} L_{nm}(\lambda). \tag{2.23}$$

n is the atom density, g_n the statistical weight of energy level n, Z the partition function (cf. Sect. 2.1.2). The factor $L_{nm}(\lambda)$ describes the line broadening of the transition as described in "Line Broadening" (see later). Integrating $L_{nm}(\lambda)$ over all wavelengths yields 1. The density of atoms in excited states (and thus the intensity of line radiation) is proportional to the atom density, i.e., the line radiation is proportional to the pressure. Moreover, the density of excited states (and thus the intensity of line radiation) increases with increasing temperature. Equation (2.23) is valid for so-called optically thin radiation, i.e., radiation which leaves the discharge without being absorbed by other atoms. Optically thick lines must be treated differently as described in "Self-absorption and Optically Thick Lines" (see later).

In HID lamps of very high pressure (e.g., 20 MPa = 200 bar in UHP lamps, cf. Sect. 1.3.6), the density of molecules gets high enough to contribute a considerable portion of molecular band radiation to the spectral power distribution. The partial pressure of Hg_2 molecules in the UHP lamp with 20 MPa of mercury can be estimated to be 1 MPa. Since molecules have energy bands instead of energy levels, the transitions between different bands result in a broad spectral power distribution, the overlap of multiple band systems gives a quasicontinuum radiation. This radiation is not reabsorbed like the resonance lines of atoms and can thus contribute to the light emitted from the discharge. In the UHP lamp with a pressure of 20 MPa, the share of the continuum radiation can be estimated to be more than 50%. New investigations indicate that a considerable part of the continuum radiation might be due to electron-atom bremsstrahlung [44, 45, 75, 105, 138].

Line Broadening

The broadening of spectral lines has already been visualized in Figs. 1.23 and 1.29 for high-pressure mercury and sodium discharge lamps. Even in low-pressure discharge lamps, the spectral lines are very sharp but still have a certain line width. This natural line width is coupled to the length of the wave train via Fourier analysis. Only an infinitely long wave train would have an absolute sharply defined wavelength or frequency. The energy levels for an electron in an atom are not infinitely sharp, they have an energy width ΔE couple by Heisenberg's uncertainty principle to the lifetime τ of this level, i.e.,

$$\Delta E \approx \frac{h}{2\pi\tau}. \tag{2.24}$$

Only the ground state with infinite lifetime is sharp, all other energy levels have a certain width. This determines the natural line width of radiation coming from an isolated atom. Typical lifetimes are of the order of 10^{-8} s, the resulting natural line width is thus of the order of 10^7 s^{-1}. This natural line width can be broadened by different mechanisms which depend on the velocity of the emitting atom (Doppler broadening), collisions between the emitting atom and other particles (collision broadening), and the change of energy levels by the surrounding particles (pressure broadening).

If the atom emits radiation while having a nonzero velocity component in the direction of the observer, the observed frequency shifts to lower (atom is flying away from observer) or higher values (atom is flying in the direction of the observer). This so-called Doppler line broadening is not coupled to the atom but to the velocity between the atom and an observer, i.e., it has nothing to do with the energy levels of the atom. This line broadening is proportional to the average velocity (Maxwell velocity distribution) and thus proportional to the square root of the temperature (cf. Maxwell velocity distribution in Sect. 2.1.3 and (2.7)).

The natural lifetime of an excited level is shortened by collisions with other particles. Since high-pressure discharge lamps are collision-dominated plasmas, this is an important effect. The average time between two collisions can be estimated by the mean free path and the mean velocity of the atoms (cf. Sects. 2.1.3 and 2.1.6). This collision broadening is proportional to the pressure p of the discharge, because the number of collisions increases with increasing pressure. Thus, in many HID lamps the collision broadening is larger than the Doppler broadening.

The energy levels for an electron in an atom shown in Figs. 2.9–2.11 are valid for an isolated atom. The surrounding atoms, ions, and electrons change these energy levels depending on the distance from the emitting atom. The more other particles are in the neighborhood, the stronger the occurrence of this effect, i.e., this so-called pressure broadening increases with increasing pressure. This line broadening mechanism is of particular importance for the resonance lines, since the ground level changes differently compared to excited energy levels depending on the surrounding particles, so that transitions between ground level and excited levels are notably altered. Transitions between excited states are broadened less, since the excited states change their energy levels in a similar way depending on the surrounding particles. The energy difference between excited energy levels and ground level decreases with decreasing distance from other atoms, thus the broadening becomes asymmetrical, i.e., the broadening is toward the longer wavelength side. Especially in high-pressure sodium lamps, pressure broadening is the main broadening mechanism (cf. Figs. 1.27–1.29 with the asymmetrical broadening of the 589.0 and 589.6 nm resonance lines). Depending on the surrounding particles, one can distinguish between Stark broadening (interaction with surrounding electric field of surrounding electrons and ions), van der Waals broadening (interaction with surrounding neutral particles), and resonance interaction between excited atoms and atoms of the same species in the ground state. The extreme case of the pressure broadening is the continuous black body radiation of a solid state, where all atoms are extremely close together and the electrons have no discrete energy levels [48, 72, 80, 116, 132, 172].

Self-Absorption and Optically Thick Lines

Photons resulting from a radiative transition from energy level *n* to energy level *m* of an atom ($E_n > E_m$) can be absorbed by an atom being in energy level *m* (E_m). This is the inverse process to the radiative de-excitation and will thus result in an excited atom with excitation energy E_n. The photon is transporting energy from one position to another position, i.e., the reabsorption or self-absorption of line radiation contributes to the energy transport in discharge lamps, which is discussed in Sect. 2.3.4.

Not only energy is transported during this process, but also the reabsorbed photons are absent in the spectral power distribution. The absorption of photons is proportional to the density of atoms with energy level *m*, i.e., atoms having the excitation energy E_m. Since most atoms are in the ground state, the self-absorption is of particular importance for resonance radiation originating from transitions from excited states to the ground level. The probability for an atom to absorb radiation depends on the energy of the photons. The closer the energy of the photons to the center of the emitted line radiation, the higher the probability for the reabsorption. This results in a flattening of the emitted line profile, i.e., the half-value width increases. In HID lamps, the emission of radiation takes place in the center of the discharge, where plasma temperature is high (cf. Fig. 1.35). To escape from the lamp, the emitted photons have to pass through a colder plasma region with high atom density (cf. (2.13)). Most of the atoms are in the ground state, i.e., they are able to absorb resonance radiation. This can result in the so-called self-reversal of resonance lines, i.e., the measured intensity in the center of the line is smaller than the intensity in the wings of the line. It is possible that practically all photons near the center of the line are absorbed, resulting in the spectral power distributions shown, e.g., in Figs. 1.27–1.29.

In a high-pressure mercury plasma, the mercury resonance lines at 185 and 254 nm are normally reabsorbed within the plasma and are thus not included in the net radiation emission coefficient u. These lines can be treated separately for radiative energy transport, e.g., using the approximation of radiation diffusion (see Sect. 2.3.4). For spectral lines which are partially absorbed within the plasma, a radiation transport calculation yields spectral line weakening factors χ [160]. According to [160], χ is set to 0.4 for the mercury atomic lines at 365.02, 365.48, and 546.08 nm, and 0.5 for the line at 435.83 nm. For xenon infrared lines at 820.86, 828.24, 823.39, and 834.91 nm, the spectral line weakening factor χ is set to 0.7 [161].

Recombination and Bremsstrahlung

The optically thin line radiation is the most important, but not the only way by which radiation is generated in the plasma. Besides the molecular radiation, which is for example important in the super high-pressure mercury lamp UHP (see earlier), there are two other sources of radiation: radiation due to recombination and the bremsstrahlung. Both mechanisms produce photons which leave the plasma without absorption and thus contribute to the net radiation emission coefficient.

Radiation due to recombination might be emitted when a free electron recombines with an ion[6]. Since the electron recombining with the atom might have arbitrary kinetic energy, the energy of the emitted photon is the sum of the kinetic energy of the electron and the ionization energy. The spectral power distribution of such a recombination is thus a continuous one with a sharp boundary at the long wavelengths corresponding to the ionization energy of the atom. Since the recombination of an electron with an ion must not result in a ground state atom, there are also recombinations resulting in an excited atom. The corresponding energy of the emitted photon is then the result of the kinetic energy of the electron plus the ionization energy minus the excitation energy of the resulting excited atom. The radiation due to recombination is proportional to the product of ion and electron density and thus it starts being important at higher temperatures. The product of ion and electron density is proportional to the neutral particle density (cf. (2.21)), i.e., the intensity of the radiation due to recombination is proportional to the pressure (as line radiation).

A further contribution to the net emission radiation coefficient u is bremsstrahlung. This type of radiation occurs when electrons are decelerated in the Coulomb field of ions or atoms. Accelerated or decelerated charges give off electromagnetic radiation. Bremsstrahlung is characterized by a continuous spectrum of radiation which becomes more intense and shifts toward higher

[6] The energy being released during recombination might also be passed to a third particle in the form of kinetic energy. This process is called three-body collisional recombination.

frequencies when the energy of the accelerated or decelerated electrons is increased. One can distinguish between electron-atom and electron-ion bremsstrahlung, i.e., the electron might be decelerated in the Coulomb field of atoms or ions. New investigations concerned with high-pressure mercury lamps indicate, that a considerable part of the continuum radiation might be due to electron-atom bremsstrahlung [75, 105, 160].

Net Radiation Emission Coefficient

Taking all emission and absorption processes together, we get the net radiation emission coefficient u, which determines the radiation energy per unit time and volume as a function of plasma temperature in local thermal equilibrium:

$$u = e - a = \int_{\lambda} (e_{\lambda} - a_{\lambda})\,\mathrm{d}\lambda = \int_{\lambda} \chi_{\lambda} e_{\lambda}\,\mathrm{d}\lambda. \tag{2.25}$$

e is the radiation emission coefficient, a the absorption coefficient, the index λ indicates that both coefficients depend on the wavelength. χ is a spectral line weakening factor for partially absorbed spectral lines [160]. This avoids the complex calculation of the absorption coefficient and includes the absorption in the emission. χ is calculated using radiation transport calculation. According to [160], χ is equal to 0.4 for the mercury atomic lines at 365.02, 365.48, and 546.08 nm, and $\chi = 0.5$ for the line at 435.83 nm. For xenon infrared lines at 820.86, 828.24, 823.39, and 834.91 nm, the spectral line weakening factor χ is set to 0.7 [161].

Summarizing, all photon emission processes are induced by electrons changing their energy level. Figure 2.12 shows the most important processes. On the left-hand

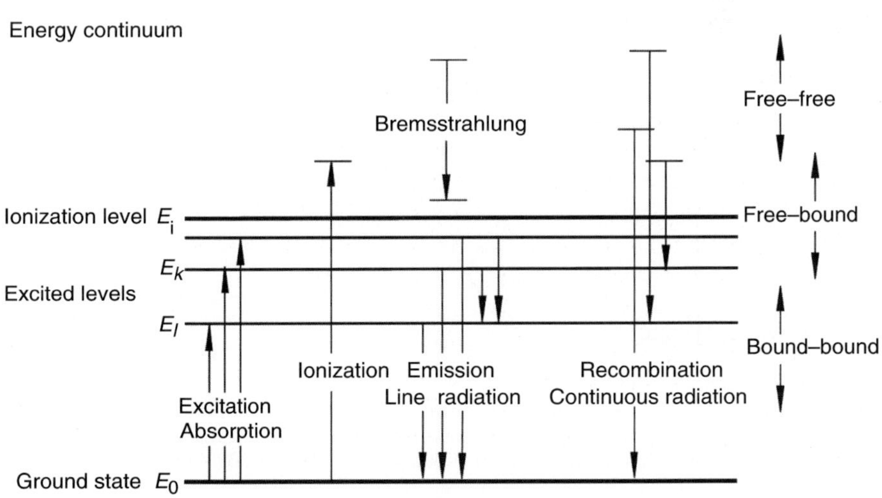

Fig. 2.12. Summary of main processes involving transitions of electrons to different energy levels [95]

side, excitation of electrons from the ground state to an excited state and ionization (setting free of an electron) are shown (arrows upwards). The excitation of an electron from an excited state is not included in Fig. 2.12 since it is rather unlikely. The electrons on the right-hand side of Fig. 2.12 (arrows downwards) make transitions to lower energy levels which is (might be) accompanied by photon emission. Depending on the state of the electron (free or bound to an atom/ion), one distinguishes between bound–bound transitions (line radiation), free–bound transitions (recombination), or free–free transitions (bremsstrahlung).

An example of the net radiation emission coefficient *u* for a mercury and xenon plasma of different pressures is given in Fig. 2.13. The intensity of the radiation is proportional to the pressure and strongly increases with temperature (logarithmic scale in Figs. 2.13–2.16). A detailed breakdown of the different contributions to the net radiation emission coefficient is shown in Figs. 2.14–2.16. Figure 2.14 depicts the different contributions to the net radiation emission coefficient for a 4 MPa mercury plasma, showing that the main light emitting process is the emission of spectral lines. In the case of xenon (Fig. 2.15), the line radiation still clearly dominates the net radiation emission coefficient, but contributions due to recombination and bremsstrahlung are nevertheless of some importance. The line radiation of the xenon ion can be neglected.

An example of the complex situation in the plasma of a D2 automotive headlight lamp is given in Fig. 2.16. Despite the high partial pressures of xenon and mercury of 5 and 2 MPa, respectively (50 and 20 bar), the radiation of scandium and sodium dominates the net radiation emission coefficient (partial pressures of NaI: 0.03 bar, ScI_3: 0.05 bar). Only for very high temperatures of 6,000 K and more (as in the hot

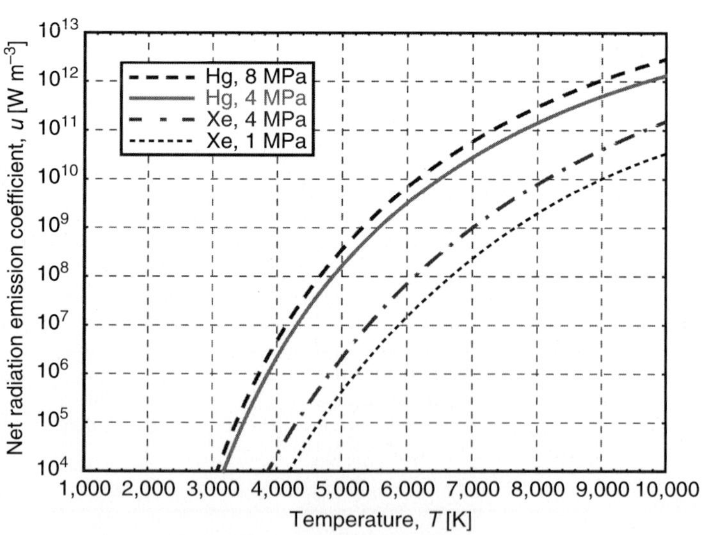

Fig. 2.13. Net radiation emission coefficient *u* of a mercury (4 and 8 MPa) and a xenon plasma (1 and 4 MPa) vs. plasma temperature [159, 160]

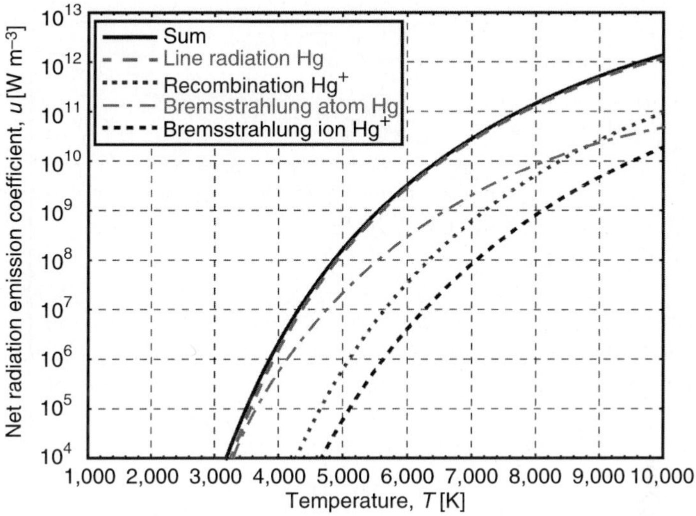

Fig. 2.14. Net radiation emission coefficient u (and components) of a 4 MPa (40 bar) mercury plasma vs. plasma temperature [159, 160]

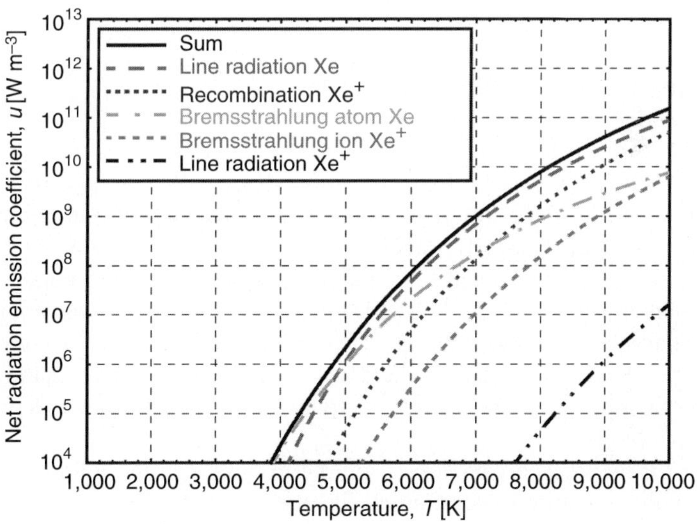

Fig. 2.15. Net radiation emission coefficient u (and components) of a 4 MPa (40 bar) xenon plasma vs. plasma temperature [159, 160]

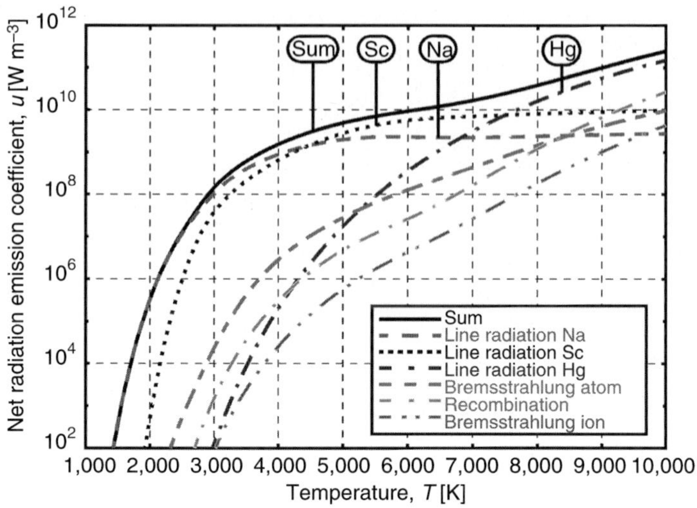

Fig. 2.16. Net radiation emission coefficient u (and components) of a D2 automotive headlight lamp plasma (Xe, Hg, NaI, ScI_3 with partial pressures of 50, 20, 0.03, and 0.05 bar, respectively) vs. plasma temperature [159, 160]

plasma spots, cf. Fig. 1.35), line radiation of mercury matters. As in the high-pressure mercury and xenon plasma, line radiation dominates all other radiation processes.

2.3 Transport Coefficients

The gas and plasma properties described in Sects. 2.1 and 2.2 are related to the local characteristics of the plasma like ionization degree or net radiation emission coefficient. Apart from these properties, transport processes play an important role in HID lamps. All transport processes are driven by spatial nonuniformities (gradients) and are proportional to some coefficient. The spatial nonuniformities lead to a net transport of, for example, particles due to a gradient of particle densities (particle diffusion), charge carriers due to a gradient of the electric potential, i.e., due to an electric field (electrical conductivity), energy due to a gradient of temperature (thermal conductivity), or momentum due to a gradient of velocities (viscosity).

Some of these transport processes will be discussed in the following sections. The corresponding transport coefficients like the diffusion coefficient, the electrical conductivity, or the thermal conductivity will be calculated or estimated using the so-called simplified kinetic theory. In this theory, the transport coefficients are calculated using simple mean free path considerations. This approach to the transport properties has the advantage of focusing on the underlying physical processes and, in addition, leads to formulas which are easy to analyze. The results from this theory are in general accurate to within a factor of 2 or 3.

Nevertheless, the transport coefficients shown in Figs. 2.18–2.26 in the following sections are based on the work of Hirschfelder, Curtiss, and Bird [90], using rigorous kinetic theory according to Chapman–Enskog theory [171]. This theory starts with the undisturbed distribution function $f_i(\vec{r},\vec{v},t)$ for particle of species i. This distribution function describes the probability to find a particles of species i at time t in an infinitesimal volume around $\vec{r}$ with velocity $\vec{v}$. The undisturbed distribution function, i.e., the distribution function in the absence of gradients of particle densities, temperature, velocity, or electrical fields, is given by the Maxwell or Maxwell–Boltzmann velocity distribution (cf. (2.5)). In the Chapman–Enskog theory, a series expansion starting from this undisturbed distribution function is used to take into account deviations from the undisturbed situation. In the LTE plasma in HID lamps, the gradients which change the undisturbed situation are sufficiently small, so that it is sufficient to consider only deviations of the first order. The Chapman–Enskog theory connects the transport coefficients to so-called collision integrals, which describe binary collision with energy or velocity dependent collision cross sections.

2.3.1 Diffusion

The diffusion of particles describes the net particle flux in a gas or plasma due to a gradient of particle density. A gradient of particle density means that there are areas (e.g., area A) where the particle density is higher than in other areas (e.g., area B). Since all particles have (approximately) a Maxwell velocity distribution, some of the particles of area A might fly to area B and vice versa. Since area A has a higher density of particles, more particles fly from A to B than from B to A. This corresponds to a net particle flux from A to B which eventually balances the unequal particle densities in A and B. Thus, transport processes tend to balance spatial nonuniformities. Let us put this into appropriate equations:

The diffusion coefficient D is defined by

$$\vec{j} = -D\,\mathrm{grad}(n). \tag{2.26}$$

Let us consider the diffusion in z direction, i.e., let us assume a particle gradient in z direction. The net particle flux through a plane at position z is determined by the particles from a plane at position $z-\ell$ flying in positive z direction and from a plane at position $z+\ell$ flying in negative z direction (cf. Fig. 2.17). ℓ is the mean free path and we assume that one-sixth of the particles at the planes at position $z-\ell$ and $z+\ell$ fly in positive z direction, one-sixth in negative z direction, one-sixth in positive x

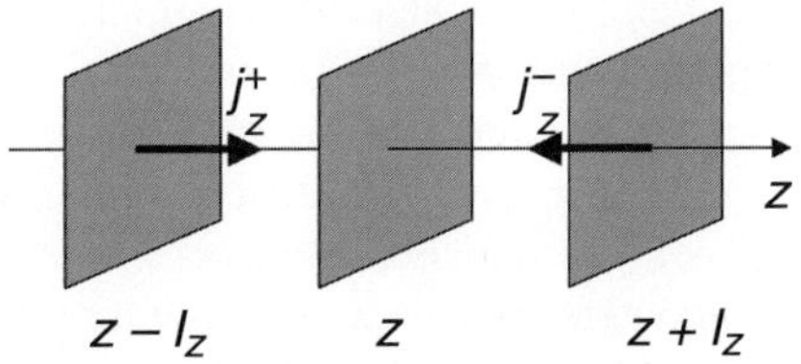

Fig. 2.17. Illustration of the net particle flux through a plane at position z

direction, etc. The distance ℓ between the planes reflects the consideration that the particles arriving at the plane at position z had their last collision at the planes at position $z-\ell$ and $z+\ell$, i.e., the particle densities at these positions are relevant for the diffusion process.

Since one-sixth of the particles fly with the mean velocity $\overline{v}$ in positive z direction and one-sixth in negative z direction, the net particle flux at the plane at position z is given by:

$$\begin{aligned} \vec{j} &= \vec{j}^{+} - \vec{j}^{-} \\ &= \frac{1}{6} n(z-\ell)\overline{v} - \frac{1}{6} n(z+\ell)\overline{v} \\ &= -\left[\frac{n(z+\ell)-n(z-\ell)}{2\ell}\right]\frac{1}{3}\ell\overline{v} \\ &= -\frac{\mathrm{d}n}{\mathrm{d}z}\frac{1}{3}\ell\overline{v} \\ &= -D\frac{\mathrm{d}n}{\mathrm{d}z}. \end{aligned} \tag{2.27}$$

The particle density at the plane at position $z-\ell$ is given by $n(z-\ell)$, at $z+\ell$ by $n(z+\ell)$. Comparing the last two lines of (2.27), the diffusion coefficient D can be expressed by $\ell\overline{v}/3$. Using the mean free path ℓ from (2.18), the mean velocity $\overline{v}$ from (2.7), and the ideal gas law ($p = nkT$, (2.13)) yields the formula:

$$D = \frac{1}{3}\ell\overline{v} = \frac{1}{3}\frac{1}{\sqrt{2}nQ}\sqrt{\frac{8kT}{\pi m}} = \frac{2}{3\sqrt{\pi}}\frac{1}{pQ}\sqrt{\frac{(kT)^3}{m}} \propto \frac{\sqrt{T^3}}{p\sqrt{m}}. \tag{2.28}$$

Q is the cross section for elastic collisions between the particles of the same species with particle density n.

Considering the case of two different types of particles, i.e., particles of species 1 (e.g., mercury ions) and 2 (e.g., mercury atoms), the diffusion of ions can be described by the binary diffusion coefficient $D_{i,n}$. According to [171], the binary diffusion coefficient $D_{i,n}$ derived with the rigorous kinetic theory is given by

$$D_{i,n} = \frac{3\sqrt{\pi}}{8}\frac{1}{p\overline{Q}_{\mathrm{in}}^{(1,1)}}\sqrt{\frac{kT^3}{2\mu}}. \tag{2.29}$$

μ is the reduced mass, i.e., $\mu = (m_i m_n)/(m_i + m_n)$, and $\overline{Q}_{\mathrm{in}}^{(1,1)}$ a temperature dependent collision integral. Comparing (2.28) (simple kinetic theory) and (2.29) (rigorous kinetic theory), the proportionality to T and p is the same, only the numerical value of the multiplier has changed but remains at the same order of magnitude (0.38 in (2.28) compared to 0.66 in (2.29)), i.e., the simple kinetic theory is capable of calculating a good approximation for the transport coefficients.

Figure 2.18 shows the binary diffusion coefficient $D_{i,n}$ between ions and atoms for a mercury and xenon plasma of different pressures calculated using rigorous

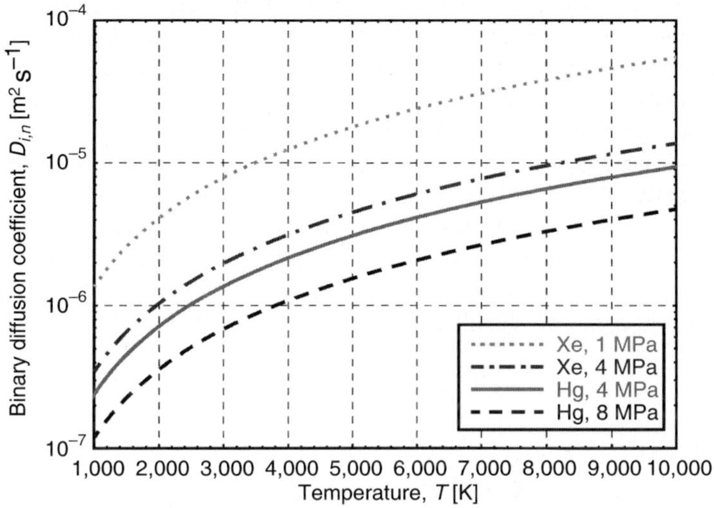

Fig. 2.18. Binary diffusion coefficient $D_{i,n}$ of a mercury (4 and 8 MPa) and a xenon plasma (1 and 4 MPa) vs. plasma temperature [171]

kinetic theory [171]. The diffusion is inversely proportional to the pressure p, because increasing pressure decreases the mean free path (logarithmic scale in Fig. 2.18). Increasing temperature increases the thermal velocity as well as the mean free path (at constant pressure), thus the binary diffusion coefficient increases with temperature. The difference between mercury and xenon (compare mercury and xenon at 4 MPa) is largely due to the mass difference: Mercury has an atomic mass of 200.6 u, xenon of 131.3 u. The higher mass of mercury decreases the thermal velocity and thus the diffusivity. The collision integrals for ion-atom collisions in mercury and xenon are very similar [72, 171].

2.3.2 Electrical Conductivity

The electrical conductivity σ determines the power injection into the plasma, which is given by Ohm's law ($jE = \sigma E^2$). For a fixed electric current I or a fixed electric current density j, a high electrical conductivity causes small electric fields E and thus small electrical power densities. In many HID lamps, a high electrical power density is required. At the same time, large electric currents stress the electrodes, leading to melting, evaporation, or sputtering of the electrodes. This determines often the end-of-life of a HID lamp. A small electrical conductivity thus provides for the small lamp currents and high electrical power densities wanted.

The electric current is carried by electrons and ions, so to calculate the electrical conductivity we need the electron and ion mobility (μ_e and μ_i) in the plasma. The mobility describes the relation between drift velocity v_d of electrons or ions in a gas or plasma and an applied electric field. The drift velocity should not be confused with the thermal velocity, which is in general larger but has no specific direction.[7] The mobility is inversely proportional to the mass of the charge carrier, as will be shown in the following. Due to the large mass difference between electrons (5.49×10^{-4} u) and ions (atomic mass of about 20 u (neon or sodium) up to 200 u (mercury) and more), the electric current is carried to more than 99% by electrons.

Let v_d be the drift velocity of an electron with mass m_e in a plasma. It can be assumed that the electron is scattered at each collision with an atom in arbitrary direction, so that the drift velocity is reset to zero after each collision. During two collisions, the constant acceleration is given by eE/m_e, the average time between two collisions is $\tau = \ell_e/\bar{v}_e$ (cf. (2.17) with ℓ_e the mean free path of electrons in a plasma). So, immediately before the next collision, the drift velocity is given by $eE\ell_e/\bar{v}_e m_e$. A more detailed derivation yields a slightly different numerical value of the multiplier [141, 171]:

$$v_d = \mu_e E = \frac{\sqrt{8}}{\pi}\frac{e\ell_e}{m_e\bar{v}_e}E. \tag{2.30}$$

The drift velocity is inversely proportional to the mass of the charge carriers, which has already been used earlier to show that the electric current is carried to more than 99% by electrons, because their mobility is approximately 1,000 times larger than the mobility of ions. The electrical conductivity σ, which is defined by

$$\vec{j}_{el} = -\sigma\,\mathrm{grad}(\varphi) = \sigma\vec{E} \tag{2.31}$$

can now be expressed by

$$\begin{aligned}\overrightarrow{j_{el}} &= \overrightarrow{j_{ion}} + \overrightarrow{j_{el}} \approx \overrightarrow{j_{el}}\\ &= -en_e\vec{v}_d\\ &= en_e\frac{\sqrt{8}}{\pi}\frac{e\ell_e}{m_e\bar{v}_e}\vec{E}\\ &= \sigma\vec{E}\end{aligned} \tag{2.32}$$

Using the mean free path of electrons ℓ_e from (2.18) with the cross section for elastic collisions of electrons with atoms[8] Q_{ea} and the mean velocity $\bar{v}_e$ from (2.7), the following formula results:

$$\begin{aligned}\sigma = en_e\frac{\sqrt{8}}{\pi}\frac{e\ell_e}{m_e\bar{v}_e} &= en_e\frac{\sqrt{8}}{\pi}\frac{ekT\sqrt{\pi m_e}}{m_e\sqrt{2}Q_{ea}p\sqrt{8kT}}\\ &= \frac{n_e e^2\sqrt{kT}}{\sqrt{2\pi m_e}Q_{ea}p} \propto \frac{n_e\sqrt{T}}{pQ_{ea}}.\end{aligned} \tag{2.33}$$

[7] The average of the thermal velocity is zero. This should not be confused with the average magnitude of the velocity given in (2.7).

[8] A weakly ionized plasma is assumed, i.e., the electron density is much smaller than the density of atoms. Then the mean free path of electrons is determined by elastic collisions between electrons and atoms.

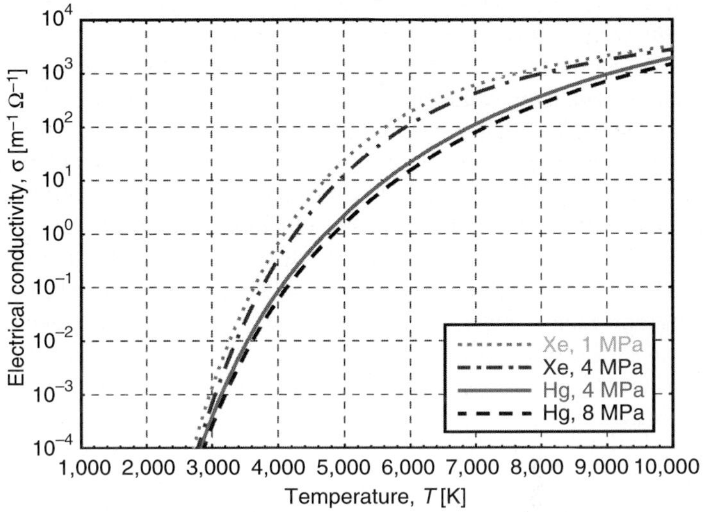

Fig. 2.19. Electrical conductivity σ of a mercury (4 and 8 MPa) and a xenon plasma (1 and 4 MPa) vs. plasma temperature [171]

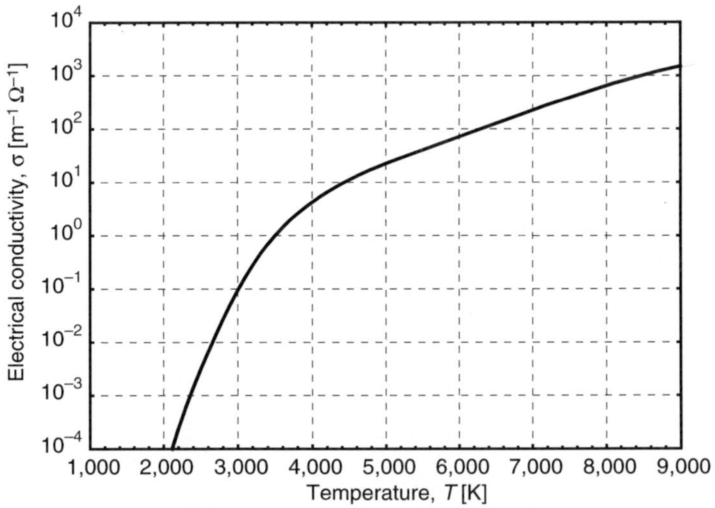

Fig. 2.20. Electrical conductivity σ of a D2 automotive headlight lamp plasma (Xe, Hg, NaI, and ScI_3 with partial pressures of 50, 20, 0.03, and 0.05 bar, respectively) vs. plasma temperature [171]

Figure 2.19 depicts the electrical conductivity for a mercury and xenon plasma of different pressures calculated using rigorous kinetic theory [171]. An example of σ for a complex plasma as in the D2 automotive headlight lamp is given in Fig. 2.20 (logarithmic scale in Figs. 2.19 and 2.20). The main influence on the electrical conductivity has the electron density, which depends strongly on the temperature (cf. Figs. 2.6–2.8). Comparing the electrical conductivity σ to the ratio n_e/n, one can deduce that the temperature dependence of the cross-section for elastic collisions of electrons with atoms Q_{ea} is of minor importance (cf. Figs. 4.18 to 4.20). Nevertheless, the absolute value of Q_{ea} is very important, since it is the only parameter depending on the plasma composition and therefore responsible for the difference of the electrical conductivity σ of a xenon or a mercury plasma (compare mercury and xenon at 4 MPa in Fig. 2.19). Mercury has a higher electron density but a lower electrical conductivity compared to xenon. This is due to the large cross section for elastic collisions of electrons with mercury atoms. For this reason, mercury is used in many high-pressure discharge lamps (high-pressure mercury and sodium discharge lamps, metal halide lamps, cf. Sects. 1.3.6–1.3.8) allowing for high electrical power input at high electric fields and small electric currents. The fact that mercury is very volatile further improves this advantage, since increasing pressure decreases the electrical conductivity.

2.3.3 Thermal Conductivity

The thermal conductivity describes the net power flux $\vec{q}$ in a gas or plasma due to a temperature gradient. The thermal conductivity λ is defined by

$$\vec{q} = -\lambda \, \mathrm{grad}(T). \tag{2.34}$$

Let us consider the power flux in z direction, i.e., let us assume a temperature gradient in z direction as in Sect. 2.3.1. The net power flux through a plane at position z is determined by the power flux from a plane at position $z-\ell$ in positive z direction and from a plane at position $z+\ell$ in negative z direction (cf. Fig. 2.21). ℓ is the mean free path and we assume that one sixth of the particles at the planes at position $z-\ell$ and $z+\ell$ fly in positive z direction and one sixth in negative z direction. The distance ℓ between the planes reflects the consideration that the particles arriving at the plane at position z had their last collision at the planes at position $z-\ell$ and $z+\ell$, i.e., the energy of the particles at $z-\ell$ and $z+\ell$ is relevant for the power flux at the position z.

Since one-sixth of the particles fly in positive z direction and one-sixth in negative z direction, the net power flux at the plane at position z is given by:

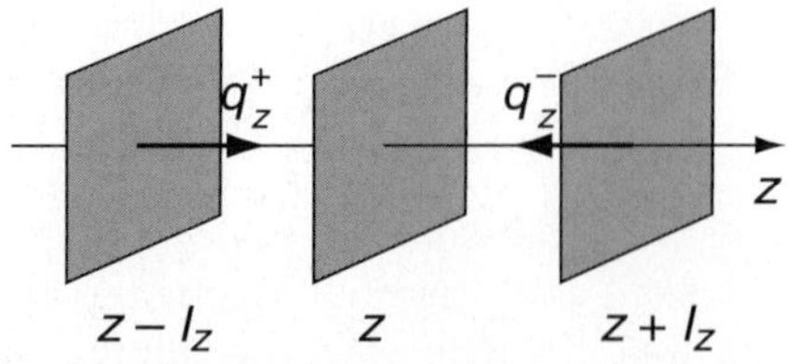

Fig. 2.21. Illustration of the net power flux through a plane at position z

$$\begin{aligned}
\bar{q} &= \bar{q}^{+} - \bar{q}^{-} \\
&= \frac{1}{6} n\bar{v}U(z+\ell) - \frac{1}{6} n\bar{v}U(z-\ell) \\
&= -\left[\frac{U(z+\ell)-U(z-\ell)}{2\ell}\right]\frac{1}{3}\ell n\bar{v} \\
&= -\frac{\mathrm{d}U}{\mathrm{d}z}\frac{1}{3}\ell n\bar{v} \\
&= -\frac{\mathrm{d}(fkT/2)}{\mathrm{d}z}\frac{1}{3}\ell n\bar{v} \\
&= -\frac{1}{6} fk\,\ell n\bar{v}\frac{\mathrm{d}T}{\mathrm{d}z} \\
&= -\lambda\frac{\mathrm{d}T}{\mathrm{d}z}.
\end{aligned} \tag{2.35}$$

The average energy U per particle at the plane at position $z-\ell$ is given by $U(z-\ell)$, at $z+\ell$ by $U(z+\ell)$. U can be expressed in terms of the temperature T by $U = 0.5fkT$, where f is the number of degrees of freedom of the atoms or molecules of the gas. In a monoatomic gas, vibrational and rotational energies are absent and all of the gas internal energy is in the form of translational kinetic energy, i.e., $f = 3$ and (2.10) results. Molecules have vibrational and rotational energies, i.e., f is higher.

Comparing the last two lines of (2.35), the thermal conductivity λ can be expressed by $fk\ell n\bar{v}/6$. Using the mean free path ℓ from (2.18) and the mean velocity $\bar{v}$ from (2.7), the following formula results:

$$\lambda = \frac{1}{6} fk\ell n\bar{v} = \frac{1}{3}\frac{fk\sqrt{kT}}{\sqrt{\pi m}Q} \propto \frac{f\sqrt{T}}{Q\sqrt{m}}. \tag{2.36}$$

Q is the cross section for elastic collisions between the particles of the same species with particle density n. The derivation of the thermal conductivity using the simple kinetic theory includes the transport of internal energy of heavy particles along a temperature gradient. According to (2.36), it is independent of the pressure p. This is due to the increase of the particle density with increasing pressure but the decrease of the mean free path with increasing pressure. In fact, the thermal conductivity can be described by (2.36) only for temperatures up to 4,000 K. At higher temperatures other contributions to the thermal conductivity become more important. Figure 2.22 shows the thermal conductivity for a mercury and xenon plasma of different pressures calculated using rigorous kinetic theory [171, 172]. Apparently, the thermal conductivity depends on the pressure p for temperatures above 4,000 K (logarithmic scale in Fig. 2.22). In the case of xenon, the thermal conductivity decreases with increasing pressure, in the case of mercury, the thermal conductivity increases with increasing pressure (cf. Sect. 2.3.4).

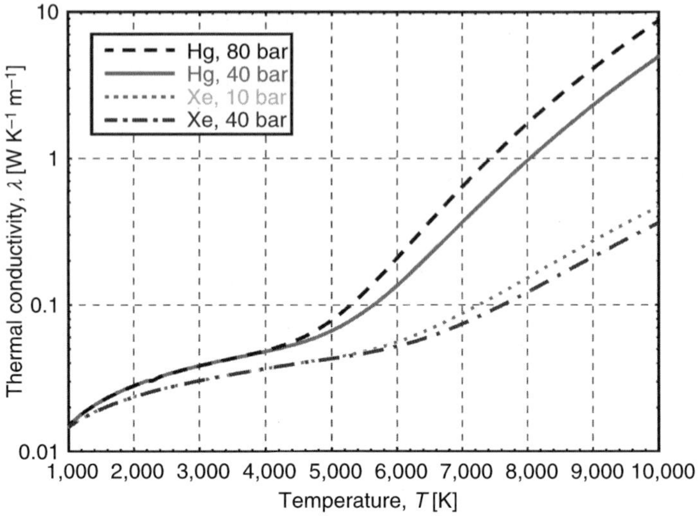

Fig. 2.22. Thermal conductivity λ of a mercury (4 and 8 MPa) and a xenon plasma (1 and 4 MPa) vs. plasma temperature [171, 172]

Due to the strongly increasing electron density with increasing temperature, the electrons contribute to the thermal conductivity for higher temperatures. Since electron density increases proportional to the square root of the pressure p and the mean free path decreases proportional to the pressure p, the thermal conductivity due to electrons is inversely proportional to the pressure, i.e., the thermal conductivity decreases with increasing pressure. This can be seen in Fig. 2.22 in the case of xenon. Another contribution to the thermal conductivity is the energy transport due to ionization and recombination or dissociation and formation of molecules. Nevertheless, this "reactive energy transport" does not contribute considerably to the total thermal conductivity.

A detailed breakdown of the different contributions to the thermal conductivity λ is shown in Figs. 2.23 and 2.24 (logarithmic scale in both figures). Figure 2.23 depicts the different contributions to the thermal conductivity for a 4 MPa mercury plasma, Fig. 2.24 shows the different contributions to the thermal conductivity for a 4 MPa xenon plasma [171, 172]. In the case of mercury, the thermal conductivity for temperatures above 6,000 K is dominated by radiative energy transport (increasing with increasing pressure), which is addressed in Sect. 2.3.4.

2.3.4 Radiative Energy Transport

Photons resulting from a radiative transition from energy level n to energy level m of an atom ($E_n > E_m$) can be absorbed by a second atom being in energy level m (E_m). This is the inverse process to the radiative de-excitation and will thus result in an excited atom with excitation energy E_n. The photon transports energy from one

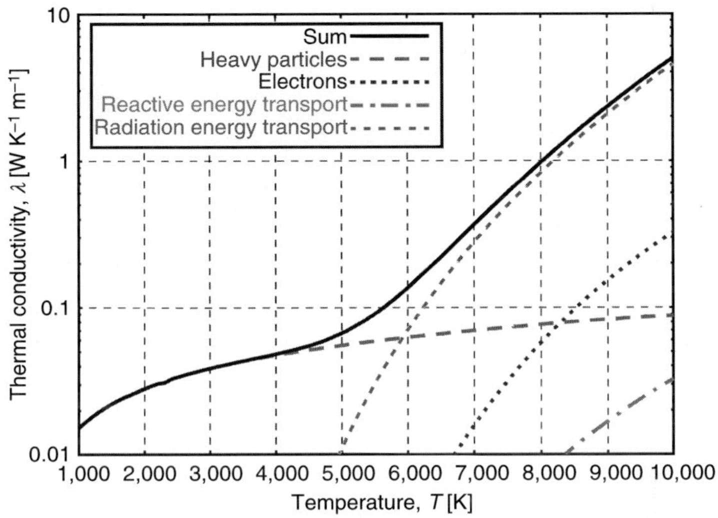

Fig. 2.23. Thermal conductivity λ (and components) of a 4 MPa mercury plasma vs. plasma temperature [171, 172]

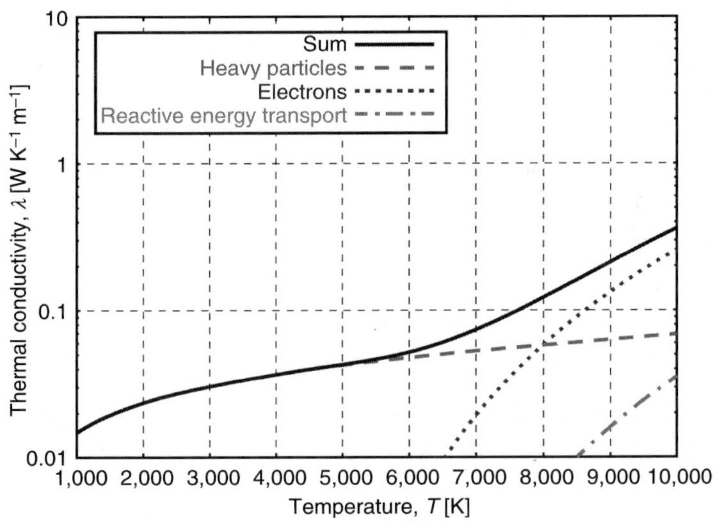

Fig. 2.24. Thermal conductivity λ (and components) of a 4 MPa xenon plasma vs. plasma temperature [171]

position to another position, i.e., the reabsorption or self-absorption of line radiation contributes to the energy transport in discharge lamps. This radiative energy transport can be calculated using radiation transport calculation.

If $L_{\lambda s}$ is the radiance at wavelength λ radiated in the direction s, then the infinitesimal change of $L_{\lambda s}$ along the optical path $\mathrm{d}s$ is given by

$$\frac{dL_{\lambda s}}{ds} = \varepsilon_\lambda - \alpha_\lambda L_{\lambda s}. \quad (2.37)$$

ε_λ and α_λ are the temperature dependent emission and absorption coefficients at wavelength λ, i.e., ε_λ describes the emitted radiation (atom and ion line radiation, molecular radiation, bremsstrahlung, radiation due to recombination) at wavelength λ and can be calculated as described earlier. The spectral distribution is important for the calculation of the reabsorption, i.e., line broadening mechanisms must be taken into account. α_λ stands for this reabsorption of radiation. The coefficient α_λ reduces the radiance $L_{\lambda s}$. The net radiation emission coefficient u, which determines the radiation energy per unit time and volume is then given by

$$u = \nabla\left(\int\limits_{\lambda,s} L_{\lambda s}\vec{e}_s \mathrm{d}\lambda \mathrm{d}\Omega_s\right). \quad (2.38)$$

This means that we have to integrate the radiance at wavelength λ radiated in the direction s ($L_{\lambda s}\vec{e}_s$) over all wavelengths λ and all directions s. The spatial change in this integral describes the total power radiated or absorbed, i.e., u can be a power sink (radiation, $u > 0$) or a power source (reabsorption, $u < 0$).

The problem with (2.38) is that u is no longer a function of plasma temperature T (as e.g., ε_λ and α_λ), but it depends on the surrounding plasma. This means that u is a local property of the discharge. If we know the temperature distribution of the plasma, we can calculate u which involves, e.g., the knowledge of the spectral data of all atomic lines, the line broadening mechanisms and the absorption coefficients. The radiation from each infinitesimal small volume in arbitrary direction has to be tracked through the whole plasma to locate possible reabsorption of this radiation. This is an enormous amount of computational work. The resulting u must then be used to calculate a new temperature distribution, making a new calculation of u necessary. The exact solution of the radiation transport equation for arbitrary lamp geometries and under consideration of, for example, convection, electrode–plasma interactions, or time-dependence in the case of ac lamp currents is beyond the ability of today's computers. Nevertheless, there are, for example, solutions taking into account only the cylindrical plasma column, i.e., reducing the problem from three to one dimension. In [84], this has been done for a 0.6 MPa mercury plasma, showing, for example, that only the line radiation at 185 and 254 nm contributes to the radiative energy transport.

To include radiative energy transport into the modeling of HID lamps, there are many different approximations [26, 29, 50, 68, 74, 84, 111], one of them will be described in the following. Instead of taking the energy transport by radiation with the help of the net radiation emission coefficient u into account, it can also be included in the thermal conductivity λ. In the case of a mercury plasma, this means that the line radiation at 185 and 254 nm is not included in the calculation of u (with (2.25)) and for other spectral lines, which are partially absorbed within the plasma, a one-dimensional radiation transport calculation yields spectral line weakening factors χ [160]. According to [160], χ is set to 0.4 for the mercury atomic lines at

365.02, 365.48, and 546.08 nm, and $\chi = 0.5$ for the line at 435.83 nm. For xenon infrared lines at 820.86, 828.24, 823.39, and 834.91 nm, the spectral line weakening factor χ is set to 0.7 [161]. For a given one-dimensional temperature profile, the atomic lines which are neither optically thin nor optically thick are included in the thermal conductivity λ giving a contribution λ_{rad}. The method is called radiation diffusion approximation, i.e., the radiation transports energy through the plasma comparable to a diffusion process with multiple emissions and absorptions of radiation (cf. [26, 51, 75, 160, 172]). The result of such a calculation can be found in Fig. 2.23 in the case of a 4 MPa mercury plasma. For this result, a typical temperature profile of a 4 MPa mercury discharge has been assumed: parabolic temperature profile with $T_{max} = 6{,}000$ K and $T_{wall} = 1{,}200$ K. Since the line emission is proportional to the pressure, the same holds for the contribution λ_{rad} by the radiation diffusion approximation to the thermal conductivity λ (cf. thermal conductivity of mercury plasma in Fig. 2.22).

Since the computation of the radiation transport with the radiation diffusion approximation is based on a given temperature profile (which of course depends on the thermal conductivity λ), the calculation of λ_{rad} is an iterative process. If the whole plasma and not only the cylindrical plasma column is taken into account, the question is which temperature profile has to be taken into account (cf. calculated temperature profile of the D2 automotive headlight lamp in Fig. 1.35). It turns out that depending on the chosen temperature profile (of the cylindrical plasma column or of the hot plasma spots in front of the electrodes), contributions of radiative energy transport to λ can be very different (compare $\lambda_{rad,col}$ and $\lambda_{rad,hs}$ in Figs. 2.25 and 2.26, logarithmic scale in both figures). In the case of the hot plasma spots with its steep temperature gradients, the emission from the center of the plasma spots can easily escape from the discharge due to the smallness of the cold plasma area, where absorption might occur. This cold plasma area is much larger in the case of the plasma column with a moderate temperature gradient compared to the hot plasma spots. To solve this problem, the author proposes a new and successful method by using $\lambda_{rad,col}$ and $\lambda_{rad,hs}$ in different temperature regions [57, 59]. For temperatures below the maximum temperature of the arc column (5,100 K in the case of the D2 automotive headlight lamp and 5,400–6,000 K in the case of a 0.6 MPa mercury plasma depending on the lamp current), the radiation transport calculation is based on the temperature profile of the arc column. For higher temperatures, the temperature profile of the hot plasma spots is taken into account to calculate radiation transport (cf. λ_{total} in Figs. 2.25 and 2.26, logarithmic scale in both figures). This method delivers good results for the temperature profile of the plasma column as well as for the temperature distribution in the hot plasma spots, i.e., the nonlocal nature of the radiative energy transport is partly accounted for. A radiation transport calculation based only on the plasma column temperature profile would result in too low plasma hot-spot temperatures, a radiation transport calculation based only on the temperature distribution within the hot plasma spots would result in too high plasma column temperature.

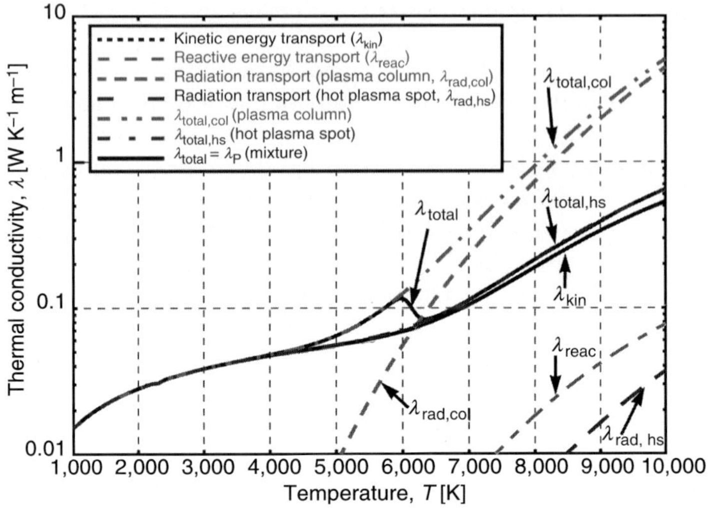

Fig. 2.25. Thermal conductivity λ (and components) of a 0.6 MPa mercury plasma vs. plasma temperature [57]

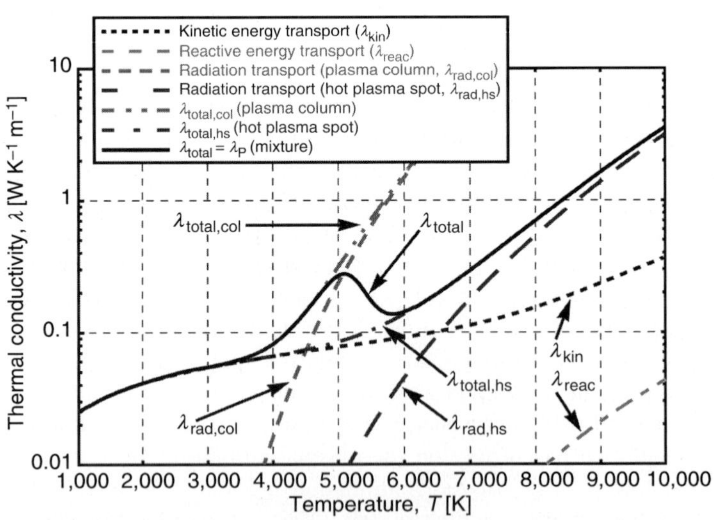

Fig. 2.26. Thermal conductivity λ (and components) of a D2 automotive headlight lamp plasma (Xe, Hg, NaI, and ScI_3 with partial pressures of 50, 20, 0.03, and 0.05 bar, respectively) vs. plasma temperature [59]

2.4 Electrode Properties

Apart from the plasma, the electrode properties are important for the operation and understanding of HID lamps. The electrodes (typically tungsten electrodes, sometimes in combination with an emitter which enhances electron emission) are responsible for the current transfer into the plasma, i.e., the cathode emits electrons and might be bombarded by an ion current, the anode collects electrons. Moreover, the electrodes are relatively cold compared to typical plasma temperatures in HID lamps. The electrode tips have typically temperatures around 2,000–3,500 K, the typical plasma temperatures in the plasma column are around 4,000–7,000 K, at the hot plasma spots the plasma temperature might be as high as 10,000 K. Thus, the electrodes act as heat sinks for the plasma, i.e., a power flux from the plasma to the electrode tips maintains the electrode temperature. This heat is passed through the electrodes to the outside connections of the electrodes (electrode root) and power is radiated from the electrode surfaces according to the Stefan–Boltzmann law. This constant drain of power out of the plasma must be balanced by an increased electrical power input into the near electrode plasma compared to the electrical power input into the plasma column. This additional power demand is the reason for the so-called electrode fall voltages (anode fall voltage and cathode fall voltage), i.e., the additional voltage drop in the near electrode plasma to balance electrode losses (cf. Sect. 3.2.2 and Chap. 4).

The electrodes are also important for the lifetime and maintenance of the lamp. A hot electrode might evaporate tungsten, which blackens the walls by condensing on the inside of the quartz vessel if no halogen cycle prevents this (cf. Sect. 1.3.6). This process can be enhanced by an ion current at the cathode, sputtering tungsten from the cathode. Wall blackening affects the light output and increases the thermal heat load to the lamp walls. Decreasing luminous efficacy (lumen depreciation) and softening or breaking of the quartz vessel might be the result of such a process. Moreover, the electrode shape might change during lifetime of the lamp due to melting of the electrode tips, tungsten transport by evaporation and condensation of tungsten at the electrodes, or sputtering due to an ion current at the cathode. Important examples are the modification of the electrode tips in UHP lamps [45, 119, 136], cf. Sect. 3.1.13), different electrode tip geometries in operating D2 automotive headlight lamps and other HID lamps [59, 85, 86], Sects. 4.4.4 and 4.4.6), cold end attacks (electrode root erosion due to tungsten evaporation, cf. [75]), or theoretical studies ([57] and Sect. 4.4.4). The change of the electrode shape modifies the electrode properties (e.g., electrode fall voltages, arc attachment at electrode tip) or might lead to total failure of the lamp.

2.4.1 Thermal Conductivity of Tungsten

The thermal conductivity of metals can be derived analogous to the derivation of the thermal conductivity of plasmas. In metals, the bulk of the thermal current is carried by the conduction electrons, which can be treated as an electron gas in metals. Thermal conductivity by the atoms is much less important than thermal conduction by the conduction electrons [3, 72]. Nevertheless, since most electrodes of HID lamps are made of tungsten (sometimes in combination with an emitter which enhances electron emission), it is sufficient and more precisely to use experimentally determined thermal conductivities. The thermal conductivity of tungsten in the temperature range from 300 to 3,500 K is given in Fig. 2.27 [94]. Compared to the thermal conductivity of the plasma (0.01–10 W K^{-1} m^{-1}), tungsten has a much higher thermal conductivity, i.e., the power flux from the plasma to the electrode tips is very efficiently passed through the electrode to the outside connections of the electrodes (electrode root).

The higher the thermal heat flux inside the electrodes, the cooler the electrode tips and the higher the power loss of the plasma (electrode fall voltages) due to the electrodes. Since the thermal conductivities of tungsten is fixed, a variation of the diameter of the electrodes can modify the heat flux inside the electrodes. A larger electrode diameter corresponds to lower electrode temperatures and higher electrode fall voltages, a smaller electrode diameter increases the electrode temperature and decreases the electrode fall voltages. The electrodes must be designed to have a low temperature (large diameter) to prevent evaporation of tungsten from the hot electrode. At the same time, the electrodes should have a small diameter to restrict the power loss of the plasma through the electrodes and the cathode should be hot enough to emit sufficient electrons to sustain the lamp current (cf. Sect. 2.4.4). If the electron emission current is too low (cathode too cold), a substantial part of the lamp current is carried by ions near the cathode surface, which might be a problem due to sputtering of electrode material. The optimal electrode geometry is thus a question of many parameters like lamp current, type of plasma, temperature-sensitivity of emitter material, work function of electrode (cf. Sect. 2.4.4), etc.

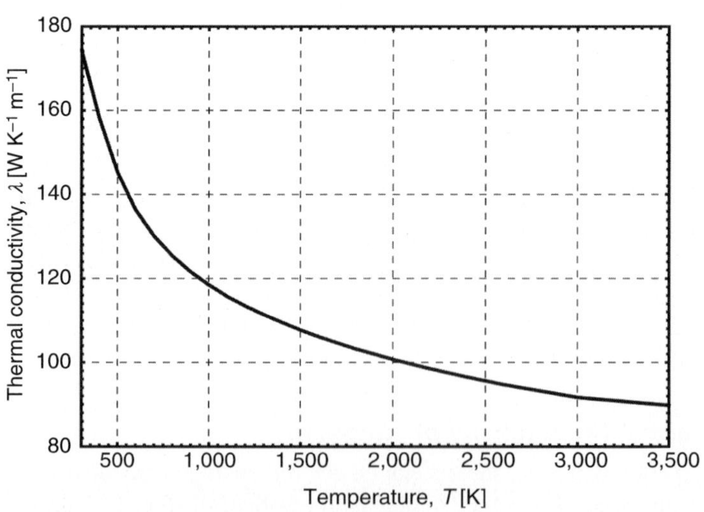

Fig. 2.27. Thermal conductivity λ of tungsten vs. plasma temperature [94]

2.4.2 Mass Density and Specific Heat of Tungsten

The mass density ρ_E of tungsten is 19,300 kg m^{-3}, the specific heat c_E is described by 129 + 0.0175T according to [31] (T in K, c_E in J kg^{-1} K^{-1}). The index E in ρ_E and c_E refers to the electrodes which are normally made of tungsten.

2.4.3 Black Body Radiation

Planck's Radiation Formula

The electromagnetic radiation from a solid state can be described by Planck's law of black body radiation. A black body is an object that absorbs all electromagnetic radiation (e.g., light) that falls onto it: no light passes through it nor is reflected. Despite the name, black bodies do radiate electromagnetic radiation. The ideal black body is a cavity with a small hole: The hole absorbs practically no energy from the outside due to its smallness, and the radiation escaping through the hole does not disturb the energy balance in the inside of the cavity noticeably. A careful analysis of the radiation inside this cavity leads to Planck's law of black body radiation. The assumptions are that the cavity walls are in thermal equilibrium with the radiation, no power input into or output from the cavity is present. All walls have the same temperature. Furthermore, it is assumed that the electromagnetic modes in the cavity are quantized in energy with the quantum energy equal to Planck's constant times the frequency ($h\nu$). The average energy per electromagnetic mode is the energy of the mode (i.e., $h\nu$) times the probability that it will be occupied. This probability is given by the Bose–Einstein distribution function, which is different to the Boltzmann or Maxwell–Boltzmann distribution discussed in Sect. 2.1.1. The Boltzmann distribution, which is proportional to $\exp(-E/kT)$ is valid for identical but distinguishable particles, like atoms or molecules. The Bose–Einstein distribution function f_{BE} is valid for identical but indistinguishable particles with integer spin (bosons), e.g., photons:

$$f_{BE}(E) = \frac{1}{e^{\frac{E}{kT}} - 1}. \tag{2.39}$$

The average energy per electromagnetic mode, i.e., the energy of the mode $h\nu$ times the probability that it will be occupied ((2.39)), must be multiplied with the number of modes per unit frequency and unit volume[9] to get the energy density per unit frequency $\rho(\nu,T)$ inside the cavity:

[9] A mode for an electromagnetic wave in a cavity must satisfy the condition of zero electric field at the wall. If the mode is of shorter wavelength, there are more ways it can be fit into the cavity to meet that condition. Careful analysis by Rayleigh and Jeans showed that the number of modes is proportional to the frequency squared ($8\pi\nu^2/c^3$, e.g., [72, 127]).

$$\mathrm{d}\rho(\nu,T) = \frac{8\pi h\nu^3}{c^3}\frac{1}{\mathrm{e}^{h\nu/kT}-1}\mathrm{d}\nu. \tag{2.40}$$

When transforming (2.40) into the energy density per unit wavelength $\rho(\lambda,T)$ inside the cavity, we have to use the relation $c = \lambda\nu$, i.e., $\mathrm{d}\nu = -c\lambda^{-2}\,\mathrm{d}\lambda$, so that we get

$$\mathrm{d}\rho(\lambda,T) = \frac{8\pi hc^3}{\lambda^5}\frac{1}{\mathrm{e}^{hc/\lambda kT}-1}\mathrm{d}\lambda. \tag{2.41}$$

We are not interested in the energy density per unit frequency $\rho(\nu,T)$ or the energy density per unit wavelength $\rho(\lambda,T)$ inside the cavity, but the radiated power per unit area $L(\lambda,T)$ from a surface at temperature T. Therefore, we have to multiply (2.41) by the velocity of the photons, i.e., the speed of light c. Moreover, the energy density given in (2.41) is for thermal equilibrium, so setting inward = outward radiation gives a factor of 1/2 for the radiated power outward. Furthermore, the averaging over all angles gives another factor of 1/2 for the angular dependence which is the square of the cosine [48, 72, 127]. The resulting radiated power per unit area $L(\lambda,T)$ is thus:

$$L(\lambda,T) = \frac{2\pi hc^2}{\lambda^5}\frac{1}{e^{hc/\lambda kT}-1}. \tag{2.42}$$

So, by measuring the radiation of a black body at a given (arbitrary) wavelength λ (filter), we can calculate the temperature T of this black body. In Fig. 2.28 the radiated power per unit area $L(\lambda,T)$ is drawn as a function of λ for various temperatures between 2,000 and 3,200 K. Even for temperatures as high as 3,200 K, the emitted radiation is mainly in the infrared part of the electromagnetic spectrum and only a small part of the radiation is emitted in the visible (cf. Figs. 1.8–1.10). This explains, for example, the poor luminous efficacy of incandescent lamps (cf. Sect. 1.3.2).

Wien's Law and Wien's Displacement Law

If we are interested in the radiation in the visible part of the electromagnetic spectrum, e.g., for temperature measurements at the electrodes using wavelengths in the visible (cf. Sects. 3.1.1–3.1.7), we can use an approximation: For short wavelengths λ with $\lambda \ll hc/kT$, the exponential function in (2.42) is much greater than 1, i.e., the 1 in the denominator of (2.42) can be neglected, leading to Wien's law:

$$L(\lambda,T) = \frac{2\pi hc^2}{\lambda^5}\mathrm{e}^{-hc/\lambda kT} \tag{2.43}$$

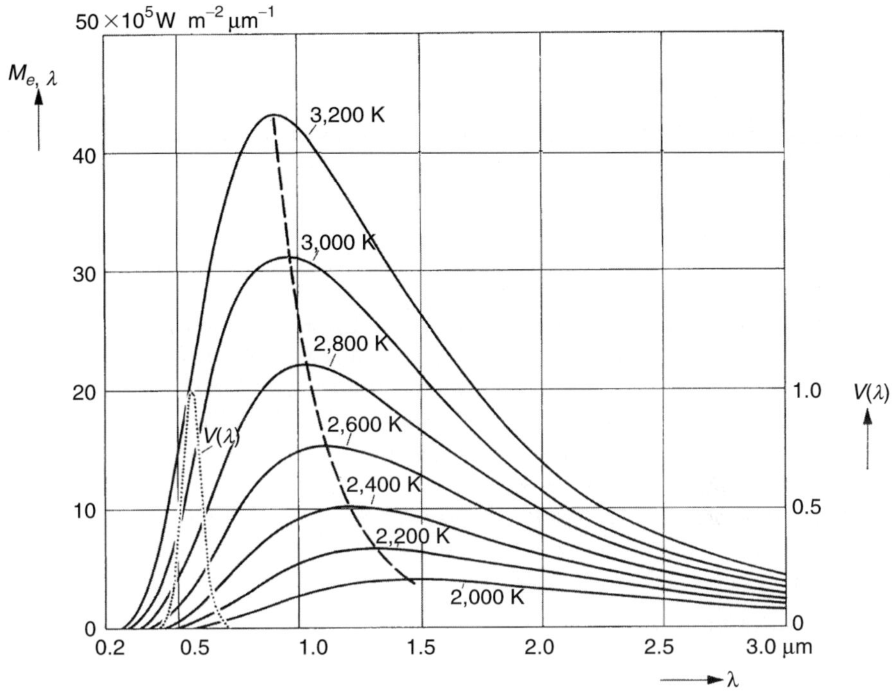

Fig. 2.28. Radiated power per unit area $L(\lambda,T)$ for different temperatures vs. wavelength λ superimposed by the sensitivity curve $V(\lambda)$ of the light adapted human eye [48]

Wien's law is applicable for $hc/\lambda kT \gg 1$, which is fulfilled for wavelengths in the visible part of the electromagnetic radiation and temperatures below 3,700 K (cf. Sect. 3.1.3). Equation (2.43) is easier to use than (2.42), especially the conversion of (2.43) into an expression for temperature T as a function of (the measured) $L(\lambda,T)$ is straightforward. The Rayleigh–Jeans law, which can be regarded as an approximation to Planck's law for $hc/\lambda kT \ll 1$, is not important in the context of this book. Historically, the approximations to Planck's law for short and long wavelengths (Wien's law and Rayleigh–Jeans law, respectively) have been known before Planck postulated (2.42), later they turned out to be approximations to the general Planck radiation formula.

Planck's radiation formula has a maximum at a certain wavelength $\lambda_{max}(T)$, which is unique for a given temperature (cf. Fig. 2.28). By measuring the emitted radiation of a black body of unknown temperature one can determine the temperature by identifying $\lambda_{\max}(T)$. $\lambda_{\max}(T)$ can be calculated by setting the first derivative of $L(\lambda,T)$ from (2.42) zero with respect to λ, which yields:

$$T\lambda_{\max}(T) = 2.898\times10^{-3}\,\mathrm{m\,K}. \tag{2.44}$$

This relationship is called Wien's displacement law. When the temperature T of a black body radiator increases, the overall radiated energy increases (cf.

"Stefan Boltzmann Law" described later) and the peak of the radiation curve moves to shorter wavelengths $\lambda_{max}(T)$ (cf. Fig. 2.28). Since the transformation from wavelength λ to frequency ν cannot be simply done by $c = \lambda\nu$ (see earlier), (2.42) cannot be transformed straightforward into an expression involving ν_{max}. $\nu_{max}(T)$ can be calculated, e.g., by setting the first derivative of $\rho(\nu,T)$ zero with respect to ν, which yields:

$$\frac{\nu_{max}(T)}{T} = 5.88\times10^{10}(\mathrm{s\,K})^{-1}. \tag{2.45}$$

Considering the black body radiation at $T = 5{,}000$ K, the peak of the radiation curve plotted as a function of wavelength λ is at $\lambda_{max}(T = 5{,}000\text{ K}) = 579$ nm, i.e., very close to the maximum of the human eye sensitivity curve at 555 nm (photopic vision, cf. Fig. 1.3). Plotting the black body radiation at $T = 5{,}000$ K as a function of frequency, the maximum is at $\nu_{max}(T = 5{,}000\text{ K}) = 294\times10^{12}$ Hz, corresponding to $\lambda = 1.019$ µm, which is in the infrared, i.e., the position of the maximum of Planck's radiation formula depends on the variable used on the horizontal axis. This should be kept at the back of one's mind if the maximum of Planck's radiation formula is discussed.

Stefan–Boltzmann Law

Integrating the radiated power per unit area $L(\lambda,T)$ from (2.42) from $\lambda = 0$ to ∞, the total power flux of a black body radiator can be obtained:

$$P = \varepsilon\sigma_{SB}AT^4. \tag{2.46}$$

This relationship is called Stefan–Boltzmann law[10]. ε is the emissivity of the object ($\varepsilon = 1$ for a black body, cf. "Kirchhoff's Law and Emissivity" later), σ_{SB} the Stefan–Boltzmann constant, and A the surface area. If the object under consideration is radiating power to its cooler or hotter surrounding of temperature T_s, the net radiation loss rate takes the form

$$P = \varepsilon\sigma_{SB}A(T^4 - T_s^4). \tag{2.47}$$

Because of the fourth power of T in (2.47), T_s can be neglected if $T \gg T_s$. The Stefan–Boltzmann law can be used to calculate the power balance of electrodes in HID lamps. The power input into the electrode tip by the plasma is radiated from the surface and conducted through the inside of the electrode. The missing factor is the emissivity ε, which is described in "Kirchhoff's Law and Emissivity".

---

[10] The Boltzmann or Maxwell–Boltzmann distribution and Boltzmann's constant originate and are named after Ludwig Boltzmann (1844–1906), who was an Austrian physicist famous for his important contributions to statistical mechanics. The Stefan–Boltzmann law was found in 1879 by Josef Stefan (1835–1893) and theoretically explained by Ludwig Boltzmann in 1884. Planck's radiation formula was formulated in 1900.

Kirchhoff's Law and Emissivity

The ideal black body, which absorbs all electromagnetic radiation that falls onto it without letting it passing through or reflecting it, is a very theoretical object. The absorption and the emission of a real surface is smaller than that of an ideal black body. The radiation of a real surface can be characterized by the spectral emissivity $\varepsilon(\lambda,T)$ describing the ratio of the radiation of the surface to that of an ideal black body of the same area and the same temperature. The spectral emissivity $\varepsilon(\lambda,T)$ normally depends on the wavelength λ under consideration and the temperature T of the surface, but it is always smaller or equal to 1.[11] Kirchhoff's law (also Kirchhoff's law of thermal radiation) connects the spectral emissivity $\varepsilon(\lambda,T)$ of an object to its spectral absorptivity (or absorbance or absorption coefficient) $\alpha(\lambda,T)$, which is the fraction of incident radiation that is absorbed by the surface. An ideal black body has an absorptivity equal to 1, i.e., the spectral emissivity and the spectral absorptivity are equal in the case of a black body. According to Kirchhoff's law, this is true for all surfaces, i.e.,

$$\varepsilon(\lambda,T)=\alpha(\lambda,T). \tag{2.48}$$

Kirchhoff's law can be derived by simple considerations: If an ideal black body and a "real" surface face each other and no radiation can escape (e.g., by mirrors), i.e., the black body radiates on the "real" surface and vice versa; thermal equilibrium states that both surfaces will have the same temperature if no additional power is brought into or taken out of the system. If the emissivity ε of the "real" surface would be higher than its absorptivity α, the resulting net power transfer to the black body would violate thermal equilibrium. The same is true for the case $\alpha > \varepsilon$. Kirchhoff's law is sometimes recapitulated by "poor reflectors are good emitters and good reflectors are poor emitters." This is the reason why thermos flask or emergency thermal blankets are typically made of reflective material: They lose little heat by radiation.

In the case of electrodes, the spectral emissivity is given by the spectral emissivity of tungsten. We can distinguish between two types of emissivities: For the power balance of the electrodes (2.46) or (2.47) we need an integral emissivity $\varepsilon(T)$, i.e., the emissivity does not depend on the wavelength λ. If the temperature of the electrodes shall be determined by measuring the radiation at a certain wavelength (pyrometry, cf. Sects. 3.1.1–3.1.7), we need the wavelength-sensitive spectral emissivity $\varepsilon(\lambda,T)$. The integral emissivity $\varepsilon(T)$ can be deduced from the spectral emissivity $\varepsilon(\lambda,T)$ but not the opposite way around. Both ($\varepsilon(T)$ and $\varepsilon(\lambda,T)$) depend in general on the temperature T of the surface. Planck's law for the radiated power per unit area $L(\lambda,T)$ from (2.42) adapted for a real surface is given by:

---

[11] There are some exception, for example if the object is fluorescent. Then, incident power at one wavelength is "reflected" at another wavelength, i.e., the spectral emissivity at some wavelengths can exceed unity, but not the emissivity integrated over all wavelengths.

$$L(\lambda,T)=\varepsilon(\lambda,T)\frac{2\pi hc^2}{\lambda^5}\frac{1}{e^{hc/\lambda kT}-1}. \quad (2.49)$$

Accordingly, Wien's law from (2.43) is given for a real surface by

$$L(\lambda,T)=\varepsilon(\lambda,T)\frac{2\pi hc^2}{\lambda^5}e^{-hc/\lambda kT}. \quad (2.50)$$

The integral emissivity $\varepsilon(T)$ of tungsten given in Fig. 2.29 can be expressed by a simple formula [174]:

$$\varepsilon(T)=-0.0266+1.8197\times10^{-4}\frac{T}{K}-2.1946\times10^{-8}\frac{T^2}{K^2}, \quad (2.51)$$

where the surface temperature T is given in Kelvin [K].

The spectral emissivity $\varepsilon(\lambda,T)$ of tungsten is shown in Fig. 2.30 (logarithmic scale). As already mentioned in Sect. 1.3.2, the spectral emissivity $\varepsilon(\lambda,T)$ of tungsten is high in the visible part[12] of the spectral power distribution (about 0.4–0.5), but low in the infrared part of the spectral power distribution (0.15–0.4). This enhances the efficacy of an incandescent lamp in comparison to a theoretical black body radiator of the same temperature (cf. Fig. 1.9). Increasing temperature decreases the spectral emissivity for wavelengths in the visible but increases the spectral emissivity $\varepsilon(\lambda,T)$ of tungsten for long wavelengths. The integral emissivity $\varepsilon(T)$ increases with increasing temperature.

The emissivity ε might be a problem, because it depends on the surface conditions of the electrodes [9, 43, 130]. A rough surface has a considerably higher emissivity ε than a polished one (black body effect). This problem arises especially at the electrode tips, where the surface conditions might change with time. This is an important point for the analysis of pyrometric measurements, where the exact value of $\varepsilon(\lambda,T)$ is often important (cf. Sect. 3.1.3). If the front surface of an electrode gets deeply structured with large bumps as shown in [86], not only the emissivity ε might change, but also the emitting surface itself is increased.

2.4.4 Electron Emission

In addition to photon emission, the surface of a solid state also emits electrons. The photon emission is described by Planck's law as described earlier. The electron emission current is composed of three parts: thermal electron emission, field-enhanced thermal electron emission, and field emission. In addition to these electron emission processes, electrons can also be released from electrodes

[12] λ between 380 (violet) and 780 nm (red) or between 0.38 and 0.78 μm.

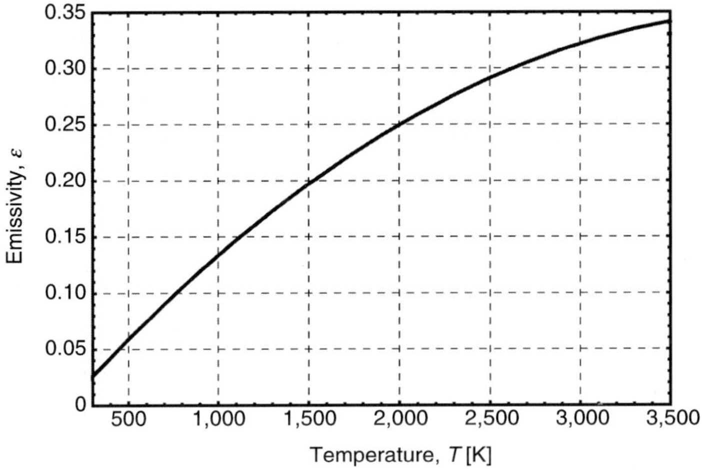

Fig. 2.29. Integral emissivity $\varepsilon(T)$ of tungsten vs. temperature [174]

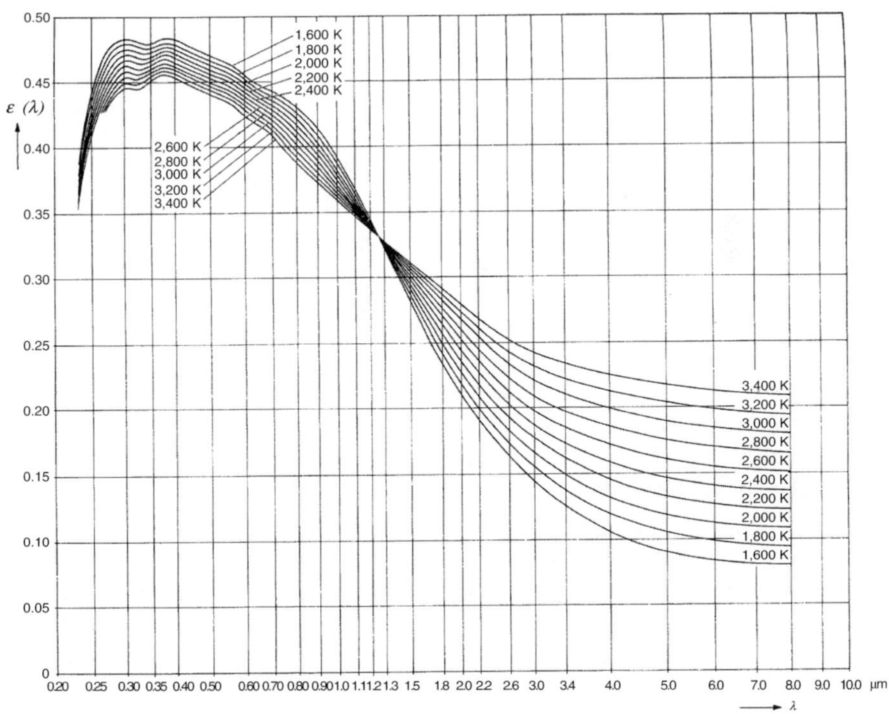

Fig. 2.30. Spectral emissivity $\varepsilon(\lambda,T)$ of tungsten as a function of wavelength for several temperatures [48]

by photons (photoemission), electrons (secondary emission), ions (γ-effect or secondary emission), or excited atoms. All these mechanisms will be discussed in the following.

A sufficiently high electron emission current is of particular importance at the cathode: If the electron emission current is less than the total lamp current, the difference is sustained by an ion current. The generation of this ion current in the near-cathode plasma consumes energy, which is reflected in a high cathode fall voltage. The ions deposit their energy (minus the energy necessary to release an electron to neutralize the ion) at the cathode, heating thus the cathode and enhancing electron emission, which depends on the local cathode temperature (see later). The enhanced electron emission current reduces the ion current, so that the interplay between ion current and electron emission current adjusts the ratio of electron to ion current until a stable power balance of the cathode is established (cf. Chap. 4, especially Sect. 4.4). Besides the power consumption for the generation of the ion current, the ion current at the cathode sputters tungsten from the cathode. This effect might lead to wall blackening or unwanted alteration of electrode geometry.

Thermal Electron Emission

To understand the emission of electrons from tungsten, we have to understand how electrons can be described in a metal. The starting point is the Drude model of electrons in a metal. In a metallic crystal lattice, the outer electron orbits (valence electrons) overlap and are shared by all the atoms in the solid. These electrons are not bound and are free to conduct current. Typical free electron densities in a metal are between 10^{28} and 10^{29} m^{-3}. The electrons form a "gas". At absolute zero temperature (0 K), all electrons pack into the lowest available energy states and build up a so-called "Fermi sea" of electron energy states. The reason for this is that electrons are fermions and by the Pauli exclusion principle cannot exist in identical energy states. The distribution function $f(E)$, which is the probability that an electron is in energy state E, is given by the Fermi–Dirac distribution.

The already discussed Boltzmann distribution (Sect. 2.1.1), which is proportional to $\exp(-E/kT)$ is valid for identical but distinguishable particles, like atoms or molecules. The Bose–Einstein distribution function, which was discussed in the context of Planck's law (Sect. 2.4.3), is valid for identical and indistinguishable particles with integer spin (bosons). The distribution function $f_{\mathrm{F}}(E)$ applicable to fermions (identical and indistinguishable particles with half-integer spin), is the Fermi–Dirac distribution given by

$$f_{\mathrm{F}}(E) = \frac{1}{\mathrm{e}^{E-E_{\mathrm{F}}/kT}+1}, \tag{2.52}$$

where E_{F} is the so-called Fermi energy or Fermi level, which depends on the metal under consideration (tungsten: $E_{\mathrm{F}} \cong 7$ eV). The Fermi energy is the maximum energy occupied by an electron in a metal at 0 K, i.e., the Fermi energy is the surface

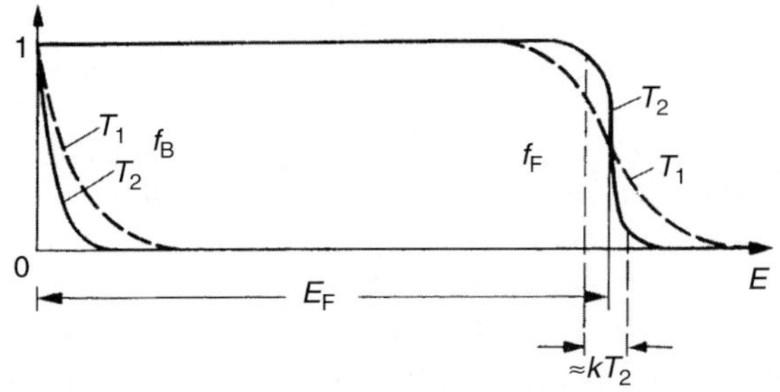

Fig. 2.31. Visualization of Fermi–Dirac distribution f_F in comparison with Boltzmann distribution f_B for two different temperatures $T_1 > T_2$ vs. electron energy E [72]

of that "Fermi sea" at absolute zero temperature where no electron has enough energy to rise above this surface. Figure 2.31 compares the Fermi–Dirac distribution f_F with the Boltzmann distribution f_B for two different temperatures $T_1 > T_2$. For both temperatures, the energy states below the Fermi energy E_F have a probability of essentially 1, those above E_F essentially zero. Only in a small energy band around E_F, the distribution differs from 1 and 0. The width of this energy band is proportional to kT. Compared to typical values of E_F of 1–10 eV, kT is relatively small ($kT = 0.026$ eV for $T = 300$ K and $kT = 0.259$ eV for $T = 3{,}000$ K).

The Fermi energy can be calculated by considering the possible energy states electrons can occupy in a metal: Fermions are not allowed to be in the same state, i.e., they must differ in their characterizing properties. The electrons in a metal are not localized, i.e., their uncertainty of position is given by the dimensions of the metal L. According to Heisenberg's uncertainty principle, the momentum p of an electron can then not be given more precisely than with an uncertainty of $\Delta p > h/L$, with h being Planck's constant. Replacing ">" by "=" we get

$$\Delta p = \frac{h}{L}. \tag{2.53}$$

So, every volume $(\Delta p)^3$ in the momentum space can only be occupied by two electrons (spin up and spin down). If we want to find the configuration "consuming" the least energy, we have to "construct" a sphere in momentum space with radius p_F containing all electrons from the electron "gas" (the so-called Fermi sphere).

The density of electrons can be calculated from the density of the metal (one electron from each atom), i.e., we know the number of electrons N per volume $V = L^3$. If we assume the lowest energy distribution and bear in mind the minimum volume of each electron in momentum space given by $0.5\,(\Delta p)^3$ (spin up and spin down), then the volume in momentum space is given by $N\,0.5\,(\Delta p)^3$. This volume must be identical to the volume of the Fermi sphere in momentum sphere, i.e.,

$$N\frac{\Delta p^3}{2} = \frac{4\pi}{3}p_F^{\ 3}. \tag{2.54}$$

Combining (2.53) and (2.54) we get for p_F:

$$p_F = 3\sqrt{\frac{3n}{8\pi}}\,h. \tag{2.55}$$

Because $E_F = (p_F)^2/2m_e$, we get for the Fermi energy E_F:

$$E_F = \frac{h^2}{2m_e}\left(\frac{3n}{8\pi}\right)^{2/3}. \tag{2.56}$$

Following these considerations, we can calculate the density of possible states for electrons as a function of energy by expressing n from (2.56) as a function of E and differentiating n by E:

$$\rho(E) = \frac{\mathrm{d}n}{\mathrm{d}E} = 4\pi\left(\frac{2m_e}{\mathrm{h}^2}\right)^{3/2}\sqrt{E}. \tag{2.57}$$

The Fermi–Dirac distribution given in (2.52) determines the probability that a given energy state will be occupied by an electron. This probability must be multiplied by the density of states function (2.57) to weight the probability by the number of states available at a given energy. Thus we get the number of conduction electrons per unit volume per unit energy:

$$\rho(E) f_F(E) = 4\pi\left(\frac{2m_e}{h^2}\right)^{3/2}\frac{\sqrt{E}}{e^{E-E_F/kT}+1}. \tag{2.58}$$

To release electrons from a metal, they must have enough energy to escape from the solid state. This energy is given by the so-called work function φ, which is defined as the distance from the Fermi energy to the ionization potential of the metal. Typical values for φ are between 2 and 6 eV (cf. Fig. 2.32).

The thermal electron emission current is the electron current due to the energy distribution of the electrons in the metal (2.58). Depending on temperature T, some electrons have sufficient energy to escape from the metal, i.e., they have a kinetic energy greater than the work function φ plus the fermi energy E_F. To calculate the thermal electron emission current, one must integrate (2.58) over all electrons having a velocity component perpendicular to the metal surface with kinetic energy greater than the work function φ plus the fermi energy E_F:

$$j_{\text{Richardson}} = A_R T^2 e^{-\varphi/kT}, \tag{2.59}$$

where A_R is the Richardson constant (theoretical value: $A_R = 1.2017\times10^6$ A m^{-2} K^{-2}). Equation (2.59) is called Richardson or Richardson–Dushman equation. Since only electrons with sufficiently high energy escape from the metal, the metal is cooled by this electron emission. The remaining electrons will redistribute their energy resulting in a lower average energy of the remaining electrons. At room temperature, there is virtually no thermal electron emission current, only at temperatures above 1,000 K there is a notable current. This thermal emission current is not only relevant in discharge lamps, it is also important, e.g., for television sets and other monitors for computers, automated teller machines, video game machines, etc. which are based on cathode ray tubes (CRT) and for vacuum tubes (diodes, triodes, magnetrons, etc.).

Values for A_R and φ for different metals can be found, e.g., in [130] and [168] (quoting [2]). For pure tungsten, we find values for φ between 4.2 and 6.0 eV with A_R equal to 0.75×10^6 A K^{-2} m^{-2} [130] or 600×10^6 A K^{-2} m^{-2} [167]. For thoriated tungsten, φ is 2.6 eV [130, 168] with A_R equal to 0.5×10^6 A K^{-2} m^{-2} [130] or 0.03×10^6 A K^{-2} m^{-2} [168]. In [145] and [146] the work function φ in an operating mercury HID lamp was measured to be 4.52 eV (pure tungsten electrode) and 3.7 eV (thoriated tungsten electrode), respectively. Moreover, φ might depend, for example, on the surface state (crystal orientation of tungsten, polycrystalline tungsten), for doped tungsten the local surface concentration of the dopant is important, the ambiance influences φ (in general measurements are done in vacuum, but the real electrodes are operated in different plasmas), etc. So, although the electron emission is very important for the numerical simulation of electrodes in HID lamps, we have to live with poor data concerning φ and A_R (cf. Sect. 4.4.6). Within this book, $\varphi = 4.5$ eV for pure tungsten electrodes, and 2.6 or 3.5 eV for thoriated tungsten is chosen. A_R is set to the theoretical value of approximately 1.2×10^6 A K^{-2} m^{-2} (as e.g., in [30, 32, 53–65, 162, 168].

Field-Enhanced Thermal Electron Emission

The work function φ depends on the material as described earlier. Moreover, the work function can be lowered as shown in Fig. 2.32: The potential energy curve of an electron escaping from the metal is bend down by the superposition of the undisturbed potential energy curve for the electron and the potential for the electron in an external electric field. The resulting lowering of the work function $\Delta\varphi$ (in eV) is given by:

$$\Delta\varphi = e\sqrt{\frac{eE_c}{4\pi\varepsilon_0}}, \tag{2.60}$$

where E_c is the electric field at the metal surface and ε_0 the permittivity constant ($\varepsilon_0 = 8.85418\times10^{-12}$ A s V^{-1} m^{-1}). This lowering of the work function is called the Schottky correction formula, the resulting emission current is given by the Richardson–Schottky equation:

$$j_{\text{Richardson–Schottky}} = A_R T^2 \exp\left(-\frac{\varphi - e\sqrt{\frac{eE_C}{4\pi\varepsilon_0}}}{kT}\right). \tag{2.61}$$

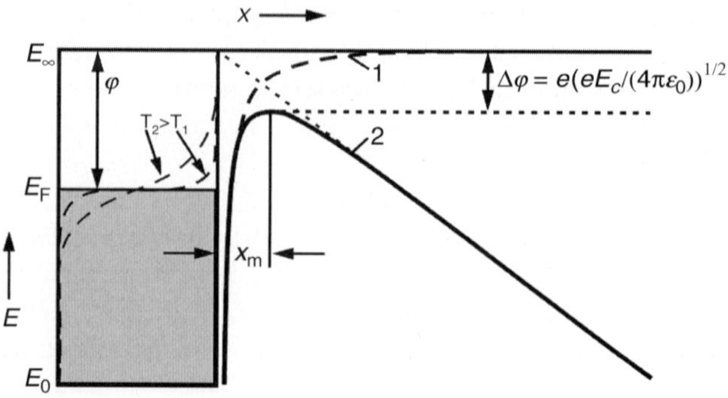

Fig. 2.32. Visualization of the lowering of the work function φ by superposition of the potential for the electron without external electric field (curve 1) with the potential for the electron in an external electric field without the metal (straight broken line) to the resulting potential energy curve (curve 2) as a function of distance x to metal surface

Field Emission

The Schottky effect is accompanied by a tunneling of electrons through the finite width of the barrier (cf. Fig. 2.32). Increasing electric fields decrease the width of the barrier and thus increase the probability for electron tunneling. Nevertheless, the electric field must be above 10^9 V m^{-1} to yield a considerable contribution of pure field emission to the total emission current j_{emission} (the sum of thermal emission, field-enhanced thermal emission, and field emission). Typical electric fields at the cathode of HID lamps are normally below 10^9 V m^{-1}. For completeness, j_{emission} is always calculated within this book taking thermal emission, field-enhanced thermal emission, and field emission into account. The governing equations taking thermal emission, field-enhanced thermal emission, and field emission into account are given by Dyke and Dolan in [47]:

$$j_{\text{emission}} = \frac{A_R T}{k} \int_{-\infty}^{\infty} N(T,\varepsilon) D(E_c,\varepsilon,\varphi) \, d\varepsilon, \tag{2.62}$$

$$N(T,\varepsilon) = \ln\left(1 + e^{-\varepsilon/kT}\right), \tag{2.63}$$

$$D(E_c,\varepsilon,\varphi) = \exp\left\{-\frac{8\pi}{3h}\sqrt{2m_e e}\,\frac{(\varphi-\varepsilon)^{3/2}}{E_e} f\left(\sqrt{\frac{e}{4\pi\varepsilon_0}}\frac{\sqrt{E_c}}{\varphi-\varepsilon}\right)\right\}. \tag{2.64}$$

$N(T,\varepsilon)$ is a measure for the number of electrons in the metal having kinetic energy ε (perpendicular to the metal surface), $D(E_c,\varepsilon,\varphi)$ gives the probability for an electron with energy ε to tunnel through the finite width of the barrier if the work function φ is reduced by the electric field E_c. The function f (in the exponent of (2.64)) is a dimensionless function with values between 0 and 1. This function is tabulated by Nordheim[13] and can be found, e.g., in [47].

For high electric fields well above 10^9 V m^{-1}, (2.62) passes over to pure field emission, for lower electric fields it yields the same results as (2.61). Figure 2.33 compares three different approaches for the calculation of the electron emission current as a function of the electric field for two different work function φ (double logarithmic scale in Fig. 2.33). The Richardson–Schottky approach ((2.61), solid line in Fig. 2.33) includes thermal electron emission and field-enhanced thermal electron emission. The approaches of Dyke and Dolan ([47], used within this book, dashed line in Fig. 2.33) and Paulini et al. ([135], dotted line in Fig. 2.33) include additionally pure field emission. All three approaches agree on the electron emission current for a given temperature (here 3,000 K) and an electric field up to 10^8 V m^{-1}. Only for electric fields above 10^9 Vm^{-1}, the effect of field emission comes into play.

Figure 2.34 shows the electron emission current as a function of electrode temperature and electric field (double logarithmic scale in Fig. 2.34). The electron emission current is calculated using (2.62), i.e., thermal emission, field-enhanced thermal emission, and field emission are taken into account. In the case of low electric fields ($E_c < 10^6$ V m^{-1}), the electron emission is dominated by thermal emission, i.e., the electrode temperature determines the emission current. For higher electric fields in the range between 10^6 and 10^9 V m^{-1}, both electrode temperature and electric field determine the emission current, i.e., we have field-enhanced thermal emission. For very high electric fields (e.g., 10^{10} V m^{-1}), the emission current depends only on the electric field, i.e., we have pure field emission. The work function φ is (besides the temperature and the electric field) the most important factor determining the electron emission current (compare Figs. 2.33 and 2.34).

Other Electron Emission Processes

In addition to these electron emission processes, electrons can also be released from electrodes by photons (photoemission, photoelectric effect), electrons (secondary emission), ions (γ-effect or secondary emission), or excited atoms.

The photoemission current requires photons of sufficiently high energy, i.e., sufficiently small wavelength to release an electron from the electrode (e.g., $\varphi = 4.5$ eV corresponds to $\lambda = 275$ nm). During normal lamp operation, the photoemission current can be neglected due to the small photoelectric yield, but the effect can be used for an in situ measurement of the work function (cf. Sect. 3.1.11,

[13] The electron emission in the case of pure field emission is described by the so-called Fowler–Nordheim equation. Equation (2.62) includes thermal emission, field-enhanced thermal emission, and field emission.

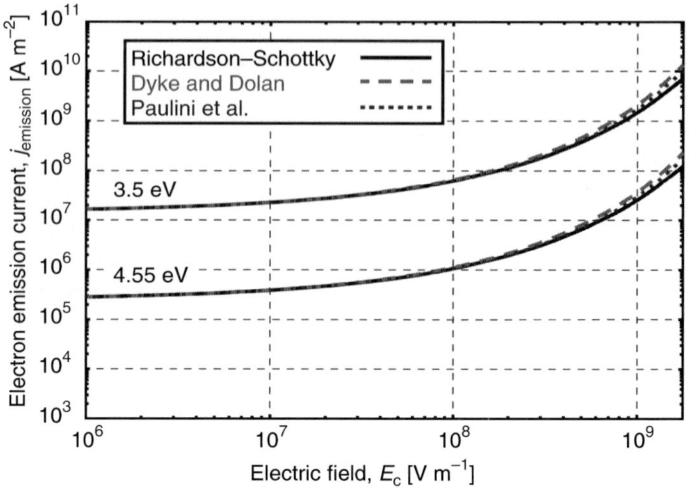

Fig. 2.33. Electron emission current j_{emission} for $T = 3{,}000$ K vs. electric field for two work functions $\varphi = 3.5$ eV and 4.55 eV computed with three different models [59]

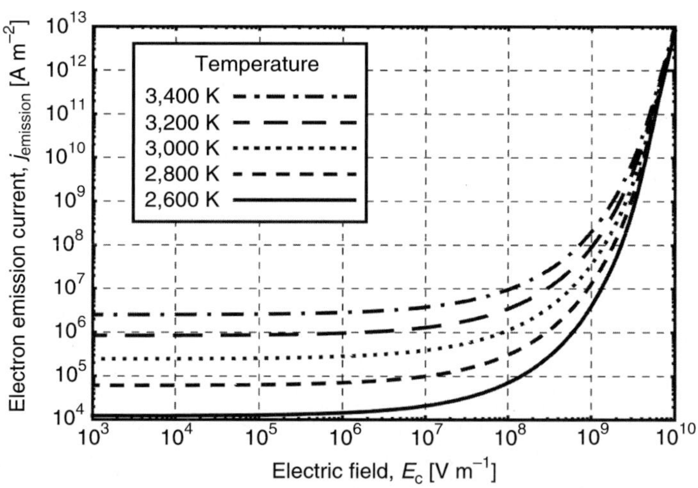

Fig. 2.34. Electron emission current j_{emission} vs. electric field for several temperatures including thermal emission, field-enhanced thermal emission, and field emission

[145, 146]). Moreover, the photoeffect can be used to reduce the ignition voltage as demonstrated in [136] for an UHP lamp (cf. Sect. 1.3.6).

Secondary electron emission by electrons and ions does not play an important role in HID lamps, but especially the secondary electron emission due to ions falling on the cathode surface is important for low-pressure discharge lamps and during the starting of HID lamps [48, 168].

3

Experimental Investigations

In Chap. 1, high-intensity discharge (HID) lamps were introduced, the basic principles explained, and the differences compared to low-pressure discharge lamps were pointed out. Chapter 2 gave a deeper insight into the gas or plasma properties and the characteristics of the electrodes. In the present chapter, this knowledge will be used to discuss experimental investigations of electrodes and plasma in HID lamps. Experimental investigations are necessary to understand, improve, or develop new high-pressure discharge lamps. Moreover, the overview of the manifold experimental methods to explore electrodes and plasma given within this chapter will improve the principle understanding of HID lamps and their components. The focus of this chapter will be on the experimental determination of the electrode temperature. This is to explain the challenges and difficulties of experimental methods and to visualize the possible accuracy of experimental investigations using this example. In Sect. 3.1, the investigation of electrode properties like temperature, power balance, or work function will be addressed, Sect. 3.2 is dedicated to the research into plasma properties like temperature, plasma potential, or operating pressure. It will be shown that plasma and electrodes are strongly coupled: The radiation coming from the plasma interferes with the temperature measurements on the electrodes, the plasma potential and especially the electrode fall voltages give information about the power balance of the electrodes, or the plasma composition might influence the work function of the electrodes. This coupling of electrodes and plasma will be further analyzed and studied in Chap. 4, dealing with the numerical investigation of HID lamps. Many other results of the experimental investigations will also appear in Chap. 4 to validate numerical models or to deliver deeper insight into the experimental results by explanations deduced from numerical results.

3.1 Electrode Diagnostics

The electrode properties are important for the operation and understanding of HID lamps. The electrodes (typically tungsten electrodes, sometimes in combination with an emitter) are responsible for the current transfer into the plasma, i.e., the cathode emits electrons and might be bombarded by an ion current, the anode collects electrons. Moreover, the electrodes are relatively cold compared to typical plasma temperatures in HID lamps. The electrode tips have typically temperatures around 2,000–3,500 K, the typical plasma temperatures in the plasma column are around 4,000–7,000 K, at the hot plasma spots the plasma temperature might be as high as 10,000 K. Thus, the electrodes act as heat sinks for the plasma, i.e., a power flux from the plasma to the electrode tips maintains the electrode temperature. This heat is passed through the electrodes to the outside connections of the electrodes (electrode root) and power is radiated from the electrode surface according to the Stefan–Boltzmann law. This constant drain of power out of the plasma must be balanced by an increased electrical power input in the near electrode plasma compared to the electrical power input into the plasma column. This additional power demand is the reason for the so-called electrode fall voltages (anode fall voltage and cathode fall voltage), i.e., the additional voltage drop in the near electrode plasma to balance electrode losses.

The electrodes are also important for the lifetime and maintenance of the lamp: A hot electrode might evaporate tungsten, which blackens the walls by condensing on the inside of the quartz vessel if no halogen cycle prevents this (cf. Sect. 1.3.6). This process can be enhanced by an ion current at the cathode, sputtering tungsten from the cathode. Wall blackening affects the light output and increases the thermal heat load to the lamp walls. Decreasing luminous efficacy (lumen depreciation) and softening or recrystallization of the quartz vessel might be the result of such a process. Moreover, the electrode shape might change during lifetime of the lamp due to melting of the electrode tips, tungsten transport by evaporation and condensation of tungsten at the electrodes, or sputtering due to an ion current at the cathode. Important examples are the modification of the electrode tips in UHP lamps ([45, 119, 136], cf. Sect. 3.1.13), different electrode tip geometries in operating D2 automotive headlight lamps and other HID lamps ([59, 85, 86], cf. Sect. 3.1.13), cold end attacks (electrode root erosion due to tungsten evaporation, cf. [75]), or theoretical studies ([57] and Sect. 4.4.4). The change of the electrode shape modifies the electrode properties (e.g., electrode fall voltages, arc attachment at electrode tip) or might lead to total failure of the lamp.

Thus, the understanding of the electrodes is important for developing new and further improving existing HID lamps. A substantial interest is in the temperature distribution of the electrodes. Especially the electrode tip temperature is a matter of particular interest, since it determines, for example, the electron emission current and the evaporation rate. Moreover, the electrode might melt and change its shape or an emitter on the cathode surface might evaporate and thus forfeit its duty to reduce the work function.[1] The latter would result in even higher electrode tip temperatures (cf. e.g., Sects. 3.1.12 or 4.4.7). Another interesting question concerns the work function (which is normally measured in a vacuum) in the operating lamp (Sect. 3.1.11). The determination of the electrode fall voltages is another subject of experimental investigations. They reflect the additional power demand in the near electrode plasma and thus effect the luminous efficacy of the lamp. Other questions addressed by experimental investigations are concerned with the electrode tip geometry and its alteration with time and the type of arc attachment to the electrodes (spot or diffuse arc attachment, Sect. 3.1.13).

Unfortunately, experimental investigations of electrodes in HID lamps are rather difficult. Their smallness and optical distortions caused by the quartz bulb are not the only challenges experimental setups have to face (cf. Fig. 3.1). Determining the electrode temperature by pyrometry is impeded, for example, by reflections of the plasma radiation especially at the electrode tip, where the temperature measurement is of particular importance. Nevertheless, in the following sections many interesting experimental methods for the investigation of electrodes in HID lamps are described, resulting in a well-founded knowledge. Where experimental methods fail

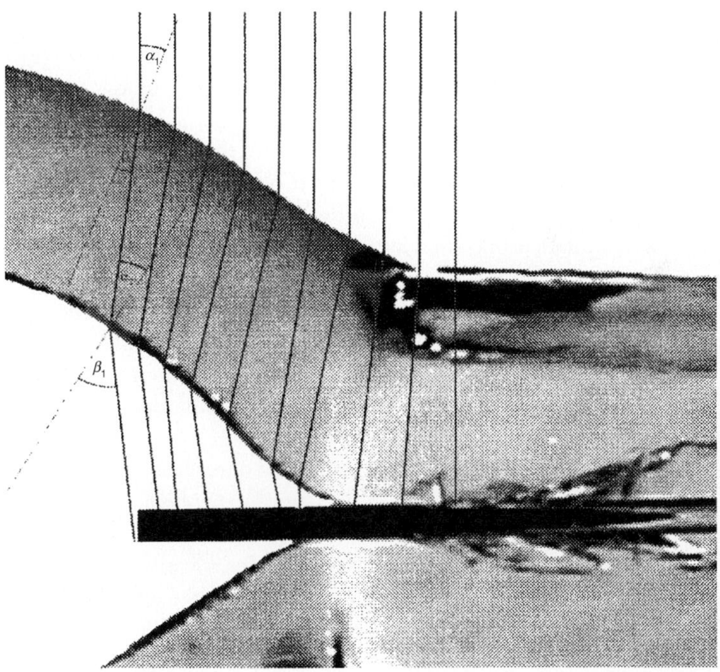

Fig. 3.1. Optical path through the quartz wall of a D2 automotive headlight lamp [85]

[1] If a thoriated electrode is operated at temperatures higher than roughly 3,000 K, the ThO_2 starts to evaporate and the work function increases [123].

due to principle problems, the smallness of the lamps, short time scales, etc., numerical simulations are often able to contribute further results to improve the understanding of HID lamps. They will be described in Chap. 4.

3.1.1 Electrode Pyrometry

As stated earlier, electrode temperature distribution is of substantial interest. Because the electrodes are very hot (typically around 2,000–3,500 K) and enclosed in the discharge tube, it is necessary to measure the electrode temperature from a distance without making contact. The method used for carrying out these non-contacting temperature measurements is known as pyrometry. Electrode pyrometry is based on the radiation coming from the electrodes, which is described by Planck's formula.

The entire radiation of an object is used by a bolometer to determine its temperature using the Stefan–Boltzmann law (cf. (2.47) and (3.1), respectively). In electrode pyrometry, however, the radiation emitted by the electrode at a certain wavelength or frequency is normally used for the determination of the temperature. The relationship between electrode temperature and radiation at a certain wavelength is given by Planck's law (cf. (2.49) and (3.3), respectively) or Wien's law (cf. (2.50) and (3.4), respectively) in combination with Kirchhoff's law (emissivity). Unfortunately, there are some problems and challenges using pyrometry: For example, the spectral emissivity $\varepsilon(\lambda,T)$ of the electrode surface, one of the parameters of the Planck equation, must be known, before the actual temperature may be computed. The emissivity ε might be a problem because it depends on the surface conditions of the electrodes [9, 43]. A rough surface has a considerably higher emissivity ε than a polished one (black body effect). This problem arises especially at the electrode tips, where the surface conditions might change with time. Moreover, transmission losses due to wall blackening might influence the results. Furthermore, the radiation of the plasma is reflected especially at the electrode tips, where the temperature measurement is of particular importance. Even the radiation of one electrode can be reflected at the surface of the other electrode, manipulating the temperature measurement by pyrometric methods.

If the radiation of the electrode is measured at one wavelength, the method is referred to as 1–λ pyrometry (Sect. 3.1.3). 1–λ pyrometry bears some problems due to uncertainties concerning the emissivity, possible transmission losses, and reflections from plasma radiation at the electrodes. Ratio, 2–λ, or two-color pyrometry eliminates the problems due to uncertainties in the emissivity and possible transmission losses by dividing the measured radiation intensities at two wavelengths (Sect. 3.1.5). However, the problems with reflections of plasma radiation at the electrode tips is not solved by 2–λ pyrometry. Thus, a further advancement is the elimination of the error due to reflections of plasma radiation at the electrodes by pyrometry with plasma correction. This method is referred to as (1+1)–λ pyrometry in the case of 1–λ pyrometry with plasma correction and (2+1)–λ or 3–λ pyrometry in the case of 2–λ or ratio pyrometry with plasma correction (Sect. 3.1.6).

3.1.2 Bolometer

A bolometer consists basically of a usually black surface of a thin film of area A_B absorbing the radiation of the surface whose temperature shall be measured (area A_X). The temperature of the thin film changes its resistance due to the heating, which can be related to the temperature of the surface under consideration. The speed of response of the bolometer is limited by the mass of the absorbing surface, thus thin films are usually used to increase this speed.

The bolometer is based on the Stefan–Boltzmann law, which has been derived in in the section "Stefan–Boltzmann Law" in Sect. 2.4.3:

$$P = \varepsilon \sigma_{SB} A \left(T^4 - T_S^4 \right), \tag{3.1}$$

where T_S is the temperature of the surrounding.

In the case of the bolometer, the surface A_X of the object under consideration must be known. If the bolometer absorbs all the radiation of this surface A_X, the total power heating the thin bolometer film of surface area A_B (often a platinum strip) is given by:

$$P_{heat} = \varepsilon_B \varepsilon_X \sigma_{SB} A_X T_X^4 - \varepsilon_B \sigma_{SB} A_B T_B^4. \tag{3.2}$$

$\varepsilon_B(T_B)$ is the (integral) emissivity of the bolometer material, $\varepsilon_X(T_X)$ is the (integral) emissivity of the object under consideration, T_X and T_B are the temperatures of object and bolometer, respectively. Equation (3.2) assumes that no other heating or cooling mechanisms exist at the bolometer, i.e., the bolometer material is not cooled by, e.g., convection. Due to their large reaction time, bolometers are normally not used for electrode pyrometry.

3.1.3 1–λ Pyrometry

1–λ pyrometry is based on Planck's law of black body radiation in combination with Kirchhoff's law as discussed in Sect. 2.4.3. The radiated power per unit area at wavelength λ of an electrode having temperature T and spectral emissivity $\varepsilon(\lambda,T)$ is given by $L(\lambda,T)$ (cf. (2.49)):

$$L(\lambda,T) = \varepsilon(\lambda,T) \frac{2\pi hc^2}{\lambda^5} \frac{1}{e^{\frac{hc}{\lambda kT}} - 1}. \tag{3.3}$$

If the exponent $hc/\lambda kT$ in (3.3) is much larger than one, the 1 in the denominator of (3.3) can be neglected, leading to Wien's law:

$$L(\lambda,T) = \varepsilon(\lambda,T) \frac{2\pi hc^2}{\lambda^5} e^{-\frac{hc}{\lambda kT}}. \tag{3.4}$$

In the case of optical pyrometry, i.e., pyrometry using wavelengths in the visible part of the electromagnetic spectrum, the largest error induced by this approximation can be calculated: The exponent $hc/\lambda kT$ in (3.3) and (3.4) has its smallest value (largest error) for large wavelengths λ and large temperatures T. Taking for example $\lambda = 700$ nm and $T = 3{,}700$ K (melting point of tungsten: 3,680 K), the fractions in (3.3) and (3.4) evaluate to

$$\frac{1}{e^{\frac{hc}{\lambda kT}} - 1} = \frac{1}{e^{5.555} - 1} = 3.883 \times 10^{-3} \tag{3.5}$$

and

$$\frac{1}{e^{\frac{hc}{\lambda kT}}} = \frac{1}{e^{5.555}} = 3.868 \times 10^{-3}. \tag{3.6}$$

That means, using Wien's law, the radiated power per unit area at wavelength $\lambda = 700$ nm of an electrode having temperature $T = 3{,}700$ K is undervalued by only 0.4%. For smaller wavelengths and smaller electrode temperatures, this error reduces further. If the radiated power is measured and T is calculated using (3.3) or (3.4), the error in the resulting temperature is even smaller. This is due to the exponential dependence of the radiated power on the temperature: A small temperature change results in a large change in the radiated power. If the radiated power per unit area at wavelength $\lambda = 700$ nm of an electrode having temperature $T = 3{,}700$ K is calculated with Planck's law (see (3.3)) and this radiated power is used to calculate the temperature using Wien's law (see (3.4)), the resulting "Wien-temperature" T_W is given by:

$$T_W = \frac{hc}{\lambda k} \frac{1}{\ln\left(\exp\left(\left(\frac{hc}{\lambda kT}\right) - 1\right)\right)}. \tag{3.7}$$

Equation (3.7) can be derived using Wien's law solved for T (see (3.8)) with $L(\lambda,T)$ from Planck's law (see (3.3)). For $\lambda = 700$ nm and $T = 3{,}700$ K the "Wien-temperature" calculates to $T_W = 3{,}697$ K, i.e., the error is less than 0.1%. Again, for smaller wavelengths and smaller electrode temperatures, the error induced by using Wien's law instead of Planck's law is further reduced. Thus, in the following Wien's law will be used without indicating this by the index "W", i.e., we set $T = T_W$. The advantage of using Wien's law is the easier analytic handling of (3.4) compared to (3.3).

An easy to use pyrometer is shown in Fig. 3.2: the optical pyrometer. The emitted radiation of the source S at a defined distance is compared with the radiation of the filament of an incandescent lamp L through an optical interference filter F. The interference filter F is thus responsible for selecting the desired wavelength λ. The temperature and thus the radiation of the filament can be adjusted with the help of the resistor R, i.e., by changing the current through the filament. Before using the pyrometer it has to be calibrated. This can be done by using standardized radiation sources. The resistor R can then be related to the temperatures of the standardized radiation sources by equalizing the radiation of the filament and the radiation of the standardized radiation sources. Thus, the

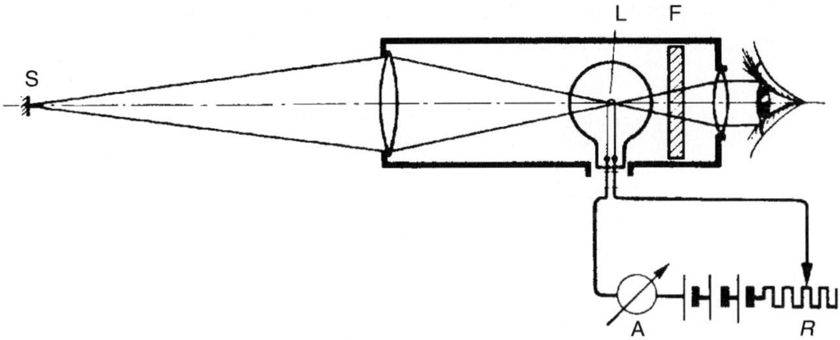

Fig. 3.2. Optical pyrometer measuring the temperature of source S by comparing the emitted radiation with the radiation of a filament of an incandescent lamp L through an optical interference filter F [72]

temperature of any source can be read directly from the scale of the adjusted resistor. So far, no formula has to be used. As soon as the emissivity of the object or transmission losses shall be taken into account, a formula as, for example, Wien's law has to be used, as described in the following. The emissivity of the filament does not matter, because the calibration has automatically taken the emissivity of the filament into account. Because of the manual operation and the errors and problems incidental thereto, the optical pyrometer is not commonly used in modern electrode pyrometry.

A more elegant and objective way of measuring electrode temperatures is by using electronic devices to measure the radiation coming from the electrodes directly, instead of manually comparing this radiation with the radiation of a reference source as in the case of the optical pyrometer. The thus measured radiated power $L(\lambda,T)$ at wavelength λ of an electrode having temperature T and spectral emissivity $\varepsilon(\lambda,T)$ can be used to calculate the electrode temperature T using Wien's law (3.4):

$$\begin{aligned}
L(\lambda,T) &= \varepsilon(\lambda,T)\frac{2\pi hc^2}{\lambda^5}\mathrm{e}^{-\frac{hc}{\lambda kT}} \\
\Leftrightarrow \quad \mathrm{e}^{-\frac{hc}{\lambda kT}} &= \frac{L(\lambda,T)\lambda^5}{2\varepsilon(\lambda,T)\pi hc^2} \\
\Leftrightarrow \quad \frac{hc}{\lambda kT} &= \ln\left(\frac{2\varepsilon(\lambda,T)\pi hc^2}{L(\lambda,T)\lambda^5}\right) \\
\Leftrightarrow \quad T &= \left[\frac{\lambda k}{hc}\ln\left(\frac{2\varepsilon(\lambda,T)\pi hc^2}{L(\lambda,T)\lambda^5}\right)\right]^{-1} .
\end{aligned} \tag{3.8}$$

λ as fixed by the interference filter, $L(\lambda,T)$ is the measured value, and $\varepsilon(\lambda,T)$ is the spectral emissivity of the electrode. As in the case of the optical pyrometer, the pyrometer has to be calibrated. Using a standardized radiation source of known black body temperature T_S, (3.8) can be transformed into the following expression using the black body temperature T_S instead of the measured radiated power $L(\lambda,T)$:

$$T = \left[\frac{\lambda k}{hc} \left(\ln \left(\frac{2\pi hc^2}{L(\lambda,T)\lambda^5} \right) + \ln(\varepsilon(\lambda,T)) \right) \right]^{-1}$$
$$\Leftrightarrow \quad T = \left[\frac{1}{T_S} + \frac{\lambda k}{hc} \ln(\varepsilon(\lambda,T)) \right]^{-1} . \tag{3.9}$$

The replacement of the measured radiated power $L(\lambda,T)$ by the black body temperature T_S from the calibration means that the measured radiated power $L(\lambda,T)$ corresponds to the same radiated power of an ideal black body ($\varepsilon(\lambda,T) = 1$) of temperature T_S. Because the electrode has a spectral emissivity $\varepsilon(\lambda,T)$ less than 1, its temperature T will be higher than the black body temperature creating the same radiation, i.e., $T > T_S$. Mathematically spoken, $\ln(\varepsilon)$ is negative for $\varepsilon < 1$, resulting in $T > T_S$ in (3.9). Because the spectral emissivity $\varepsilon(\lambda,T)$ is itself a function of temperature T, (3.9) has to be solved iteratively.

Equation (3.9) can be used to estimate the error resulting from the wrong choice of $\varepsilon(\lambda,T)$ or uncertainties in the spectral emissivity $\varepsilon(\lambda,T)$. In Fig. 3.8, the spectral emissivity $\varepsilon(\lambda,T)$ of tungsten for different temperatures as a function of wavelength λ is shown. Selecting a wavelength λ with the help of an optical interference filter results in a spectral emissivity $\varepsilon(\lambda,T)$ changing with temperature. For $\lambda = 550$ nm, for example, the spectral emissivity $\varepsilon(\lambda,T)$ is varying from 0.43 to 0.46 in the case of a polished tungsten surface. Assuming radiation corresponding to a black body temperature of $T_S = 3{,}000$ K and using the wavelength $\lambda = 550$ nm, the electrode temperature computes to $T = 3{,}259$ K for $\varepsilon(\lambda,T) = 0.5$ and $T = 3{,}352$ K for $\varepsilon(\lambda,T) = 0.4$. So, for an uncertainty of 10% in $\varepsilon(\lambda,T)$, an uncertainty of 50 K in the electrode temperature results. For higher temperatures, $1/T_S$ is smaller and thus the correction term due to the spectral emissivity $\varepsilon(\lambda,T)$ has a larger influence, i.e., produces a larger error in the temperature T. Larger wavelengths λ also increase the error due to uncertainties in $\varepsilon(\lambda,T)$ because of the factor $\lambda k/hc$ in front of the logarithm term. The exponential dependence of the radiated power on the temperature avoids larger errors due to the uncertainties in $\varepsilon(\lambda,T)$ because small temperature changes result already in large changes in the radiated power, whereas $L(\lambda,T)$ depends only linearly on $\varepsilon(\lambda,T)$. Nevertheless, as soon as the surface conditions of the electrode change, $\varepsilon(\lambda,T)$ might vary much more than can be seen in Fig. 3.8. A rough surface has a considerably higher emissivity ε than a polished one (black body effect). For example, the integral emissivity ε near the melting point of tungsten is about 0.39 for a polished surface, but 0.65–0.8 in the case of a rough surface [9, 43, 130].

Another problem for electrode pyrometry are transmission losses due to wall blackening caused by evaporated tungsten of the electrodes (cf. Fig. 3.3). The

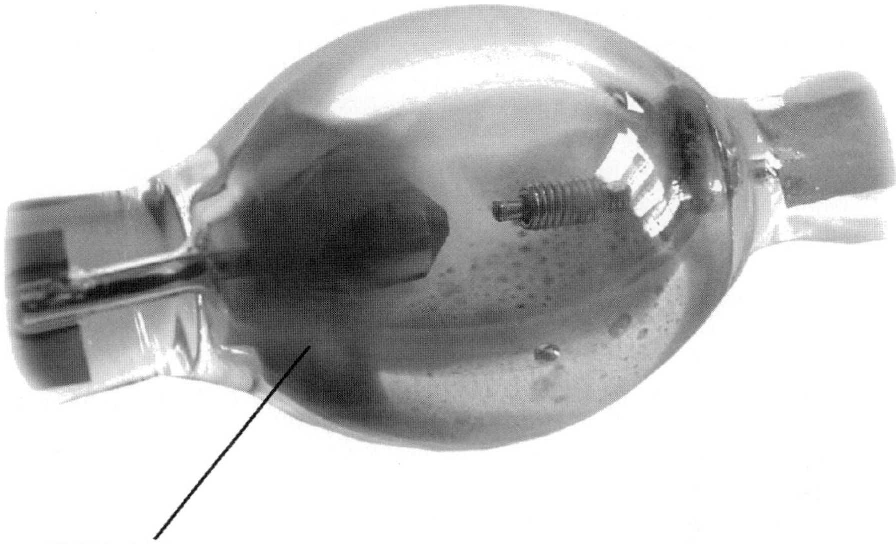

Fig. 3.3. DC mercury HID lamp demonstrating wall blackening effect due to evaporated tungsten [8]

transmission losses can be assumed to be independent of the wavelength λ of the light passing these regions, i.e., the wall blackening represents a gray optical interference filter. Apart from the limited transmission t of the quartz vessel of approximately 0.95, wall blackening might reduce this transmission to typical values of 0.8 [121]. Using Wien's law (3.4), the resulting radiated power $L(\lambda,T)$ per unit area at wavelength λ of an electrode having temperature T and spectral emissivity $\varepsilon(\lambda,T)$ is given by

$$L(\lambda,T)=t\varepsilon(\lambda,T)\frac{2\pi hc^2}{\lambda^5}\mathrm{e}^{-\frac{hc}{\lambda kT}}. \tag{3.10}$$

The resulting temperature is thus, by analogy to (3.8), given by:

$$T=\left[\frac{\lambda k}{hc}\ln\left(\frac{2t\varepsilon(\lambda,T)\pi hc^2}{L(\lambda,T)\lambda^5}\right)\right]^{-1}. \tag{3.11}$$

The correction due to these transmission losses can be illustrated by comparing the measured electrode temperature without taking transmission losses into account (T) with the measured electrode temperature including transmission losses (T_{new}):

$$T_{\text{new}}=\left[\frac{1}{\mathrm{T}}+\frac{\lambda k}{hc}\ln(t)\right]^{-1}. \tag{3.12}$$

The derivation of (3.12) is analogous to the derivation of (3.9). Taking as an example again $T = 3{,}000$ K measured at $\lambda = 550$ nm without taking transmission losses into account, a correction due to transmission losses ($t = 0.8$) results in a corrected electrode temperature of 3,079 K, i.e., the neglect of transmission losses might result in a systematic error or undervalued temperature measurements.

In summary, using 1–λ pyrometry one has to pay attention to the spectral emissivity $\varepsilon(\lambda,T)$ and the surface conditions of the electrodes (rough or polished surface) and transmission losses have to be taken into account. The radiation of the plasma which is reflected especially at the electrode tips where the temperature measurement is of particular importance (and even the radiation of one electrode being reflected at the surface of the other electrode) manipulates the temperature measurements using 1–λ pyrometry.

3.1.4 Examples of 1–λ Pyrometry

The relative spectral sensitivity of a commercial pyrometer (Impac IS10) is shown in Fig. 3.4. This pyrometer measures in the wavelength range from 650 to 1,200 nm (infrared), where black body radiation of the electrodes has its maximum or is at least very intense. The advantage of this pyrometer is the strong signal, which reduces possible noise to a minimum. The pyrometer can be focused on the electrode with a target spot of 0.3 mm or greater. The average temperature within this target spot is given by the pyrometer directly, i.e., it is already calibrated by the manufacturer. The only parameter which can be adjusted is the emissivity ε. The pyrometer assumes the radiation to be gray, i.e., ε is assumed to be independent of the wavelength λ. In [43, 121, 123], where this pyrometer is used, ε is set to 0.35. The correction due to the wavelength and temperature dependence of ε and due to

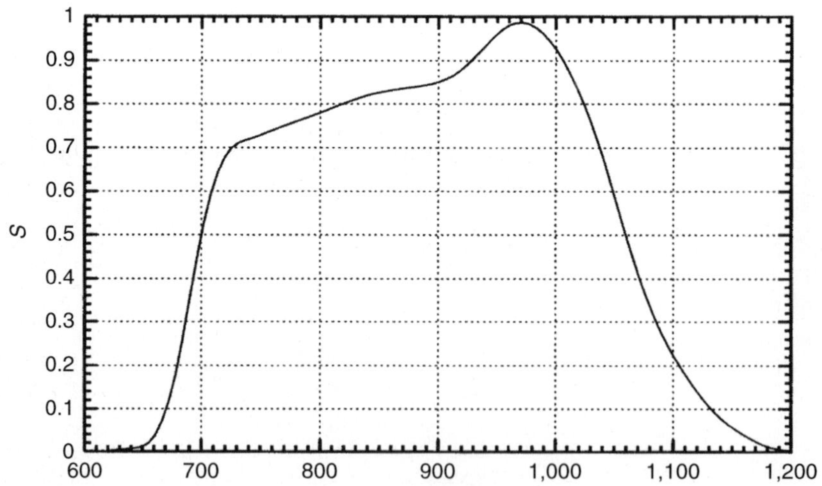

Fig. 3.4. Relative spectral sensitivity of a commercial pyrometer (Impac IS10) used in [121]

transmission losses is done during the postprocessing of the measured data according to (3.9) and (3.12). Measurements are possible within the temperature range from 1,073 to 3,773 K [43].

In [43, 121, 123], the pyrometer Impac IS10 is successfully applied to the pyrometry of electrodes in a model lamp.[2] A sketch of the experimental setup[3] is shown in Fig. 3.5 (see also Sect. 3.1.9, 3.2.2, and 3.2.3 for a description of this lamp). The electrodes are slowly moved along the target spot of the pyrometer by stepping motors to achieve the spatial resolution (cf. Fig. 3.6). This scan takes about 20s. Figure 3.6 shows the result of such a temperature scan on a cathode with a diameter of 1 mm and a length of 14 mm. The discharge consists of a 0.26 MPa argon plasma, the lamp current is 3 A. At the cathode tip, a region of approximately 1 mm shows a significantly higher temperature gradient compared to the rest of the temperature distribution. This higher gradient is caused by an increased surface roughness due to the thermal stress of the arc attachment. This higher surface roughness leads to a higher emissivity, causing an overestimation of the electrode tip temperature. Moreover, reflections of plasma radiation disturb the measured temperature distribution near the electrode tip. Thus, the region of confidence for the measured temperature is $z > 1$ mm.

The measured cathode temperature in Fig. 3.6 is superimposed by calculated electrode temperatures. They are the solution of the following heat conduction equation:

$$\pi R^2 \frac{\mathrm{d}}{\mathrm{d}z}\left(\lambda_{\mathrm{elec}} \frac{\mathrm{d}}{\mathrm{d}z} T\right) = 2\pi R \varepsilon \sigma_{\mathrm{SB}} \left(T^4(z) - T_{\mathrm{S}}^4\right). \tag{3.13}$$

Right-hand side of (3.13) represents the Stefan–Boltzmann law (cf. (2.47)), i.e., the power loss due to black body radiation of an infinitesimal small slice of the electrode: R is the radius of the electrode, ε the emissivity (depending on the wavelength λ and electrode temperature T), σ_{SB} the Stefan–Boltzmann constant, and T_{s} the temperature of the surrounding. Left-hand side of (3.13) represents the change in the power flux due to heat conduction (cf. (2.34)): $q = -\lambda_{\mathrm{elec}} \cdot \mathrm{d}T/\mathrm{d}z$ is the heat flux per unit area through an infinitesimal small slice of the electrode, where λ_{elec} is the thermal conductivity of tungsten (cf. Fig. 2.27). Thus, (3.13) traces the losses of the power flux inside the electrode due to black body radiation at the electrode surface. By assuming the power flux through the electrode end, the temperature distribution from the electrode end up to the electrode tip can be calculated[4]. In Fig. 3.6 solutions of (3.13) are shown for different power fluxes through the electrode end. The best fit is found for 16.13 W, i.e., one can expect a

[2] The lamp is also used for various other experiments described in Sects. 3.1.9, 3.2.2, and 3.2.3.

[3] This model lamp is restricted to pressures up to 1 MPa (10 bar) and the usage of rare gases. It enables fundamental research on the electrode behavior; nevertheless, properties in real lamps containing e.g., mercury might be different as described in [155].

[4] The electrode end is kept at constant temperature ($T = 293$ K) with the help of a cooling system (cf. Sect. 3.1.9).

power transfer of 16.13 W from the electrode end to the cooling fluid (cf. Sect. 3.1.9). Nevertheless, the calculated electrode tip temperatures might be wrong because the different processes of power transfer (and their spatial variation) from the plasma to the electrode tip are not taken into account (see also Sect. 4.4.3, especially footnote 40). The possible power fluxes from the plasma to the electrode tip are heating due to ion current and due to heat conduction from the plasma to the electrode tip and cooling due to electron emission.

To be able to estimate the reproducibility of the temperature measurements, a single electrode has been used for successive temperature measurements on different days. The electrode was cleaned after each measurement, the quartz vessel was replaced, and the lamp was reignited after each measurement. The electrode had a diameter of 2 mm and a length of 27 mm. The discharge was a 0.2 MPa argon plasma, the lamp current 5 A. The average of the resulting five temperature profiles and the deviations of the five individual measurements from this average are shown in Fig. 3.7 as a function of the distance z to the cathode tip. Especially the large errors near the electrode tip (top 1 mm) of ±150 K show that the problems due to a possible higher surface roughness and reflections of plasma radiation restrict the region of confidence for the measured temperature to $z > 1$ mm. Moreover, the general error of approximately ± 40 K shows that an increase of the accuracy beyond ± 40 K might not be useful because of natural variations of the electrode temperature. This conclusion is confirmed by pyrometric measurements using several identical D2 automotive headlight lamps described in [85] and pyrometric measurements using (1+1)–λ and (2+1)–λ pyrometry using high-pressure mercury model lamps (cf. Sect. 3.1.7, Fig. 3.24, [9, 128]).

The commercial pyrometer Impac IS10 is also used for temperature measurements on electrodes in model lamps with 1 MPa xenon as described in [56] and Sect. 4.4.3. Moreover, it is used for temperature measurements on electrodes in

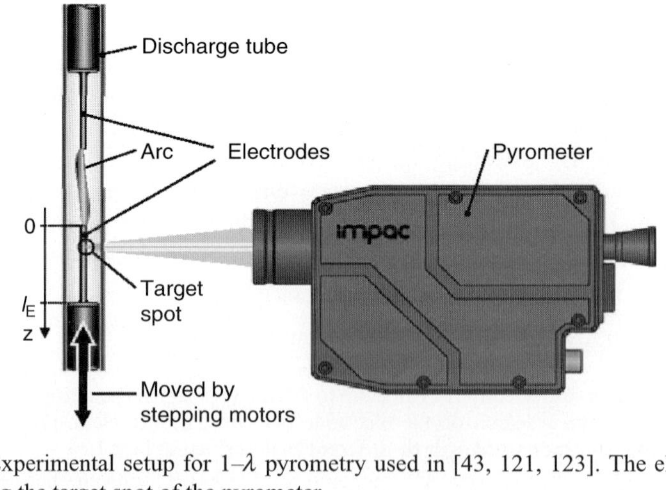

Fig. 3.5. Experimental setup for 1–λ pyrometry used in [43, 121, 123]. The electrodes are moved along the target spot of the pyrometer

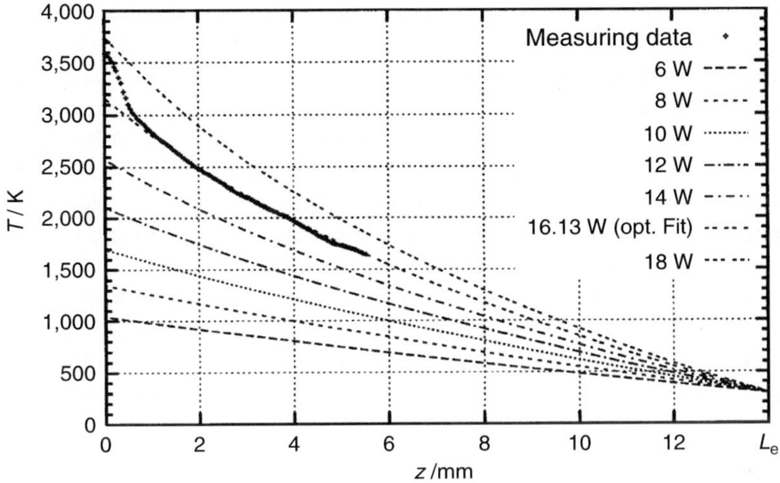

Fig. 3.6. Measured cathode temperature and solutions of a heat conduction equation with variable power output at electrode end as a function of distance z to cathode tip. The region of confidence of the measured data is $z > 1$ mm (electrode diameter: 1 mm, electrode length: 14 mm, gas: argon 0.26 MPa, lamp current: 3 A, [121])

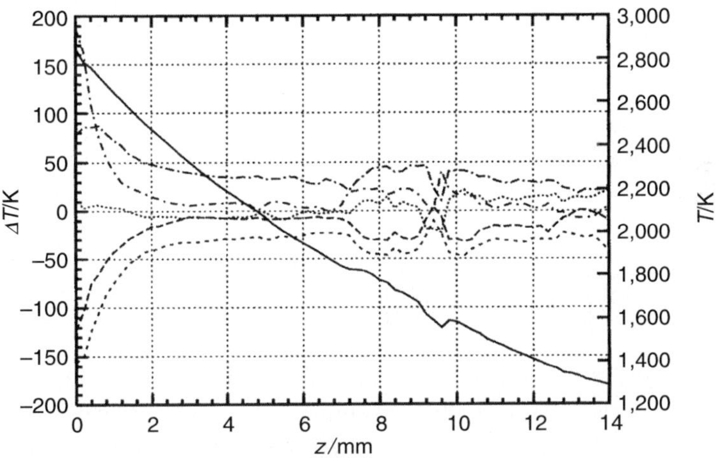

Fig. 3.7. Measured average cathode temperature (*solid line, right scale*) and deviations of the five individual measurements from average as a function of distance z to cathode tip (electrode diameter: 2 mm, electrode length: 27 mm, gas: argon 0.2 MPa, lamp current: 5 A, [121])

model lamps with 0.6 MPa mercury as described in [63, 156]. These measurements are described in more detail in Sects. 3.1.10, 4.4.4, and 4.4.5, dealing with the external laser heating of electrodes and different electrode geometries.

Pyrometric measurements on the electrodes of high-pressure sodium and metal halide lamps (D2 automotive headlight lamp, cf. Sect. 1.3.8) described in [1, 85, 86] have been carried out using an indium–antimony detector with sensibility between 500 and 2,800 nm (infrared). Depending on the spectral power distribution of the discharge, a wavelength range up to 60 nm is selected with the help of a monochromator for the temperature measurement. Within this selected wavelength range, the ratio of plasma radiation to electrode radiation should be as small as possible, reducing thus the interfering plasma radiation. In the case of a high-pressure sodium lamp, the measurements were performed using wavelengths around 1.32 and 1.58 μm, for a D2 automotive headlight lamp a wavelength range between 1.45 and 1.50 μm has been selected. To be able to record the temperature distribution along the electrode axis, the lamp was moved during the measurements using a stepping motor. This method is too slow to measure time dependent temperature profiles (current frequency in the case of a D2 automotive headlight lamp: 400 Hz), so that the measurements yield time-averaged electrode temperature profiles. They are shown in Sect. 4.4.6 dealing with the numerical simulation of D2 automotive headlight lamps.

3.1.5 2–λ Pyrometry

Ratio, 2–λ, or two-color pyrometry eliminates the problems due to uncertainties of the spectral emissivity $\varepsilon(\lambda,T)$ and possible transmission losses by dividing the measured radiation intensities at two wavelengths λ_1 and λ_2. Using Wien's law with transmission correction (3.10), the following transformations can be made:

$$\frac{L(\lambda_1,T)}{L(\lambda_2,T)}=\frac{t\varepsilon(\lambda_1,T)}{t\varepsilon(\lambda_2,T)}\left(\frac{\lambda_2}{\lambda_1}\right)^5\exp\left\{\frac{hc}{kT}\left(\frac{1}{\lambda_2}-\frac{1}{\lambda_1}\right)\right\}$$

$$\Leftrightarrow\quad \frac{L(\lambda_1,T)}{L(\lambda_2,T)}\frac{\varepsilon(\lambda_2,T)}{\varepsilon(\lambda_1,T)}\left(\frac{\lambda_1}{\lambda_2}\right)^5=\exp\left\{\frac{hc}{kT}\left(\frac{1}{\lambda_2}-\frac{1}{\lambda_1}\right)\right\} \tag{3.14}$$

$$\Leftrightarrow\quad T=\frac{hc}{k}\frac{\lambda_1-\lambda_2}{\lambda_1\lambda_2}\left[\ln\left(\frac{L(\lambda_1,T)}{L(\lambda_2,T)}\frac{\varepsilon(\lambda_2,T)}{\varepsilon(\lambda_1,T)}\left(\frac{\lambda_1}{\lambda_2}\right)^5\right)\right]^{-1}.$$

The transmission losses (factor t) have been assumed to be independent of the wavelength λ of the light, thus they are eliminated by 2–λ pyrometry. Instead of the absolute value of the spectral emissivity $\varepsilon(\lambda,T)$ as in 1–λ pyrometry, the ratio of $\varepsilon(\lambda_1,T)$ to $\varepsilon(\lambda_2,T)$ enters (3.14). This ratio still depends on the temperature T in the case of tungsten, but the variations with temperature T are much smaller compared to the variations of the absolute value of the spectral emissivity $\varepsilon(\lambda,T)$. Figure 3.8 demonstrates the selection of two wavelengths $\lambda_1 = 532$ nm and $\lambda_2 = 750$ nm and the corresponding spectral emissivities $\varepsilon(\lambda_1,T)$ and $\varepsilon(\lambda_2,T)$. Even if the ratio $\varepsilon(\lambda_1,T)/\varepsilon(\lambda_2,T)$ is taken to be constant (e.g., at its value at $T = 3{,}000$ K), the error in the temperature measurement is less than 15 K for temperatures between 2,000 and 3,400 K [9, 11, 128, 147]. Moreover, changes in the spectral emissivity $\varepsilon(\lambda,T)$ due to surface roughness can also be assumed to be independent of the wavelength λ of the light, thus they are also eliminated by 2–λ pyrometry.

Equation (3.14) can be further transformed resulting in

$$T = \frac{hc}{k}\frac{\lambda_1 - \lambda_2}{\lambda_1 \lambda_2}\left[\ln\left(\frac{L(\lambda_1,T)}{L(\lambda_2,T)}\right) - \ln\left(\frac{\varepsilon(\lambda_1,T)}{\varepsilon(\lambda_2,T)}\right) + 5\ln\left(\frac{\lambda_1}{\lambda_2}\right)^5\right]^{-1}. \tag{3.15}$$

Because the spectral emissivity $\varepsilon(\lambda_i,T)$ is itself a function of temperature T, (3.15) has to be solved iteratively.

The disadvantage of the 2–λ pyrometry is the need for the measurement at two different wavelengths. The radiation coming from the electrode of which the temperature shall be measured must either be split into two light beams (cf. Sect. 3.1.7) or the optical interference filters for choosing wavelengths λ_1 and λ_2 must be exchanged during the measurement. Both methods have their advantages and disadvantages: Splitting the radiation into two light beams involves experimental efforts but allows for simultaneous measurement of $L(\lambda_1,T)$ and $L(\lambda_2,T)$. Exchanging the interference filters increases the time needed for the measurement and one has to make sure that the electrode conditions do not shift during the exchange of the interference filters. As a possible solution the author proposes to use a rotating wheel in the optical path with different interference filters at different segments, comparable to the color wheel in certain types of video projectors. A further problem of 2–λ pyrometry is that it is more sensitive to noise.

In summary, using 2–λ or ratio pyrometry the problems due to the uncertainties in the spectral emissivity $\varepsilon(\lambda,T)$, the surface conditions of the electrodes (rough or polished surface), and transmission losses are solved. However, the problems with reflections of plasma radiation or radiation from the opposite electrode is not solved by 2–λ pyrometry. Furthermore, 2–λ pyrometry is more sensitive to noise compared to 1–λ pyrometry. The problems due to reflections of plasma radiation especially at the electrode tips (where the temperature measurement is of particular importance) can be solved using pyrometry with plasma correction, which is described in Sect. 3.1.6.

3.1.6 Pyrometry with Plasma Correction

Both, 1–λ and 2–λ pyrometry have problems due to interfering reflections of plasma radiation at the electrode tips, where the temperature measurement is of particular importance, since the electrode tip temperature determines, for example, the electron emission current and the evaporation rate. The radiation of the electrode necessary for the temperature measurement is superimposed by the plasma radiation reflected at the electrode surface[5] as shown in Fig. 3.9. Thus, a

---

[5] Direct plasma radiation, i.e., plasma radiation originating from a location on the optical path from the electrode surface to the pyrometer might be of importance as well, but since most plasma radiation is produced between the electrode tips (cf. Fig. 3.9) and the pyrometer is normally side-on directed on the electrode, it plays mostly a minor role. Nevertheless, for the plasma correction method described here, the origin of the interfering plasma radiation does not matter.

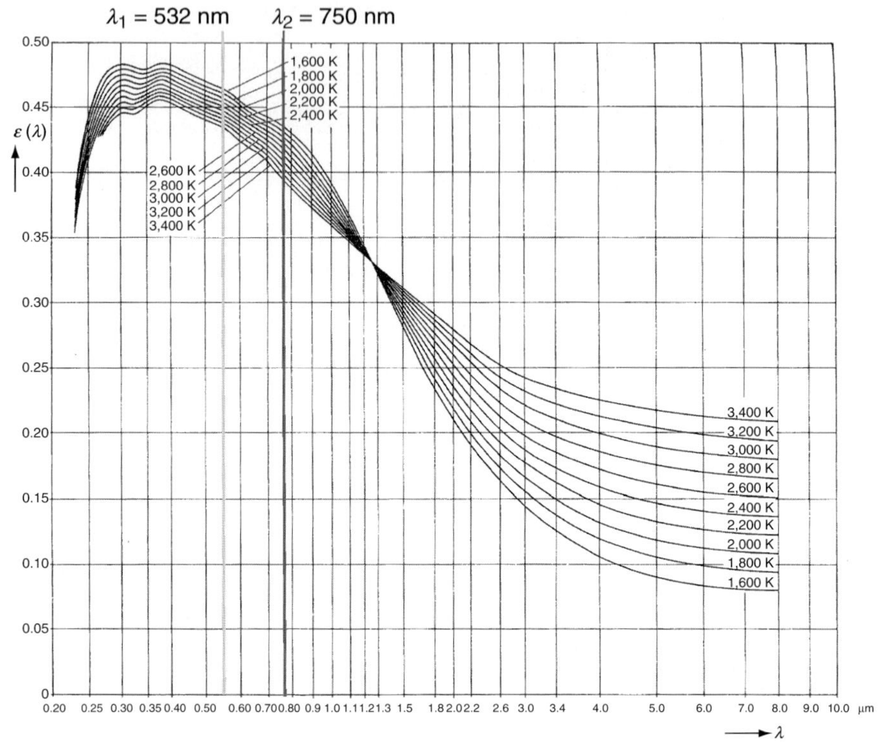

Fig. 3.8. Spectral emissivity $\varepsilon(\lambda,T)$ of tungsten vs. wavelength λ for several temperatures superimposed by two wavelengths λ_1 and λ_2 selected with the help of optical interference filters for 2–λ pyrometry [48]

further advancement of electrode pyrometry is the elimination of the error due to reflections of plasma radiation at the electrodes by pyrometry with plasma correction. To achieve this plasma correction, a measurement at a further wavelength λ_0 is necessary. This method is referred to as (1+1)–λ pyrometry in the case of 1–λ pyrometry with plasma correction and (2+1)–λ or 3–λ pyrometry in the case of 2–λ or ratio pyrometry with plasma correction [9, 11, 128, 147].

The principal problem of the reflections of plasma radiation in the case of electrode pyrometry is shown in Fig. 3.9. The measured total radiation $L(\lambda,T)$ is a superposition of the radiation emitted by the electrode $L_{el}(\lambda,T)$ and the interfering radiation from the plasma $L_{pl}(\lambda)$. The corresponding equation reads as follows:

$$L(\lambda,T) = L_{el}(\lambda,T) + L_{pl}(\lambda). \tag{3.16}$$

The plasma radiation $L_{pl}(\lambda)$ does not depend on the electrode temperature T and it is assumed that the intensity of the plasma radiation reflected at the electrodes is

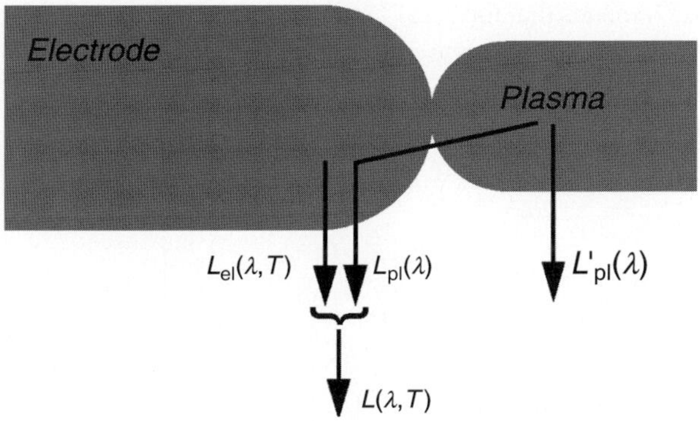

Fig. 3.9. Superposition of the radiation emitted by the electrode $L_{el}(\lambda,T)$ and the radiation from the plasma $L_{pl}(\lambda)$ to the measured total radiation $L(\lambda,T)$

proportional to the intensity of the plasma radiation coming directly out of the plasma, i.e.,

$$L_{pl}(\lambda)=\beta L'_{pl}(\lambda), \tag{3.17}$$

where β is a constant independent of the wavelength λ. In the case of (1+1)–λ pyrometry, one uses one wavelength λ_1 for the actual measurement of the electrode temperature T and one wavelength λ_0 for the plasma correction. The resulting set of equations is:

$$\begin{aligned} L(\lambda_0,T)&=L_{el}(\lambda_0,T)+L_{pl}(\lambda_0),\\ L(\lambda_1,T)&=L_{el}(\lambda_1,T)+L_{pl}(\lambda_1). \end{aligned} \tag{3.18}$$

(2+1)–λ or 3–λ pyrometry uses two wavelengths λ_1 and λ_2 for the actual measurement of the electrode temperature T and one wavelength λ_0 for the plasma correction:

$$\begin{aligned} L(\lambda_0,T)&=L_{el}(\lambda_0,T)+L_{pl}(\lambda_0),\\ L(\lambda_1,T)&=L_{el}(\lambda_1,T)+L_{pl}(\lambda_1),\\ L(\lambda_2,T)&=L_{el}(\lambda_2,T)+L_{pl}(\lambda_2). \end{aligned} \tag{3.19}$$

The procedure is similar for both methods. The description given below is visualized in Figs. 3.10 ((1+1)–λ pyrometry) and 3.11 ((2+1)–λ pyrometry). In these two figures the successive steps leading to the unknown electrode temperature T are labeled with numbers in the format (1), (2), etc., so that they will be referred to in the following by e.g., "(cf. Fig. 3.10, step (2))."

First, it is assumed that the interfering radiation due to reflections of plasma radiation is known, i.e., the undisturbed radiation of the electrodes $L_{el}(\lambda_{1,2},T)$ is given by:

$$L_{el}\left(\lambda_{1,2},T\right)=L\left(\lambda_{1,2},T\right)-L_{pl}\left(\lambda_{1,2}\right). \tag{3.20}$$

Equation (3.20) means that the "real" radiated power $L_{el}(\lambda_{1,2},T)$ at wavelengths $\lambda_{1,2}$ coming from the electrode with temperature T can be calculated by subtracting the (not yet known) radiation at wavelengths $\lambda_{1,2}$ originating from the plasma ($L_{pl}(\lambda_{1,2})$) from the measured radiation $L(\lambda_{1,2},T)$. Using (3.11) in combination with (3.20), the electrode temperature in the case of (1+1)–λ pyrometry (cf. Fig. 3.10, step (1)) is given by:

$$T=\left[\frac{\lambda_1 k}{hc}\ln\left(\frac{2t\varepsilon(\lambda_1,T)\pi hc^2}{\lambda^5\left(L(\lambda_1,T)-L_{pl}(\lambda_1)\right)}\right)\right]^{-1}. \tag{3.21}$$

In the case of (2+1)–λ pyrometry, combining (3.14) with (3.20) yields the wanted electrode temperature (cf. Fig. 3.11, step (1)):

$$T=\frac{hc}{k}\frac{\lambda_1-\lambda_2}{\lambda_1\lambda_2}\left[\ln\left(\left(\frac{L(\lambda_1,T)-L_{pl}(\lambda_1)}{L(\lambda_2,T)-L_{pl}(\lambda_2)}\right)\left(\frac{\varepsilon(\lambda_2,T)}{\varepsilon(\lambda_1,T)}\right)\left(\frac{\lambda_1}{\lambda_2}\right)^5\right)\right]^{-1}. \tag{3.22}$$

In the next step, the radiation at wavelengths $\lambda_{1,2}$ originating from the plasma ($L_{pl}(\lambda_{1,2})$) has to be replaced. This is done by defining the ratio of the plasma radiation at wavelengths $\lambda_{1,2}$ to the plasma radiation at wavelength λ_0:

$$C_{i,0}=\frac{L_{pl}(\lambda_i)}{L_{pl}(\lambda_0)}=\frac{L'_{pl}(\lambda_i)}{L'_{pl}(\lambda_0)} \quad \text{for } i=1,2 \tag{3.23}$$

Here, the assumption that the intensity of the plasma radiation reflected at the electrodes ($L_{pl}(\lambda)$) is proportional to the intensity of the plasma radiation coming directly out of the plasma ($L'_{pl}(\lambda)$) has been used (cf. (3.17) and Fig. 3.9). Equation (3.23) means that the ratio C_{i0} can be measured using the radiation coming directly out of the discharge before the actual temperature measurement at the electrodes starts.

The plasma radiation at wavelength λ_0 can be determined using the expression for the total radiation $L(\lambda_0,T)$ given in (3.18) and (3.19) solved for $L_{pl}(\lambda_0)$:

$$L_{pl}(\lambda_0)=L(\lambda_0,T)-L_{el}(\lambda_0,T). \tag{3.24}$$

The plasma radiation at wavelength λ_1 needed for (3.21) in the case of (1+1)–λ pyrometry can be calculated using (3.23), i.e.,

$$L_{pl}(\lambda_1)=C_{10}L_{pl}(\lambda_0). \tag{3.25}$$

The unknown $L_{pl}(\lambda_0)$ can be calculated using (3.24) together with (3.10) (Wien's law with emissivity and transmission losses taken into account) corresponding to step (2) in Fig. 3.10:

$$L_{pl}(\lambda_1)=C_{10}\left(L(\lambda_0,T)-t\varepsilon(\lambda_0,T)\frac{2\pi hc^2}{(\lambda_0)^5}e^{-\frac{hc}{\lambda_0 kT}}\right). \tag{3.26}$$

The unknown in (3.26) is the electrode temperature T, which can be calculated using (3.21) as indicated in Fig. 3.10 by step (3). In (3.21), the only unknown is

$L_{pl}(\lambda_1)$, which is given by (3.26), corresponding to step (4) in Fig. 3.10. Thus, (3.21) and (3.26) have to be solved iteratively to get the electrode temperature T using (1+1)–λ pyrometry.

The procedure in the case of (2+1)–λ pyrometry is similar: The plasma radiation at wavelengths $\lambda_{1,2}$ needed for (3.22) can be calculated using (3.23), i.e.,

$$L_{pl}(\lambda_i) = C_{i,0} L_{pl}(\lambda_0) \qquad \text{for } i = 1, 2. \tag{3.27}$$

The unknown $L_{pl}(\lambda_0)$ can be calculated using (3.24) together with (3.10) (Wien's law with emissivity and transmission losses taken into account) corresponding to step (2) in Fig. 3.11:

$$L_{pl}(\lambda_i) = C_{i0}\left(L(\lambda_0, T) - t\varepsilon(\lambda_0, T)\frac{2\pi hc^2}{(\lambda_0)^5} e^{-\frac{hc}{\lambda_0 kT}} \right) \qquad \text{for } i = 1, 2. \tag{3.28}$$

Again, (3.22) and (3.28) have to be solved iteratively to get the electrode temperature T using (2+1)–λ pyrometry (cf. steps[6] (3) and (4) in Fig. 3.11).

The arrows corresponding to step (3) in Figs. 3.10 and 3.11 do not point at the T in $L(\lambda_{0,1,2},T)$. This is correct because $L(\lambda_{0,1,2},T)$ are the measured values of the radiated power at wavelengths $\lambda_{0,1,2}$ of an electrode having temperature T. The T just indicates that the measured values depend on the electrode temperature, but this T is of course not part of the iterative scheme for the determination of the electrode temperature T.

In summary, pyrometry with plasma correction proceeds as follows: First, the plasma radiation is measured directly in the plasma at the wavelengths $\lambda_{0,1,2}$ to determine the coefficients C_{10} and C_{20}. Then, the total radiation coming from the electrode is measured at the wavelengths $\lambda_{0,1,2}$. Finally, the electrode temperature is calculated using the iterative schemes given in Figs. 3.10 and 3.11 ((1+1)–λ pyrometry and (2+1)–λ pyrometry, respectively). The wavelengths $\lambda_{1,2}$ for the determination of the electrode temperature and λ_0 for the plasma correction should be chosen carefully: At the wavelengths $\lambda_{1,2}$, the black body radiation from the electrode should be strong, the plasma radiation on the other hand should be small. At the wavelength λ_0, the electrode should hardly emit radiation, whereas the plasma should radiate considerably. All chosen wavelengths $\lambda_{0,1,2}$ should be in the continuous part of the spectral power distribution of the plasma, i.e., $\lambda_{0,1,2}$ should not correspond to atomic or molecular line radiation. The continuous part of the spectral power distribution of the plasma is less sensitive to plasma temperature changes and the ratio C_{i0} is better defined. In this way, the plasma correction works best.

Apart from the method described earlier for the elimination of the interfering plasma radiation, other methods exist as well. Depending on the spectral power distribution of the radiation coming from the plasma, wavelengths can be chosen

---

[6] The arrow pointing at T in $\varepsilon(\lambda_2,T)$ in Fig. 3.11 naturally includes T in $\varepsilon(\lambda_1,T)$, the arrow pointing at $L_{pl}(\lambda_2)$ naturally includes $L_{pl}(\lambda_1)$ also.

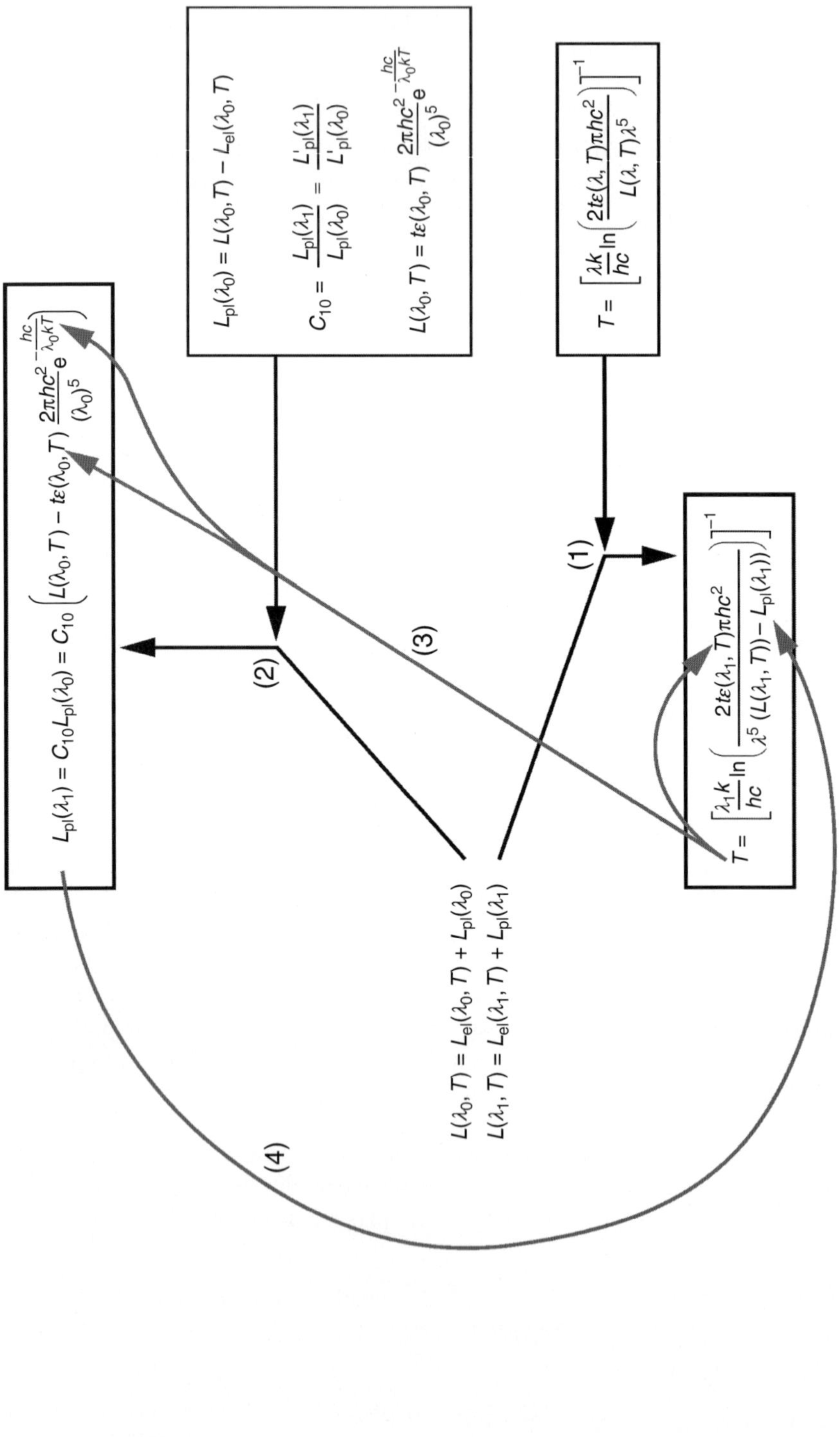

Fig. 3.10. Visualization of the iteration method used for (1+1)–λ pyrometry

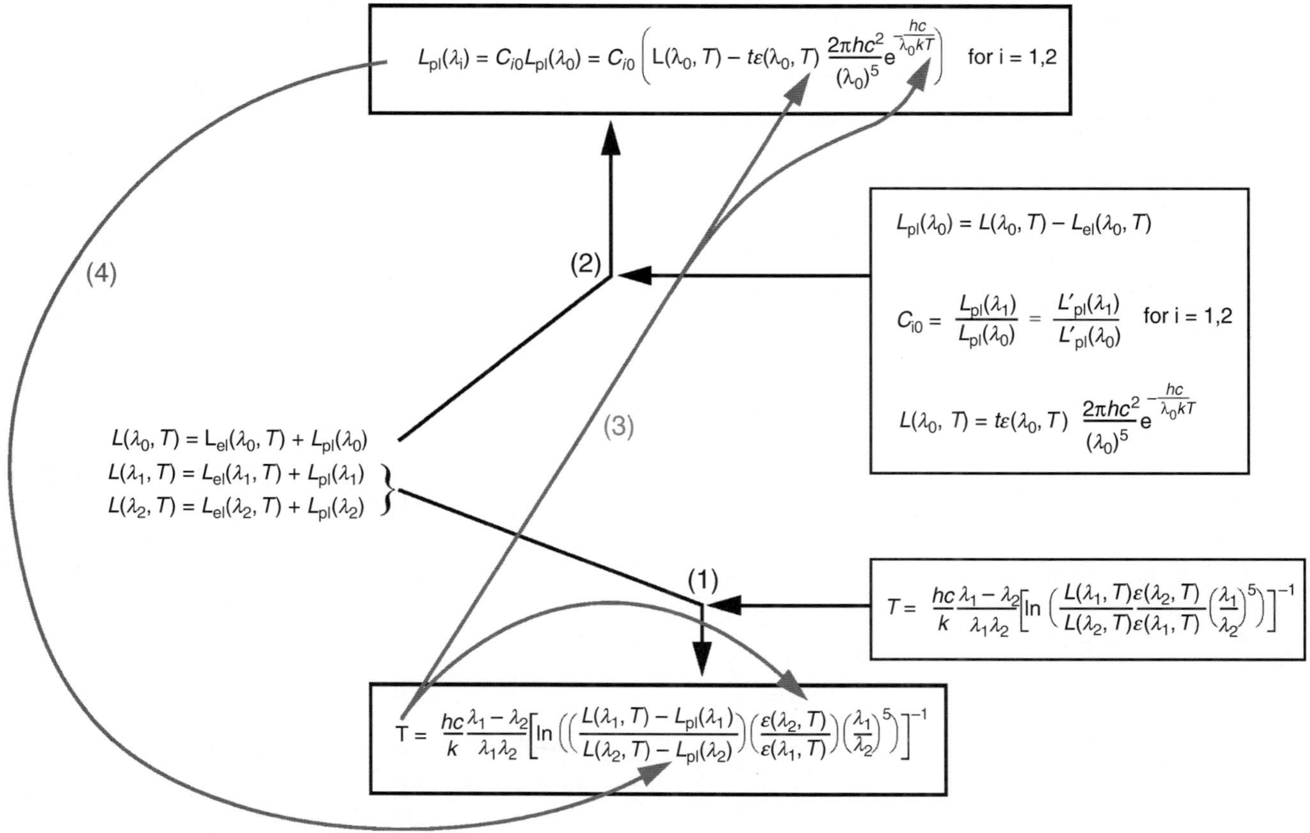

Fig. 3.11. Visualization of the iteration method used for (2+1)–λ pyrometry

where the plasma hardly emits (cf. Sect. 3.1.4). This is increasingly difficult in commercial lamps with extremely high pressure (e.g., the UHP lamp, Sect. 1.3.6) or metal halide lamps (e.g., D2 automotive headlight lamp, cf. Sect. 1.3.8). Due to the high pressure or the emission spectrum of the metals, the spectral power distribution offers no or only small gaps where plasma radiation is weak. Another possibility to reduce the interfering plasma radiation is to turn off the power just before starting with the measurement. The intensity of the plasma radiation decays faster than the radiation of the electrodes, but especially the temperature at the electrode tip also changes quite fast as soon as the lamp power is turned off. Blanking techniques switch of the lamp only during the short time necessary for the electrode temperature measurement, but even then, the heat capacity of small electrodes might be too small to ensure reliable temperature measurements.

In summary, using 3–λ or (2+1)–λ pyrometry, the problems due to the uncertainties in the spectral emissivity $\varepsilon(\lambda,T)$, the surface conditions of the electrodes (rough or polished surface), and transmission losses are solved. Moreover, the interfering reflections of plasma radiation are eliminated. Only the problem with reflections of the radiation from the opposite electrode is not solved. Because 2–λ pyrometry is more sensitive to noise than 1–λ pyrometry, (1+1)–λ pyrometry might be the better choice. (2+1)–λ pyrometry can then be used to validate the results of the (1+1)–λ pyrometry. Problems using the described pyrometry with plasma correction might occur as soon as the assumption is violated that the intensity of the plasma radiation reflected at the electrodes is proportional to the intensity of the plasma radiation coming directly out of the plasma. This might be the case in metal halide lamps, where the different kinds of atoms are not distributed evenly but depend on the local temperature (cf. Figs. 2.8 and 2.16). The resulting spectral power distribution must be analyzed carefully to decide, if the radiation can be assumed to be independent of the position within the plasma. In principle, (2+1)–λ pyrometry at the electrodes of metal halide lamps is possible as well, as shown in [129].

3.1.7 Example of (2+1)–λ Pyrometry

Pyrometry with plasma correction is used and described in [9, 11, 128, 129, 147] on electrodes of high-pressure mercury and metal halide lamps. The careful selection of the wavelengths $\lambda_{1,2}$ for the determination of the electrode temperature and λ_0 for the plasma correction is shown in Figs. 3.12 and 3.13 in the case of a high-pressure mercury lamp (here: the UHP lamp, cf. Sect. 1.3.6). At the wavelengths $\lambda_1 = 532$ nm and $\lambda_2 = 750$ nm, the black body radiation from the electrode is strong, at the wavelength $\lambda_0 = 425$ nm, the electrode hardly emits radiation (electrode temperatures between 2,400 and 3,400 K). Furthermore, the plasma radiation is stronger at λ_0, whereas the plasma radiates less at the wavelengths $\lambda_{1,2}$.

The experimental setup is illustrated in a diagram shown in Fig. 3.14. The radiation or light emitted by one electrode is imaged onto an intensified CCD array, resulting in the image of the electrode with a spatial resolution down to 4 μm. The shortest exposure time is 10 μs. In contrast to the scanning of the electrode

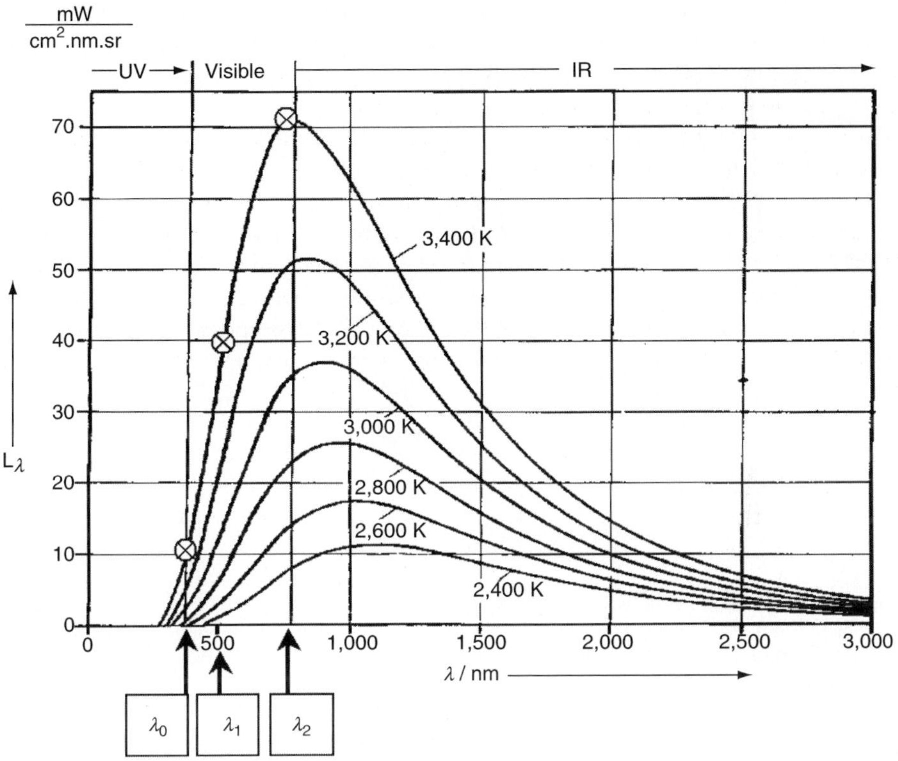

Fig. 3.12. Selection of the wavelength $\lambda_{0,1,2}$ for (2+1)–λ pyrometry: at the wavelengths $\lambda_{1,2}$, the black body radiation from the electrode is strong, at the wavelength λ_0, the electrode hardly emits radiation (electrode temperatures between 2,400 and 3,400 K [9]).

temperature with the target spot of a pyrometer, this method allows the simultaneous determination of the electrode temperature profile. Thus, time-dependent investigations on electrodes in ac operated lamps are possible.

To allow for pyrometry using the three wavelengths $\lambda_{0,1,2}$, the light beam is split into three individual light beams. This is done with the help of four dichroic mirrors, allowing for maximum light intensity at each wavelength. Three of the four dichroic mirrors shown in Fig. 3.14 are labeled with the letters A, C, and D. The corresponding labeling is used in Fig. 3.15, showing the transmission vs. wavelength λ of these different dichroic mirrors. The light which is not transmitted through the corresponding dichroic mirror is reflected with loses in the region of a couple percent. Thus, the dichroic mirror A reflects the blue light (including $\lambda_0 = 425$ nm) but lets the radiation with wavelengths greater than 520 nm through (including $\lambda_1 = 532$ nm and $\lambda_2 = 750$ nm). This remaining light beam is split a second time by the dichroic mirror D. Here, the light with wavelengths up to 610 nm is reflected (including $\lambda_1 = 532$ nm), the rest of the light is transmitted (including $\lambda_2 = 750$ nm). The resulting three light beams with different wavelength ranges are

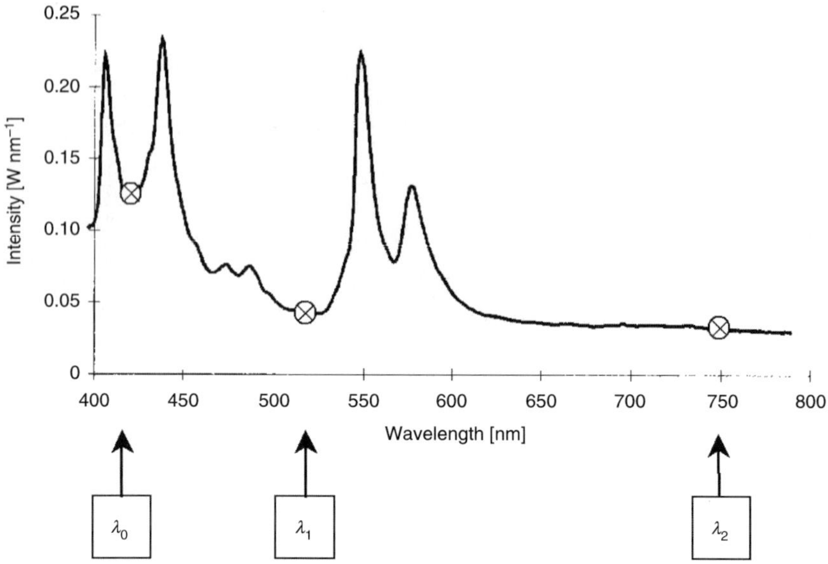

Fig. 3.13. Selection of the wavelengths $\lambda_{0,1,2}$ for (2+1)–λ pyrometry. At the wavelength λ_0, the plasma radiation from a UHP lamp is strong, at the wavelengths $\lambda_{1,2}$, the plasma radiation is smaller [9]

further reduced by three optical interference filters, selecting the wanted wavelengths $\lambda_{0,1,2}$. The three light beams with the wavelengths $\lambda_{0,1,2}$ are recombined on the CCD array with the help of two more dichroic mirrors. The light beam with wavelength $\lambda_0 = 425$ nm is reunited with the light beam with wavelength $\lambda_1 = 532$ nm using the dichroic mirror C. It lets the light beam with wavelength $\lambda_0 = 425$ nm through and reflects the light beam with wavelength $\lambda_1 = 532$ nm. The more or less wild transmission curve for wavelengths greater than 650 nm does not play a role because the light beam with wavelength $\lambda_2 = 750$ nm is only reunited with the other two light beams with the help of the next dichroic mirror (of which no transmission curve is shown in Fig. 3.15). The three images of the HID lamp electrode corresponding to the three wavelengths $\lambda_{0,1,2}$ are not mapped onto the same position on the CCD array, but they are imaged side by side onto the 1024×1024 pixels as described in the following[7] (see also Figs. 3.18 and 3.20).

The exact experimental setup used in [9, 11, 128, 129] is shown in Figs. 3.16 and 3.17. The HID lamp is hidden in a safety box with a shatterproof glass to let the light through to lens 1. The shatterproof glass transmits light in the wavelength range between 380 and 1,000 nm, the inside of the safety box is black to avoid light

[7] A design similar to this shown in Fig. 3.14 is used in many video projectors, replacing the optical interference filters by LCD panels and mapping the three resulting images on top of each other.

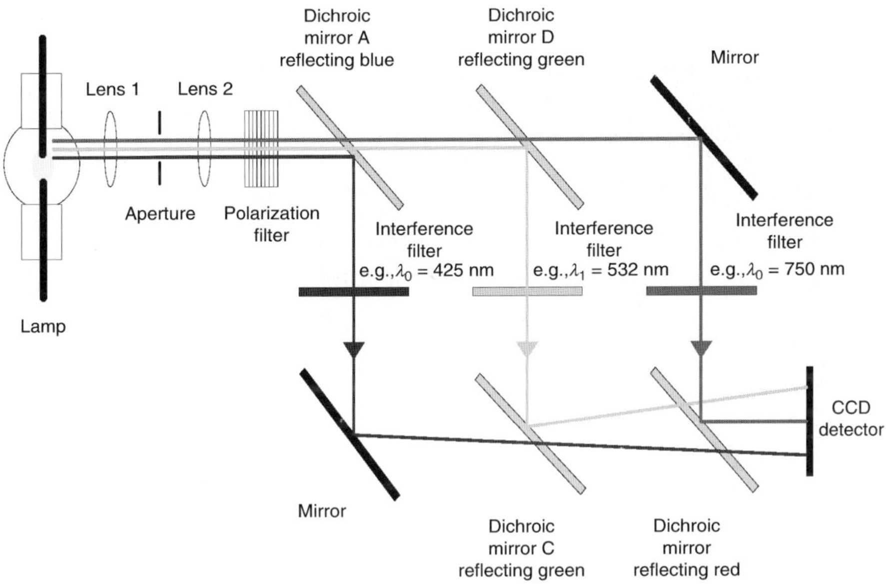

Fig. 3.14. Experimental setup for (2+1)–λ pyrometry used in [9, 11, 128, 129]

reflections. The black box shown in Fig. 3.16 contains the components for the splitting of the light beam. An insight into this box is given in Fig. 3.17, showing the six mirrors and the three optical interference filters.

A typical measurement using this experimental setup is shown in Fig. 3.18. The filament of a standardized halogen incandescent lamp (adjusted temperature: 3,100 K) is mapped onto the CCD array, resulting in three simultaneous images of the filament. Each of the single images at a given wavelength has a resolution of approximately 340×1024 pixels. From left to right, the filament is enlarged with the help of the two lenses. These test images were taken before the system was calibrated, i.e., the given intensities at the different wavelengths cannot be compared directly. Nevertheless, in accordance with the choice of the three wavelengths $\lambda_{0,1,2}$, the intensity at $\lambda_0 = 425$ nm is, as expected, the smallest. The measurement at wavelength λ_0 is not really necessary in the case of an incandescent lamp, since there is no plasma and thus no need for plasma correction at the filament of an incandescent lamp. The measured radiation intensities can be analyzed yielding the temperature distribution on the filament using one or two wavelengths pyrometry.

The radiation intensity at both sides of each wire is higher than the radiation intensity coming from the front surface. This is due to a black body effect: The radiation inside the coil is reflected at the inner surface of the coil. This is comparable to the situation in an ideal black body (a cavity), increasing the "undisturbed" emissivity considerably. Using 1–λ pyrometry, i.e., using only the

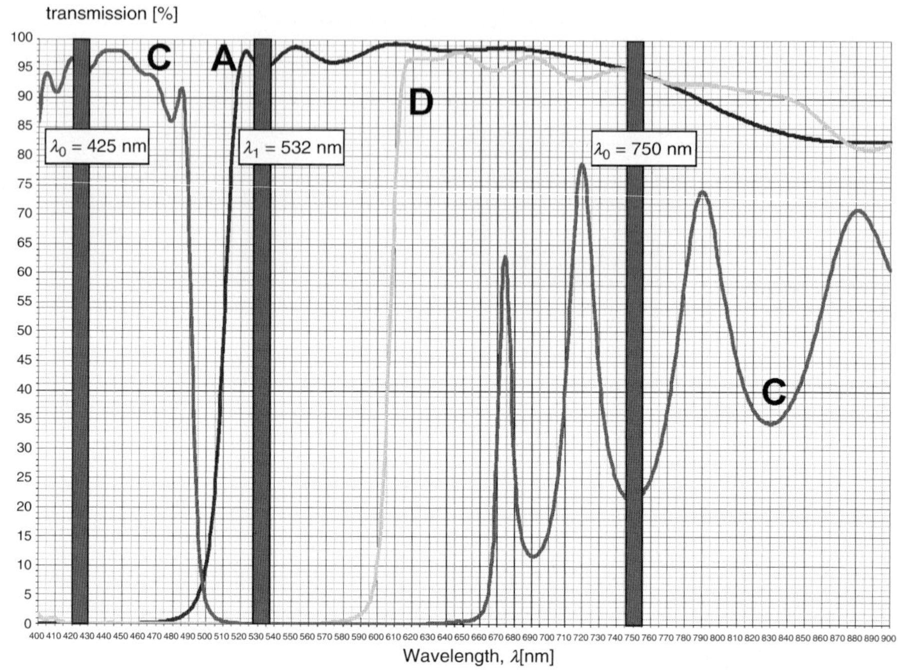

Fig. 3.15. Transmission vs. wavelength λ of different dichroic mirrors used in the experimental setup shown in Fig. 3.14 [8]

image at one wavelength of the filament for the determination of the temperature, this effect yields an overestimation of the temperature by 200–300 K. By 2–λ pyrometry, i.e., using the images at two wavelengths of the filament for the determination of the temperature by ratio pyrometry, this error can be eliminated [9, 128]. Unfortunately, this requires the nontrivial task of overlapping the two images at λ_1 and λ_2 pixel accurate. Otherwise, the ratio of two measured light intensities is not reasonable, if the two values are not from an identical position. Slight differences will enhance the noise in the calculated temperature.

A second test image, again taken before the system was calibrated, is shown in Fig. 3.20, showing the electrode of an operating UHP lamp (an image of a UHP lamp is shown in Fig. 3.19, another picture of the UHP lamp is displayed in Fig. 1.26, cf. Sect. 1.3.6). This time, the measurement at $\lambda_0 = 425$ nm is important for the plasma correction. One can clearly see the reflections of the radiation of the plasma near the electrode tip and at certain positions of the coil. This interference can be eliminated by the plasma correction with the help of the measured radiation intensity at wavelength $\lambda_0 = 425$ nm. The second problem, the enhanced emissivity between the coils can be eliminated by ratio or 2–λ pyrometry.

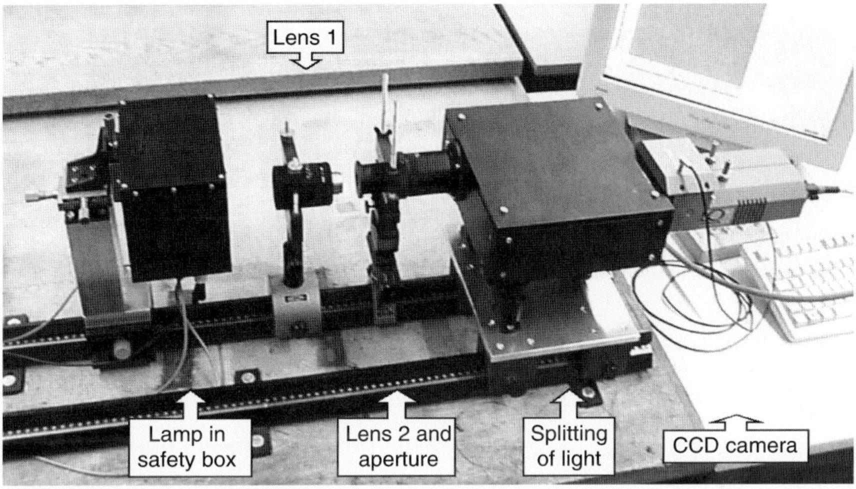

Fig. 3.16. Experimental setup for (2+1)–λ pyrometry as used in [9, 11, 128, 129]. A more detailed insight into the black box containing the components for the splitting of the light beam is shown in Fig. 3.17

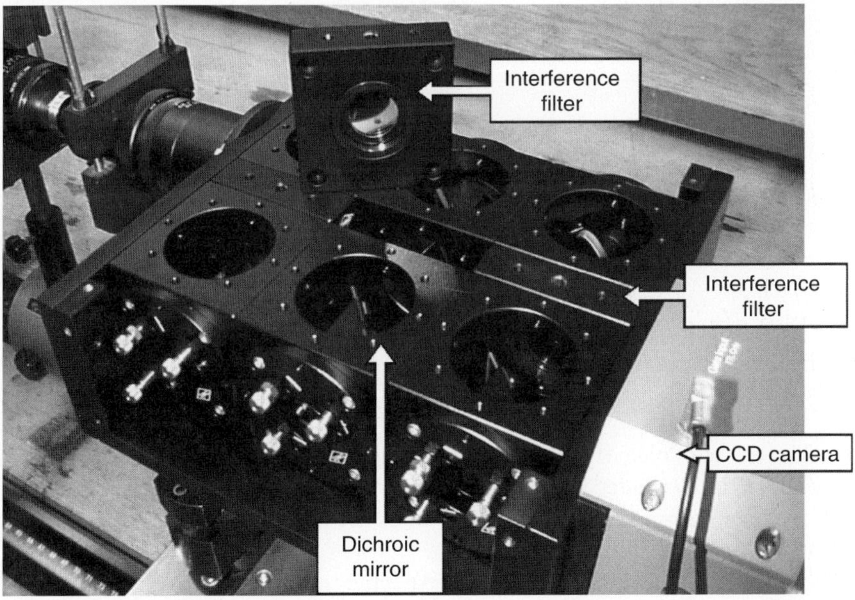

Fig. 3.17. Experimental setup for (2+1)–λ pyrometry as used in [9, 11, 128, 129]: insight into the box containing the components for the splitting of the light beam

Again, this enhanced emissivity is due to a black body effect: The radiation coming from the space between two coils is reflected one or several times at the surfaces of the two neighboring parts of the coil before it is reflected into the direction of the pyrometer and superimposed with the "direct" radiation from the coil surface.

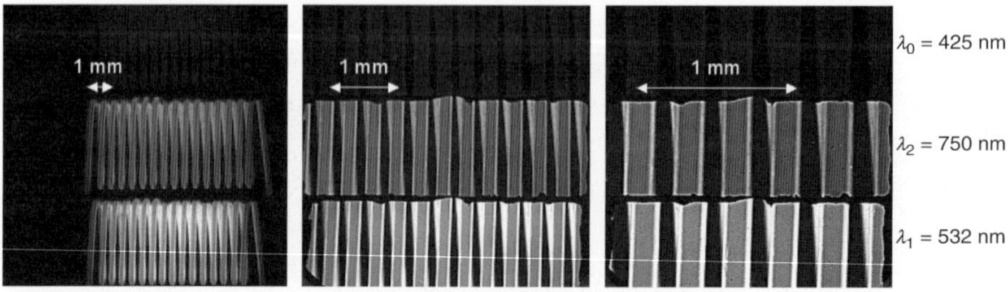

Fig. 3.18. Test images of the filament of a standardized halogen incandescent lamp for calibration purpose at three different wavelengths $\lambda_{0,1,2}$ on different scales depending on the adjustment of lenses and aperture [9, 128]

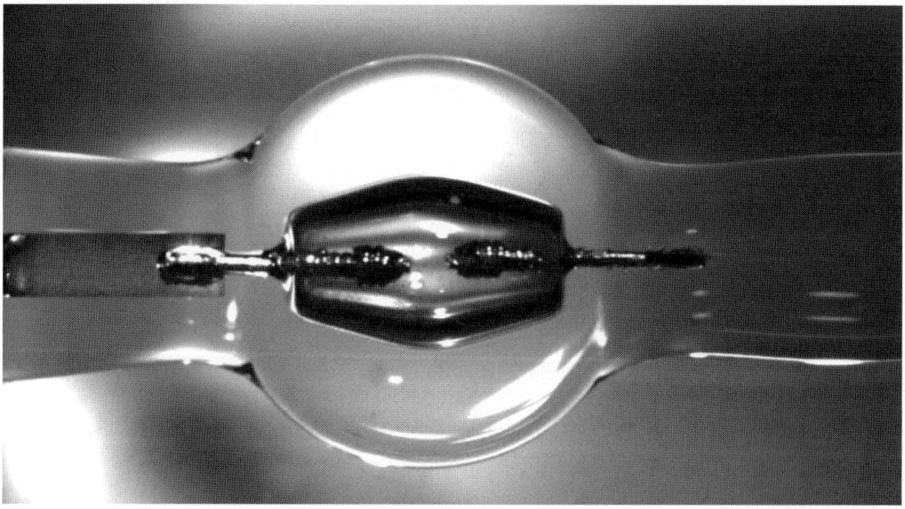

Fig. 3.19. Image of a UHP lamp in alcohol to reduce distortion on the quartz walls (courtesy of Philips, Aachen [88])

According to the choice of the three wavelengths $\lambda_{0,1,2}$, the intensity at $\lambda_2 = 750$ nm is, as expected, dominated by the radiation of the electrode, whereas the radiation of the plasma dominates the radiation intensity at $\lambda_0 = 425$ nm (cf. Figs. 3.12 and 3.13).

The intensity at $\lambda_1 = 532$ nm is, as expected, more sensible to the interfering radiation of the plasma than the measured intensity at $\lambda_2 = 750$ nm (compare Fig. 3.20 with Figs. 3.12 and 3.13). Thus, a sensible selection of wavelengths can already reduce the problems due to interfering radiation from the plasma (cf. Sect. 3.1.4). Nevertheless, the radiation intensities shown in Fig. 3.20 were taken before the system was calibrated, so that a direct assessment of the extent of the plasma interference is not possible. Moreover, even at the image of the UHP lamp electrode taken at the wavelength $\lambda_2 = 750$ nm, interfering radiation is clearly

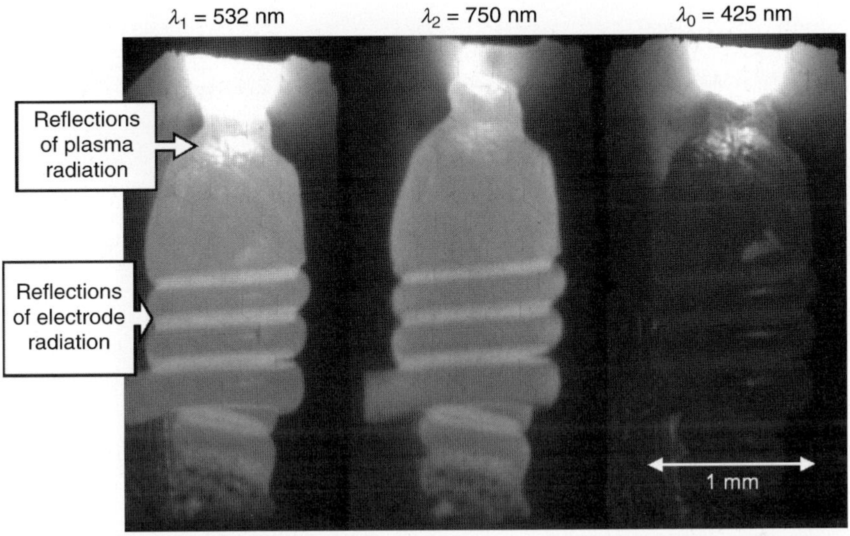

Fig. 3.20. Test image of the electrode of a UHP lamp at three different wavelengths $\lambda_{0,1,2}$ [9, 128]

visible at the very top of the electrode. This makes the determination of the electrode tip temperature difficult or even impossible. But especially the electrode tip temperature is of particular importance, since it determines, for example, the electron emission current, the evaporation rate and a possible melting of the electrode tip. (2+1)–λ pyrometry masters all these problems due to enhanced emissivity, transmission losses, and interfering radiation of the plasma, but the results are more sensible to noise than the simpler 1–λ pyrometry [9, 128]. Thus, a reasonable procedure is the following: The radiation intensities at the three wavelengths $\lambda_{0,1,2}$ are recorded with the experimental setup described earlier. Then, the resulting radiation intensities are analyzed using (2+1)–λ and (1+1)–λ pyrometry, the latter by disregarding the measurement at, e.g., λ_1. If both results are identical except for random noise, the results of the thus validated (1+1)–λ pyrometry can be used, otherwise the results from the (2+1)–λ pyrometry will be used.

Apart from the problems described earlier, which are mastered by (2+1)–λ pyrometry, another problem (amongst other problems) due to the radiation of the opposite electrode may occur. This problem was investigated using specially built dc high-pressure mercury discharge lamps, one of them shown in Fig. 3.21. Similar lamps were built with different electrode gaps (2, 3, and 4 mm), different mercury pressures (2, 4, and 6 MPa, i.e., 20, 40, and 60 bar[8]) and different electrode shapes

[8] Strictly speaking, the lamp was filled with 20, 40, or 60 mg mercury, approximately resulting in the pressures given earlier. If not the entire dose of liquid mercury is evaporated during operation, the resulting pressure is lower. Moreover, different lamp currents and different electrode gaps change the mercury pressure due to the different temperature profiles in the plasma (cf. Sect. 4.4.7).

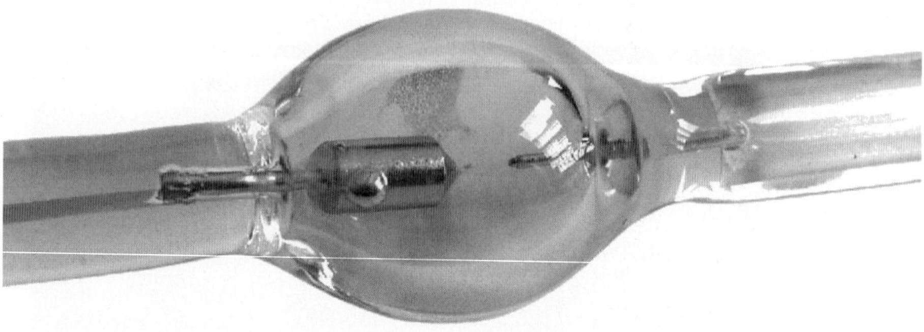

Fig. 3.21. Specially built dc high-pressure mercury discharge lamp for the investigation of the electrode temperature and the general electrode behavior at different electrode gaps, mercury pressures, electrode geometries, and lamp currents [8, 9, 11, 128]

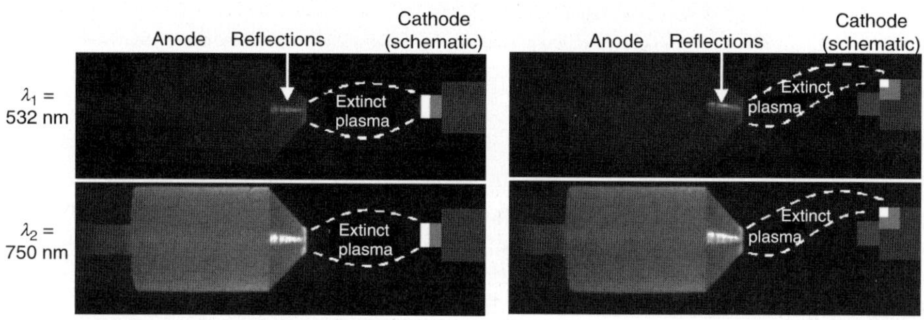

Fig. 3.22. Test image of the anode of the mercury high-pressure discharge lamp shown in Fig. 3.21 at two different wavelengths $\lambda_{1,2}$ 500 μs after extinction of the plasma. The images are different with respected to the cathodic arc attachment before the lamp was turned off [9, 128]

(Apollo-shaped or rod-shaped). Altogether, 25 different high-pressure mercury discharge lamps similar to this shown in Fig. 3.21 were built. As usual in dc high-pressure lamps, the thicker electrode acts as the anode, the thinner one as the cathode.

Figure 3.22 shows the radiation intensities at the two wavelengths $\lambda_1 = 532$ nm and $\lambda_2 = 750$ nm. Both pictures (left-hand side and right-hand side) were taken 500 μs after switching-off the lamp, i.e., the plasma radiation does not interfere with the measurements. Nevertheless, there are reflections at the conical shaped anode tip. Moreover, these reflection are different in the image on the left-hand side and the image on the right-hand side. This difference can be explained with the cathodic arc attachment before the lamp was turned off. Before the image on the left-hand side of Fig. 3.22 was taken, the arc attached at the cathode tip, as shown in Fig. 3.22, on the schematically drawn cathode. The situation prior to the record of the image on the right-hand side of Fig. 3.22 was different: The arc attached on the first winding of the coil at the cathode tip, i.e., the arc attachment was not centric, but in this case at the top of the coil at the cathode. The resulting reflection at the anode shows a higher intensity at the top of the reflection area, corresponding to the higher radiation due to the higher temperature at the top of the cathode coil. Thus, the reflections observed at the anode in Fig. 3.22 are due to the radiation coming from the cathode. Because the cathode was much hotter than the anode for the lamp investigated in Fig. 3.22, the reflections on the anode had a threefold intensity compared to the actual radiation coming from the anode.

This problem due to the reflections of one electrode at the surface of the other electrode cannot be eliminated using the techniques described earlier. A possible solution is the usage of one-dimensional temperature plots using a scan axis outside the problematic regions as shown in Fig. 3.23. The one-dimensional temperature profiles shown in Figs. 3.24 and 3.25 are obtained using this method.

Figure 3.23 shows the temperature distribution at the electrodes of an operating dc high-pressure mercury lamp, i.e., the lamp was not turned off to reduce the

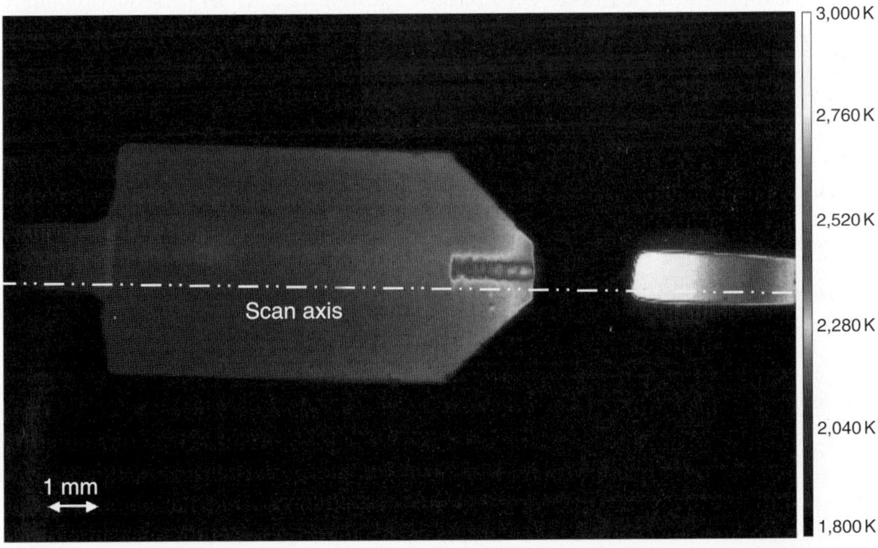

Fig. 3.23. Temperature distribution of anode (left-hand side) and cathode (right-hand side) of the mercury high-pressure discharge lamp shown in Fig. 3.21 using (2+1)–λ pyrometry. The scan axis for the determination of a one-dimensional temperature plot avoids the problems due to reflections of the radiation coming from the cathode [9, 128], mercury pressure: 2 MPa, electrode distance: 2 mm, current: 4 A)

problems due to the reflection of radiation of the plasma. The (2+1)–λ pyrometry used to produce Fig. 3.23 obviously proofs its ability to eliminate the interfering reflections of plasma radiation. The anode tip temperature is about 2,300 K, the cathode tip temperature is approximately 3,000 K. The one-dimensional temperature profile resulting from Fig. 3.23 (mercury pressure: 2 MPa, electrode distance: 2 mm, current: 4 A) is shown Fig. 3.24.

Figures 3.24 and 3.25 both show one-dimensional temperature plots of the electrodes in dc high-pressure mercury lamps as described earlier. An X-ray photograph visualizing the lamp under investigation is shown within the two figures. In Fig. 3.24, the influence of an increasing pressure on the electrode temperatures, the lamp voltage, and the lamp power is investigated. Increasing the pressure from 2 to 6 MPa nearly doubles the lamp voltage, i.e., the lamp power is nearly doubled too, since the lamp current is fixed. The voltage increase is due to the decrease of the electrical conductivity of the plasma with increasing pressure (cf. Sect. 2.3.2). The increasing power input into the plasma is accompanied by an increasing heating of the anode. The anode tip temperature rises from 2,300 K (2 MPa) to 2,550 K (6 MPa). The heating mechanisms at the anode are heating due to electrons and heating due to heat conduction from the plasma. Since the electric current is constant, the reason for the increasing anode temperature must be an increase of the heat conduction from plasma to anode.

The situation at the cathode is different: Though there is an increase in cathode tip temperature when the pressure is increased from 2 to 4 MPa, there is no

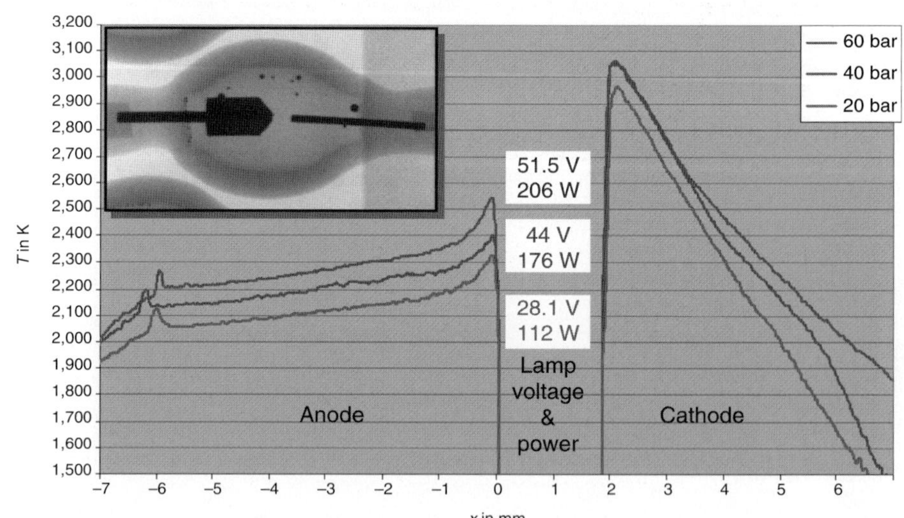

Fig. 3.24. One-dimensional temperature plot of three different dc high-pressure mercury lamps with different mercury pressures (2, 4, and 6 MPa) using (2+1)–λ pyrometry. An X-ray photograph of the lamp under investigation is shown in the top left-hand corner. The lamp voltage and lamp power corresponding to the three temperature curves are given between the temperature profiles of anode and cathode [9, 128], electrode distance: 2 mm, current: 4 A)

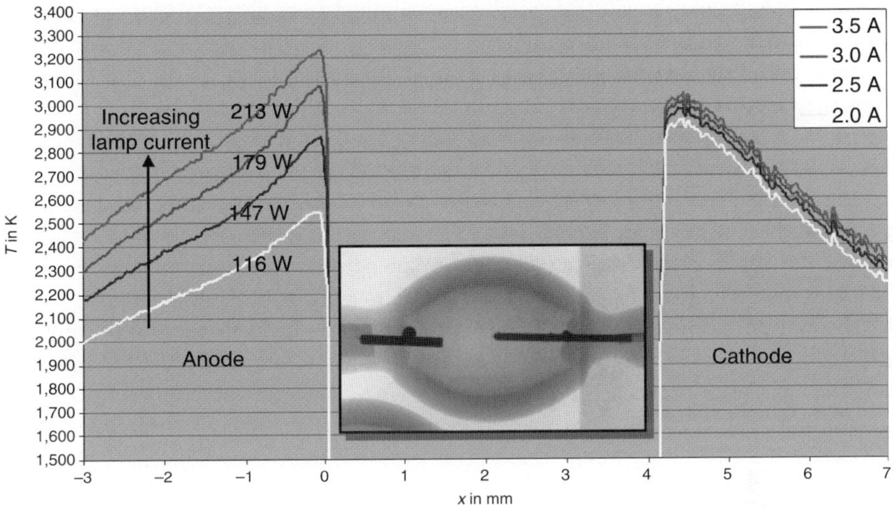

Fig. 3.25. One-dimensional temperature plot of a dc high-pressure mercury lamp at four different lamp currents (2–3.5 A) using (2+1)–λ pyrometry. An X-ray photograph visualizing the lamp under investigation is shown between the temperature profiles of anode and cathode ([9, 128], mercury pressure: 2 MPa, electrode distance: 4.23 mm)

difference in cathode tip temperature increasing the pressure further to 6 MPa. Moreover, comparing the temperature profiles of the cathode for different mercury pressures, it seems that there are problems with the measurements especially at distances more than 1 or 2 mm away from the cathode tip. The temperature profiles of the cathode for 4 and 6 MPa are identical for the first 1.5 mm starting from the cathode tip, then they start to diverge, lacking any plausible reason. This effect also emerges comparing the measurements of two dc high-pressure discharge lamps which are equal in all of their parameters (in this case, mercury pressure: 4 MPa, electrode gap: 3 mm, lamp current: 3 A). The cathode tip temperatures differ by more than 200 K and the temperature profiles are different as well [9, 128]. This means that we cannot have unquestionable confidence in the results. Unknown parameters, small differences in the lamp design, the history of the lamp, or small differences in the operating conditions might affect the measurements or the actual temperature of the electrodes. Thus, an increase of the accuracy of the determination of electrode temperature beyond ±50 K might not be useful because of these "natural" variations of the electrode temperature. This conclusion is confirmed by pyrometric measurements using several identical D2 automotive headlight lamps described in [85] and the measurements described in Sect. 3.1.4 (cf. Fig. 3.7).

The cathode seems to be more stable with respect to an increase of the mercury pressure than the anode. This statement is enhanced by the different shapes of anode and cathode investigated in Fig. 3.24: Due to the smaller dimensions of the cathode, the reaction due to changes in the pressure or the lamp current should be larger at the cathode. But the major difference between cathode and anode is that the cathode reacts more "active" to changes in, e.g., pressure or lamp current. This is due to the interplay between ion current and electron emission current at the cathode tip: When the electron emission current is less than the total lamp current, the difference is sustained by an ion current. The ions deposit their energy (minus the energy necessary to release an electron to neutralize the ion) at the cathode, thus heating the cathode and enhancing electron emission which depends strongly on the local cathode temperature (cf. Sect. 2.4.4). The enhanced electron emission current reduces the ion current, so that this interplay between ion current and electron emission current adjusts the ratio of electron to ion current until a stable power balance of the cathode is established. An increasing heat conduction from the plasma to the cathode tip increases the cathode tip temperature, thus the electron emission current is increased, leading to a decrease of the ion current. Therefore, the increasing heat load of the cathode tip is partially balanced by a decreasing heat load due to a decreasing ion current. For this reason, the temperature change at the cathode tip due to an increase in the mercury pressure is smaller compared to the temperature change at the anode tip.

The effect described earlier can also be observed in the measurements shown in Fig. 3.25: Here, both electrodes are rod-shaped, i.e., anode and cathode behavior can be compared easily. In Fig. 3.25, the lamp current is increased at constant pressure[9] and constant electrode gap. Again, the ("passive") anode tip temperature reacts more strongly on this change compared to the ("active") cathode tip temperature. Moreover, when pyrometry is used at a specific lamp, the results seem to be quite reproducible.

3.1.8 More Pyrometry

Quartz has a good transmission for radiation from 185 to 4,000 nm [118]. Within this wavelength range, pyrometry on the electrodes and plasma spectroscopy are possible[10]. Larger wavelengths, e.g., 5 μm can be used to measure the temperature of the outer quartz surface as shown in Fig. 3.26 for a UHP lamp (cf. Sect. 1.3.6). The emissivity of quartz is 0.97 at 5 μm [121], whereas the transmission is zero, i.e., no radiation of the plasma or the electrodes is interfering the measurement.

3.1.9 Calorimetric Measurements on Electrodes

The experimental setup shown in Fig. 3.5 for 1–λ pyrometry at the electrodes in a model lamp can also be used to measure the heat flow from the electrode ends to a coolant (silicone oil). The silicone oil keeps the temperature of the electrode ends at constant temperature (approximately 293 K), the temperature change of the silicone oil together with the measured mass flow of the cooling fluid yields the heat

[9] The pressure will be slightly different due to the different lamp currents and the resulting different temperature distributions in the plasma (cf. Sect. 4.4.7).

[10] The transmission is not equally high for all wavelengths; especially in the infrared region small transmissions can be observed.

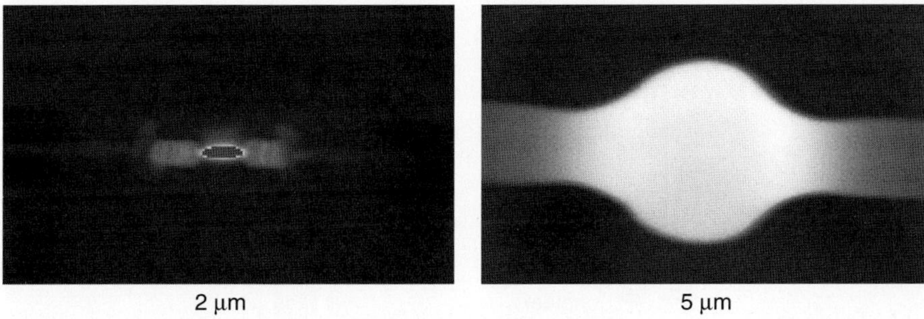

Fig. 3.26. Congruent images of a UHP lamp showing the electrodes (left-hand side, $\lambda = 2$ μm) and the quartz tube (right-hand side, $\lambda = 5$ μm) using pyrometry to determine the temperature of the electrodes and of the outer quartz tube surface (courtesy of Philips, Aachen [88])

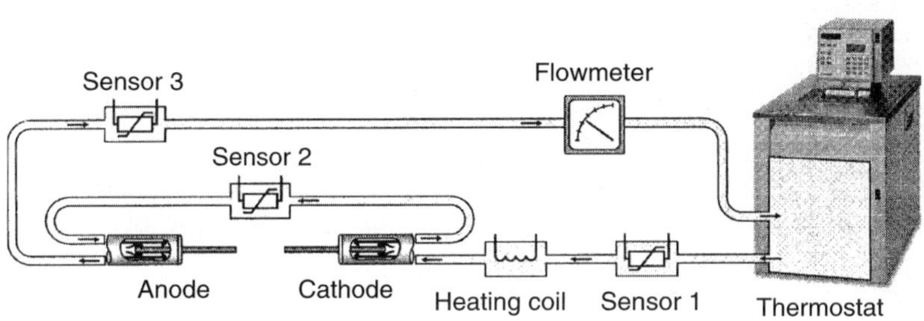

Fig. 3.27. Experimental setup for calorimetric measurements on the electrodes used in the model lamp shown in Fig. 3.5 [121]

flow from the electrode ends to the silicone oil. The temperature sensors 1, 2, and 3 shown in Fig. 3.27 enable the separate determination of the heat flow from the anode and the cathode to the coolant because the temperature sensor 2 measures the temperature of the silicone oil between anode and cathode. The calibration of this method is done with the help of a heating coil heating the silicone oil (model lamp turned off). The measured temperature change together with the measured mass flow of the cooling fluid can be associated with the known power input by the heating coil [43, 121].

The flow of the silicone oil in the electrode holder is shown in Fig. 3.28 (left-hand side). A problem of this method is the determination of the "effective" electrode length, i.e., the position of the electrode where $T = 293$ K. This position is not identical to the soldering shown in Fig. 3.28 but is situated within the electrode holder. Moreover, the temperature increase of the silicone oil is not only due to the power transfer from the electrode ends to the cooling liquid, but also due to the parasitic heat flows from the lamp tube to the electrode holders as shown in Fig. 3.28 (right-hand side). A reduction of these parasitic heat flows from the lamp tube to the electrode holders can be achieved by fixing the temperature of the lamp tube to the temperature of the cooling fluid using an additional external cooling in the region of the electrode holder [121]. Nevertheless, this additional cooling might

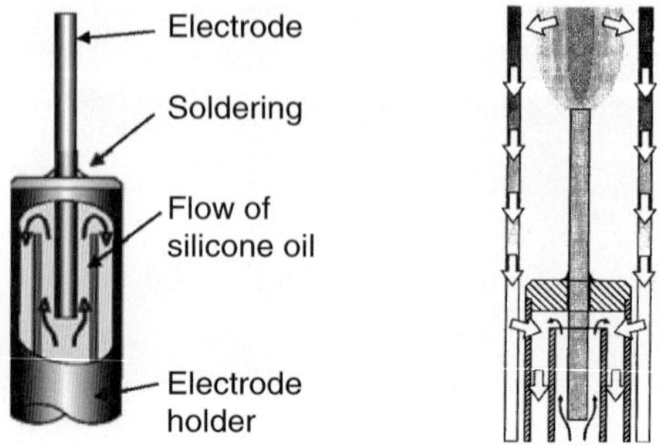

Fig. 3.28. Cooling of electrode end by a flow of silicone oil through the electrode holder (left-hand side) and parasitic heat flows from the lamp tube to the electrode holder for the model lamp shown in Figs. 3.5 and 3.27 [43, 121]

again effect the heat transfer from the electrode ends to the coolant, so that this is not an ideal solution. Moreover, due to the small increase of the temperature (some kelvin) of the silicone oil due to the heat flow through the electrode ends, the determination of the heat flow through the electrode ends is quite inexact.

The measured heat flow through the electrode ends can be compared with the calculated heat flow from the measured temperature profiles (cf. Sect. 3.1.4, Fig. 3.6, and (3.13)), thus verifying the measured temperature profiles and the resulting calculated heat flow through the electrode ends. The radiation losses (calculable from the measured temperature profiles with the help of the Stefan–Boltzmann law (2.47), (3.1), or (3.13)) together with the measured heat flow through the electrode ends specify the power balance of the electrodes, i.e., the power input into the electrode tip equals the radiation losses plus the heat flow through the electrode end. Nevertheless, instead of taking the measured value of the heat flow through the electrode ends as described in this section, the more reliable method is to use the calculated heat flow from the measured temperature profiles (cf. Sect. 3.1.4) instead of the measured values.

Such a power balance of an electrode (here, the cathode) is shown in Fig. 3.29. The radiation losses and heat flow through the cathode end are displayed for different cathode lengths from 10 to 25 mm and for different lamp currents from 1 to 5 A. The cathode material was pure tungsten, the cathode diameter 1 mm, the gas a 0.25–0.3 MPa argon plasma [121]. The radiation losses and heat flow through the cathode end were both calculated using the measured temperature profiles (using 1–λ pyrometry, cf. Sect. 3.1.4), i.e., the measured values of the heat flow through the cathode end were not used.

The temperature of the cathode tip primarily depends on the lamp current and is to a large extent independent of the cathode length. This is due to the fact that the cathode has to emit electrons to sustain the given lamp current. The electron emission is determined by the Richardson–Schottky equation (2.61) which strongly depends on the cathode tip temperature. Thus, the cathode tip temperature will adjust to a value, where electron emission current is sufficiently high, so that electron emission current and ion current equal the total lamp current (cf. Chap. 4,

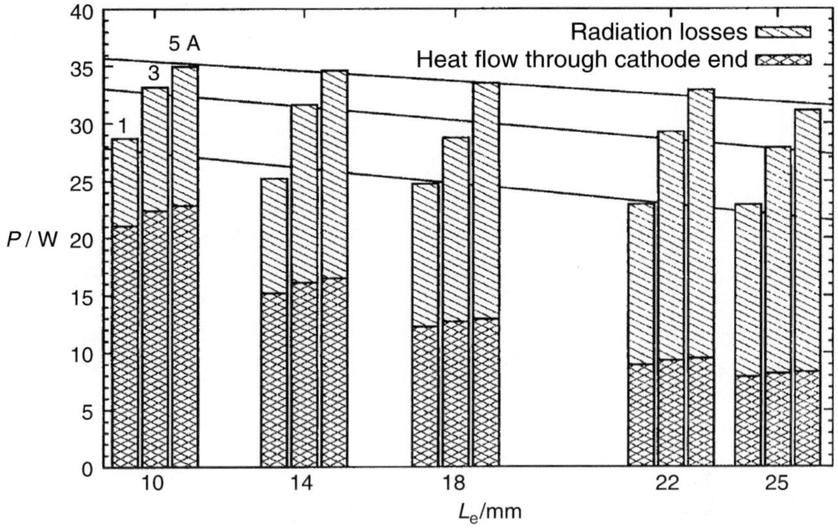

Fig. 3.29. Power balance (radiation losses and heat flow through cathode end) of pure tungsten cathode (diameter: 1 mm, gas: argon, pressure: 0.25–0.3 MPa) depending on cathode length (10–25 mm) and lamp current (1–5 A) according to [121]

especially Sect. 4.4). The temperature gradient between the cathode tip and the liquid cooled end of the cathode therefore decreases with increasing cathode length. Thus, the heat flow through the cathode end decreases with increasing cathode length (cf. Fig. 3.29). By the same token, the radiation losses increase with increasing cathode length because the cathode tip temperature is to a large extent independent of the cathode length and the radiating surface area increases with increasing cathode length.

Figure 3.30 presents the same data in a different diagram. The measured (global[11]) cathode tip temperature $T_{c,0}$ using 1–λ pyrometry (cf. Sect. 3.1.4) is given in the top left diagram. Due to the fact that the temperature of the cathode tip cannot be measured with the method applied here, the extrapolation of the measured cathode temperature according to (3.13) is used (cf. Fig. 3.6 and [121]). Because this calculation of the cathode tip temperature does not take the different processes of power transfer (and their spatial variation) from the plasma to the cathode tip into account, these values should be regarded as an approximation to the cathode tip temperature. The cathode tip temperature is to a large extent independent of the cathode length because the electron emission, which is determined by the temperature dependent Richardson–Schottky equation (2.61), adjusts the cathode tip temperature according to the total lamp current. The higher the lamp current, the higher the cathode tip temperature, so that more electrons are emitted to sustain this higher lamp current. The ratio of radiation losses P_{rad} to electrode losses P_{el} (sum of

[11] Not the temperature within the arc attachment area but the extrapolated temperature as described in the following.

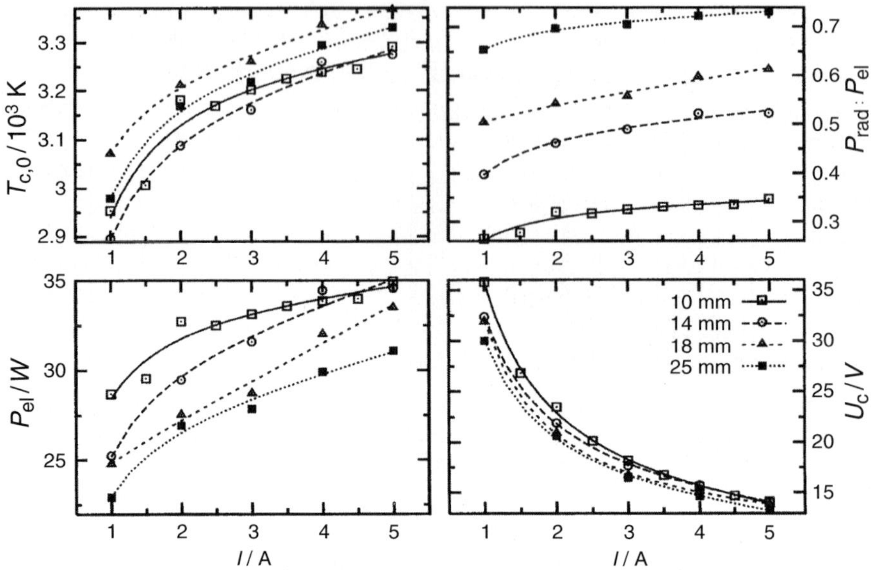

Fig. 3.30. Cathode tip temperature (top left), ratio of radiation losses P_{rad} to electrode losses P_{el} (sum of radiation losses and heat flow through cathode end, top right), electrode losses (bottom left), and calculated cathode fall voltage U_c (bottom right) of pure tungsten cathode (diameter: 1 mm, gas: argon, pressure: 0.25–0.3 MPa) vs. lamp current (1 to 5 A) as a function of cathode length (10–25 mm, legend in bottom right diagram) according to [121]

radiation losses and heat flow through cathode end) increases with increasing cathode length as described earlier (top right diagram of Fig. 3.30). The electrode losses P_{el} decrease with increasing cathode length and increase with increasing lamp current as shown in Figs. 3.29 and 3.30 (bottom left diagram).

The cathode fall voltage U_c shown in the bottom right diagram of Fig. 3.30 is calculated using the electrode losses P_{el}. The cathode fall voltage U_c is defined as the additional power consumption in the near cathode plasma due to the power transfer from the plasma to the cathode plus the increased radiation losses by the hot plasma spots near the electrode tips. It can be extracted from the electric potential on the lamp axis as shown in Fig. 3.40. If the electric potential distribution in the plasma is not available, it can be approximately calculated. The equation used for the calculation of the cathode fall voltage shown in Fig. 3.30 is:

$$U_c = \frac{P_{el}}{I} + \varphi + \frac{5kT_e}{2e}. \tag{3.29}$$

This equation means that the electrical power given by the cathode fall voltage U_c times the lamp current I equals the electrode losses P_{el} (sum of radiation losses and heat flow through cathode end) and the power $I\varphi$ necessary to "extract" the electrons with the work function φ from the cathode[12] (φ given in volt, e.g., $\varphi = 4.5$ V for pure tungsten electrodes). Moreover, the electrons are assumed to have the electron temperature T_e in the near electrode plasma. By leaving the near cathode plasma in the direction of the plasma column, they are assumed to remove the power $2.5IkT_e/e$ from the near cathode plasma.[13] Using this model and these assumptions, the resulting cathode fall voltage U_c is given in the bottom right diagram of Fig. 3.30. The cathode fall voltage decreases with increasing lamp current, but the power consumption due to the cathode fall voltage (i.e., IU_c) increases with increasing lamp current. This is due to the increase of the electrode losses P_{el} with increasing lamp current (bottom left diagram of Fig. 3.30).

3.1.10 External Laser Heating of Electrodes

The methods described earlier for the investigation of HID lamp electrodes (pyrometry and calorimetric measurements) leave the electrodes undisturbed. The radiation of the electrodes is detected or the temperature change of the cooling fluid is measured. It is also possible to learn something about electrodes by disturbing the balance between plasma and electrodes. One possibility is the external laser heating of the electrodes during the operation of the HID lamp. This experimental method will be described in this section, a comparison of measured electrode tip temperatures and total lamp voltages with the results of numerical simulations is given in Sect. 4.4.5.

The influence of the external laser heating of anode and cathode in operating HID lamps on electrode temperature and lamp voltage is presented in [156]. The lamp under investigation contains a pure mercury plasma with a working pressure of about 0.6 MPa, the electrode gap is 10 mm. The electrodes (length 13 mm, diameter 0.5 mm) are made from pure tungsten (cf. Sect. 4.4.5, Figs. 4.40[14] and 4.56). The measurements were carried out on a vertically mounted lamp driven by a rectangular-wave current (1.8 A, 0.05 Hz). The lower electrode was heated using a diode laser ($\lambda = 808$ nm; maximum laser power P_L: 37.3 W) which was turned on for 3 s each half-cycle, starting 3.5s after current reversal (half cycle length: 10 s). The laser was directed side-on at the upper part of the lower electrode just below the tip. Its beam diameter at the electrode surface was 0.5 mm. The electrode temperature near the electrode tip and along the electrode was determined using a pyrometer (Impac IS10, λ from 700 to 1,100 nm, cf. Sect. 3.1.4), the observation direction being at an angle of more than 90° compared to the laser beam. To avoid

---

[12] At the cathode surface, the total lamp current I is the sum of electron current I_e due to electron emission and ion current I_i. The emitted electrons cool the cathode with the work function φ, but the ions cool the cathode as well because they extract an electron from the cathode for their neutralization. The heating of the cathode by heat conduction from the plasma to the cathode tip and by the ionization energy of the ions arriving at the cathode surface is included in the electrode losses P_{el}.

[13] More details on electrode and plasma models can be found in Chap. 4, more details about (3.29) can be found in [20, 43, 121, 168].

[14] The electrodes shown in Fig. 4.40 have a length of 10 mm, the electrodes used for the external laser heating had a length of 13 mm.

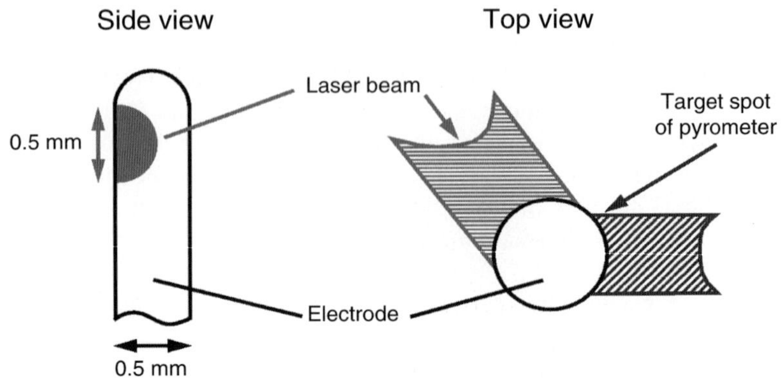

Fig. 3.31. Sketch of the experimental setup for external laser heating of HID lamp electrodes: The laser is directed side-on at the upper part of the electrode just below the tip (beam diameter: 0.5 mm). The electrode temperature is determined using the pyrometer Impac IS10, the observation direction being at an angle of more than 90° compared to the laser beam (cf. [156])

disturbance of the pyrometric measurements by the laser beam, a notch filter (10^{-4} at the laser wavelength) was used for the pyrometer (cf. Fig. 3.31).

Before investigating the influence of this external laser heating of anode and cathode in operating HID lamps on electrode temperature and lamp voltage, the "effective" laser power $P_{L,\,eff}$ (as opposed to the nominal laser power P_L, which is supposed to be provided by the laser) heating the electrode has to be determined. This has been done by heating one electrode using different laser powers without starting the discharge, i.e., the lamp was switched off. From the resulting temperature profiles of the electrode, the effective laser power $P_{L,\,eff}$ can be determined. This has been done using a simple global power balance $P_{L,\,eff} = P_{rad} + P_{HeatFlow}$, where P_{rad} is the radiation cooling and $P_{HeatFlow}$ the heat flow through the electrode end [156]. The result of this procedure is an efficiency η (ratio of effective laser power $P_{L,\,eff}$ to nominal laser power P_L) of 12–24% [153, 156].

$P_{L,\,eff}$ is an important input parameter for numerical simulations (cf. Sect. 4.4.5). The problem with the procedure described earlier is that $P_{L,\,eff}$ might be different as soon as the discharge is turned on. Then, the electrode temperature will naturally be higher, resulting in a different emissivity ε of tungsten (cf. Fig. 3.8). If ε decreases for higher temperatures ($\lambda = 808$ nm), the effective laser power should decrease as well as ε is equal to the absorption coefficient.

The results obtained by using the external laser heating revealed many interesting features of HID lamp electrodes[15]: The electrode temperature reacts strongly on the external laser heating during the anode phase, but weakly during the cathode phase. In contrast, the striking feature comparing measured and calculated

[15] The detailed results are presented in Sect. 4.4.5.

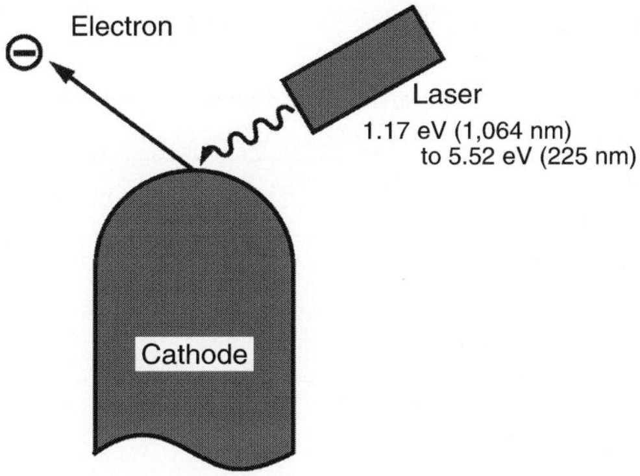

Fig. 3.32. Sketch of the experimental setup for in situ measurement of the electrode work function using the photoelectric effect with photons in the energy range from 1.17 to 5.52 eV (cf. [145, 146])

lamp voltages is a strong influence of the external laser heating during the cathode phase and a barely observable influence during the anode phase. Moreover, the time constants of the reactions to the external laser heating are different during anode and cathode phase.

Thus, the investigations described in [156] are very valuable for the verification of theoretical electrode models. Moreover, by comparing the interesting experimental findings with results from numerical simulations, one can learn much about the principles of electrode behavior and explain in detail the differences between anode and cathode phase (cf. Sect. 4.4.5).

3.1.11 In Situ Laser Diagnostics of Work Function

A different method described in [145, 146] uses a laser to determine in situ (i.e., during lamp operation) the work function of the cathode using the photoelectric effect/photoemission (cf. Sect. 2.4.4). The specific method described in [145, 146] uses short laser pulses of different wavelengths, which are focused on the electrode tip on the arc attachment area (cf. Fig. 3.32). The resulting transient electric response signals for different laser wavelengths are the basis to evaluate the electrode work function. For that, a laser system offering different photon energies in the wavelength range of 225–1,064 nm (corresponding to 5.52–1.17 eV) is necessary. In contrast to the method of electrode heating described earlier, the laser power for the determination of the work function for electron emission is high (up to 100 MW) and the pulse widths are small (order of 10 ns). This is to prevent the electrode from heating-up too much. The high power is necessary to generate sufficiently high electric signals induced by the pulse laser irradiation. Typical electric signals are about 0.1–1% of the lamp current and can be used to determine the electrode work function. This method is limited to cathodes or electrodes during cathode phase because electric fields in front of the electrode surface are needed to collect the generated photoelectrons.

Table 3.1. Results of in situ measurement of the effective electrode work function φ_{eff} [145, 146]

lamp type	cathode mode	φ_{eff} (eV)	effect	remarks
pure Hg, pure W	diffuse	4.52	–	–
	spot	3.63	Schottky	$E = 5.5\times10^8$ V m^{-1}
pure Hg, emitter	diffuse	3.67	emitter	oxides between rod and coil
	spot	3.06	Schottky	$E = 2.6\times10^8$ V m^{-1}
metal halide, pure W	nominal power	2.41	emitter	rare earth adsorption
	low power	4.45	salts condensed	pure Hg discharge

The thus determined effective work function φ_{eff} includes the lowering of the work function due to electric fields (Schottky effect, cf. (2.60) and (2.61), and Sect. 2.4.4) and the lowering of the work function due to an emitter (e.g., thoriated tungsten electrodes used in many HID lamps). The measurements can be disturbed by unwanted absorption of the laser beam in the plasma causing excitations of plasma species, thermal plasma heating, and a reduced laser power at the electrode tip. Transient thermal electrode surface heating by the absorbed laser pulse amounts to a few tens kelvin and gives rise to increased thermal emission according to Richardson's equation (2.59). Experimentally, this effect can be quantified with laser pulses of small photon energies below the work function, for example about 2 eV, which do not cause a photoelectric effect at the tungsten surfaces. These corrections are necessary for all of the laser wavelengths used.

The results of this method are summarized in Table 3.1. The measured work function for high-pressure mercury discharge lamps (modified Philips HP80W with pure tungsten electrodes instead of electrodes with emitter material) with pure tungsten electrodes depends on the type of arc attachment to the electrode tip: In the diffuse attachment mode of this lamp, a work function of 4.52 eV is determined. This value corresponds very well with literature data for polycrystalline tungsten, which confirms the applicability of this method for the determination of work functions in HID lamps. The electrode temperature was estimated to be about 3,000 K; assuming an error of ±300 K and taking other possible errors into account, the error in the calculated work function can be estimated to be ±0.25 eV. Carrying out several measurements on similar lamps yields work functions from 4.25 to 4.55 eV [145]. A reduced value of 3.63 eV is measured in the spot mode (±0.20 eV for a 10% error in the electrode temperature, 3.5–3.8 eV for similar lamps, cf. [145]). This reduction is due to the Schottky effect (cf. (2.60) and (2.61), and Sect. 2.4.4). The electric field at the cathode surface can be calculated to be about 5.5×10^8 V m^{-1} within the arc attachment spot.

The measured work function in high-pressure mercury discharge lamps (Philips HP80W with emitter material) with tungsten electrodes with emitter material (thoriated tungsten) depends also on the type of arc attachment to the electrode tip: In the diffuse arc attachment mode, a work function of 3.67 ± 0.1 eV is determined, i.e., the emitter reduces the work function by approximately 0.8 eV. In the spot mode, the work function is further reduced to 3.06 ± 0.1 eV due to the Schottky effect (calculated electric field about 2.6×10^8 V m^{-1}).

The results of the measurement of the electrode work function using a metal halide lamp with pure tungsten electrodes is also presented in Table 3.1. This lamp (Philips MHN-TD150W) always exhibited a diffuse arc attachment, but the experiments were conducted with a nominal power load and with about half of the nominal power load. Under nominal operating conditions, the measured work function was about 2.41 ± 0.2 eV. This is much smaller than the expected 4.5 eV for a diffuse arc attachment at pure tungsten electrodes (see earlier). The measured work function at the electrodes of the lamp being operated at half nominal power is 4.45 eV, which can be expected for the pure tungsten electrodes used within this lamp. The explanation given in [145, 146] is that in the metal halide lamp operated at half nominal power nearly no salts are evaporated because of the lower cold spot temperature. Thus, the discharge is essentially a mercury discharge. But in the lamp operated with nominal lamp power, the halide pressures are significant. These salts can be adsorbed onto the cathode tip and thus reduce the work function. A comparison of numerical simulations and experimental measurements on the D2 automotive headlight lamp (cf. Sect. 1.3.8) presented in Sect. 4.4.6 shows that this effect may be important as well in the D2 lamp.

The methods described here and in Sect. 3.1.10 both use a laser for the investigation of electrodes in HID lamps. Both methods in common are a careful choice of the wavelength of the laser radiation: Interactions of the laser beam with the lamp plasma have to be avoided, the energy (wavelength) of the photons must allow for the photoelectric effect (determination of work function) or must avoid the photoelectric effect (external heating), and pyrometric measurements of the electrode temperatures should not be disturbed (external heating). Moreover, both methods are used to investigate electrode behavior and electrode characteristics in situ, i.e., during normal lamp operation. Thus, they are very powerful tools for electrode diagnostics in HID lamps (see also [79]).

3.1.12 Monolayer of Sodium on Cathode

The effect of a reduced work function due to adsorption of atoms onto the cathode tip as described in Sect. 3.1.11 can also be observed in high-pressure sodium lamps, as described for example in [80]. In [1] and [85] an experimental setup is described to investigate this effect: A high-pressure sodium lamp with 16 kPa xenon as a starting gas and 4 mg pure sodium for the discharge was used to investigate the electrode temperature and thus the electrode work function as a function of the sodium coverage at the cathode tip. The sodium pressure of the lamp was adjusted by heating the cold spot of the lamp (with the condensed sodium) using the irradiation of an incandescent lamp outside the lamp bulb. Thus, the sodium pressure can be controlled independently of the lamp current. By permanently monitoring the self-reversed sodium D-lines (589.0 and 589.6 nm, cf. Sect. 3.2.4), the pressure can be adjusted to the wanted value.

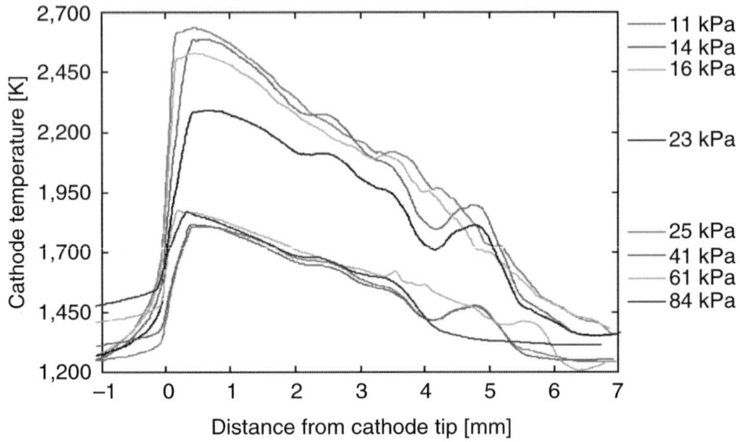

Fig. 3.33. Dependence of cathode temperature profile on sodium pressure in a high-pressure sodium lamp with pure tungsten electrodes (electrode length: 7 mm, electrode diameter: 0.7 mm, lamp power: 75 W, dc current 1.5 A, cf. [1, 85]

The cathode temperature for different sodium pressures at constant lamp current and constant lamp power (1.5 A, 75 W) was measured using 1–λ pyrometry as described earlier. Figure 3.33 shows the results of this experiment: The cathode tip temperature decreases with increasing sodium pressure. At lower sodium pressures (less than 20 kPa), no or not sufficient sodium is adsorbed onto the cathode tip to reduce the work function. Thus, the high work function of pure tungsten (about 4.5 eV) makes a high cathode temperature necessary to sustain a sufficiently high electron emission current (cf. Sect. 2.4.4). For higher sodium pressures (above 25 kPa), the sodium coverage on the cathode tip reduces the work function. Thus, a much lower cathode temperature is able to sustain a sufficiently high electron emission current. The cathode tip temperature and the calculated effective work function φ_{eff} of this lamp as a function of sodium pressure is shown in Fig. 3.34. The effective work function is reduced from well above 4 eV (pure tungsten) to less than 3 eV (sodium acts as an emitter on the cathode surface). The experimental results are confirmed by theoretical calculations of the work function as a function of sodium coverage and by the investigations described in Sect. 3.1.11. In [80], the reduced work function after the formation of a monolayer of sodium on the cathode is quoted to be 1.8–2.1 eV (referring to [66]). Further details concerning the variation of the work function for electron emission can also be found in [130] on page 77ff.

3.1.13 Deformation of Electrodes

The deformation of electrodes during lamp operation is an important subject in commercial HID lamps. This deformation can lead to different operating conditions as described for example in [85, 86] for a D2 automotive headlight lamp (for details see later). The UHP lamp used in video projectors (cf. Sect. 1.3.6) also suffers from

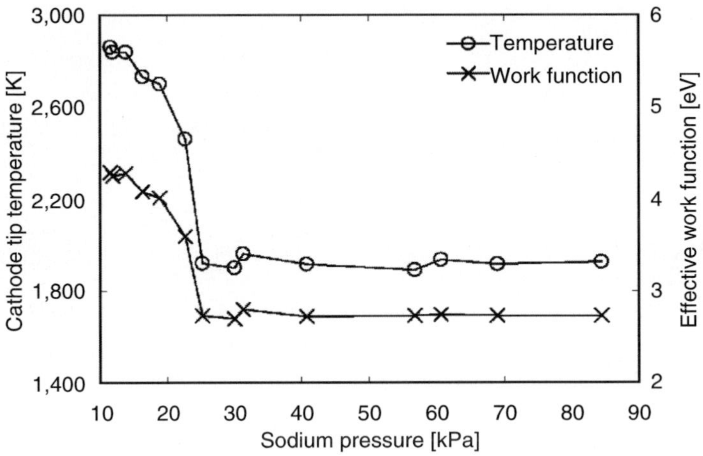

Fig. 3.34. Dependence of cathode tip temperature and effective work function φ_{eff} of pure tungsten electrodes on sodium pressure (electrode length: 7 mm, electrode diameter: 0.7 mm, lamp power: 75 W, dc current 1.5 A, cf. [1, 85]

electrode deformation during lamp operation. Problems in UHP lamps are reported in [45, 119] due to changes in the electrode geometry: Depending on the electrode shape, the arc attachment area moves over the electrode surface, leading to visible arc movements and thus unwanted artifacts in projection applications. The solution to this problem is a special pulse current as described in [45, 119]. The deformation of electrodes in various HID lamps is also reported in [43, 86, 95, 114, 121]. Besides different operating conditions as arc attachment area and electrode temperature, the possible evaporation of tungsten and a resulting wall blackening are important subjects related to the deformation of electrodes.

According to [36], there are two different pathways along which transport of tungsten (leading to the deformation of the electrode tips) could take place: First, gaseous tungsten coming from the flanks of the electrode can be transported through the plasma toward the electrode tip, and second, a change of the electrode tip structure might be induced by the flow of liquid tungsten due to several forces acting on the melt. The latter possibility presupposes the melting of the electrode tip, as it is at least observed in UHP lamps. The electrodes shown in Figs. 3.36 (D2 automotive headlight lamp) and 3.37 (xenon model lamp) exhibit as well structures, indicating that the melting temperature of tungsten (3,680 K) was reached at least locally and temporarily.

In [85] (in parts also in [86]), electrode temperature measurements at different electrodes in D2 automotive headlight lamps (cf. Sect. 1.3.8) are presented. Three different types of electrodes were investigated in operating D2 lamps: thoriated tungsten electrodes (type A), pure tungsten electrodes glowed at 1,200°C for cleaning (type B), and pure tungsten electrodes glowed at 2,400°C for cleaning (type C). For each of the three types of electrodes at least five lamps have been investigated to reduce statistical fluctuations. Thoriated tungsten electrodes should have lower

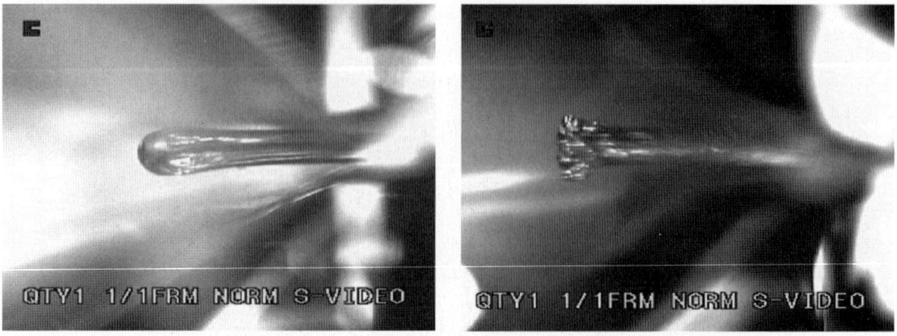

Fig. 3.35. Pure tungsten electrodes in D2 automotive headlight lamps after 1,000 h operation time. Left-hand side: electrode glowed during production at 2,400°C for cleaning (type C), right-hand side: electrode glowed during production at 1,200°C for cleaning (type B). A magnification of the electrode tip shown on the right-hand side is given in Fig. 3.36 [85, 86]

electrode temperatures compared to pure tungsten electrodes due to the enhanced electron emission by a lower work function φ for thoriated tungsten electrodes (cf. Sects. 2.4.4 and 3.1.11). This is true if one compares the electrode temperatures of type A with type C electrodes ([85], electrode temperature of up to 2,500 K (type A) and 2,900 K (type C)). But surprisingly, the pure tungsten electrodes of type B have practically the same electrode temperature as type A electrodes (thoriated tungsten). This might have many reasons including work function φ, type of arc attachment (spot or diffuse), adsorption of sodium[16] at electrode tip (cf. Sect. 3.1.11), increased front surface of deeply structured electrode tip by large bumps as shown in Fig. 3.35, etc. Figure 3.35 compares type B and C electrodes (glowed at 1,200°C and 2,400°C for cleaning before mounting) after 1,000 operating hours, Fig. 3.36 shows a magnification of the electrode tip of a type B electrode.

Due to the different "history" of the electrodes (type B and C) in the production process, the arc attachment, the electrode temperature, and the electrode shapes after operating the lamps are completely different. So, we have to keep in mind that knowing material and (initial) geometry of the electrode is not always enough to characterize electrodes and their behavior. Even small differences in the production process of the electrodes [85] or different plasma compositions or electrical power inputs (Sect. 3.1.11, [145, 146]) might change the electrode properties considerably. Especially the work function is very sensitive to plasma composition, temperature, surface structure, etc. This is important to know because the work function is essential for numerical simulations of electrodes (cf. Chap. 4 and Sect. 4.4.6).

Figure 3.37 shows different electrodes (here used as cathodes) made of pure tungsten (electrode diameter: 1 mm, electrode length: 10 mm) and used in a 0.26 MPa xenon discharge lamp[17] with a lamp current of 6 A. The cathodes were identical before operation and were all used in the same experimental setup. The

[16] Sodium is part of the plasma composition in D2 lamps, cf. Sect. 1.3.8.

[17] The lamp was the model lamp described in Sect. 3.1.4, Fig. 3.5.

Fig. 3.36. Scanning electron microscope record [86] of an electrode similar to that shown on the right-hand side of Fig. 3.35 (type B electrode). Despite the lower electrode temperature of up to 2,500 K, the electrode exhibits structures, indicating that the melting temperature of tungsten (3,680 K) was reached at least locally and temporarily

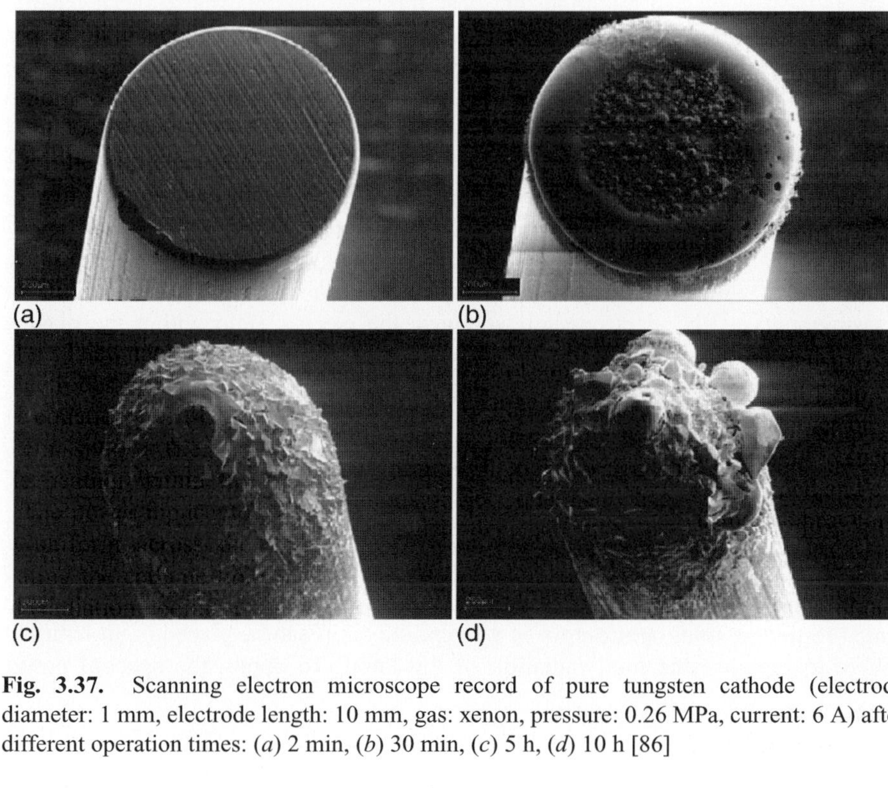

Fig. 3.37. Scanning electron microscope record of pure tungsten cathode (electrode diameter: 1 mm, electrode length: 10 mm, gas: xenon, pressure: 0.26 MPa, current: 6 A) after different operation times: (*a*) 2 min, (*b*) 30 min, (*c*) 5 h, (*d*) 10 h [86]

only difference between them is their operation time: The cathode shown in Fig. 3.37(a) was operating for only 2 min, cathode (b) for 30 min, cathode (c) for 5 h, and cathode (d) for 10 h. Not only the shape of the electrode changed during the operating time, but also the mode of arc attachment to the cathode tip when the arc was switched off was different: (a) diffuse mode, (b) spot mode, (c) and (d) so-called super spot mode. The global cathode tip temperature[18] decreases from about 3,000 K to about 1,700 K going from the diffuse arc attachment to the so-called super spot mode [86]. So, depending on the operating conditions, the deformation of the electrode tip can be quite a fast process, altering the operation condition of the whole lamp dramatically (cf., e.g., Sect. 4.4.4). The process of electrode deformation observed in operating UHP lamps also occurs on very short time scales [45, 119].

3.2 Plasma Diagnostics

The plasma is the most important part of HID lamps because the primary purpose of any lamp is the generation of light. This light is produced within the plasma, the relevant processes are described in Sect. 2.2.3. The temperature of the plasma plays a central role for the radiation processes in HID lamps. This plasma temperature can be measured using emission spectroscopy or X-ray methods as described in Sect. 3.2.1. The plasma radiation, the heat flux from the plasma to the discharge tube, the heat flow through the electrode ends, and the radiation emitted by the electrodes are loss mechanisms requiring constant electrical power input into the plasma. This electrical power input is higher in the near-electrode plasma due to the losses induced by the heat flow through the electrode ends and the radiation emitted by the electrodes. The additional electrical power input into the near-electrode plasma is revealed in the so-called electrode fall voltages, namely the cathode and the anode fall voltage. The electrical power consumption of the so-called plasma column between anode and cathode is proportional to the length of the plasma column, i.e., the electric field is constant within the plasma column. The measurements of the electrode fall voltages, the electric field within the plasma column, and the plasma potential will be discussed in Sects. 3.2.2 and 3.2.3. A further important parameter for the description of the plasma in HID lamps is the pressure. The spectral power distribution (cf., e.g., Figs. 1.23 and 1.29), the power consumption, and the plasma temperature strongly depend on the operating pressure. The experimental determination of the pressure is discussed in Sect. 3.2.4.

3.2.1 Emission Spectroscopy

The radiation escaping from a discharge lamp is the primary purpose of an HID lamp. To get the total light output of a lamp, the lamp can be placed in an Ulbricht sphere to measure the total radiation of the lamp. To obtain the spectral power

[18] See Sect. 3.1.9.

distribution of a lamp (as shown in Chap. 1 for several lamps), the light emission is measured as a function of wavelength λ. Moreover, the luminous intensity can be measured depending on the angle, resulting in the polar distribution curve of a lamp.

The most important parameter of an LTE plasma is the local plasma temperature. Together with the pressure, one can calculate the plasma composition, the ionization degree, the population density of excited atoms, and thus the plasma radiation if the local plasma temperature is known. A common technique to obtain the spatially resolved temperature profile $T(r)$ of a rotationally symmetric plasma is a line-of-sight diagnostic. Measurements of side-on intensities for various chords through the plasma with different distances y to the center of the discharge are used to calculate radially resolved emission data. From these data, the radially resolved temperature profile can be calculated (see later). The deflection of the various chords through the plasma (with different distances y to the center of the discharge) due to the refraction at the arc tube does not alter the "effective" distance of the line-of-sight from the center of the discharge, only the orientation changes (cf. Fig. 3.38).

If $\varepsilon_\lambda(r)$ denotes the emission density per steradian at wavelength λ of optically thin radiation, the measured radiation intensity at wavelength λ using a chord with distance y to the center of the discharge is given by $J_\lambda(y)$:

$$J_\lambda(y) = 2 \int_{x=0}^{x=\sqrt{R^2-y^2}} \varepsilon_\lambda(r(x))\,\mathrm{d}x. \tag{3.30}$$

The factor 2 in front of the integral accounts for the fact that the integral yields only the emission along the line-of-sight starting from $x = 0$ in the positive x direction. An illustration of this and the following equations can be found in Fig. 3.39. To transform dx to dr, the relation $\mathrm{d}r/\mathrm{d}x = r/(\sqrt{r^2-y^2})$ has to be used:

$$J_\lambda(y) = 2 \int_{r=y}^{r=R} \varepsilon_\lambda(r) \frac{r\,\mathrm{d}r}{\sqrt{r^2-y^2}}. \tag{3.31}$$

The measured value is the spatially resolved side-on intensity $J_\lambda(y)$, whereas the wanted quantity is the emission density per steradian at wavelength λ: $\varepsilon_\lambda(r)$. The so-called inverse Abel transformation or Abel inversion is used to calculate $\varepsilon_\lambda(r)$ from the line-of-sight measurements of $J_\lambda(y)$:

$$\varepsilon_\lambda(r) = -\frac{1}{\pi} \int_{y=r}^{y=R} \frac{\mathrm{d}J_\lambda(y)}{\mathrm{d}y} \frac{\mathrm{d}y}{\sqrt{y^2-r^2}} \tag{3.32}$$

To get sufficiently good values for the derivative $\mathrm{d}J_\lambda(y)/\mathrm{d}y$, the side-on intensity $J_\lambda(y)$ has to be measured at many positions y. The inverse Abel transformation not only assumes a rotationally symmetric plasma, the measured emission must also escape from the plasma without absorption, i.e., only optically thin radiation might be considered.

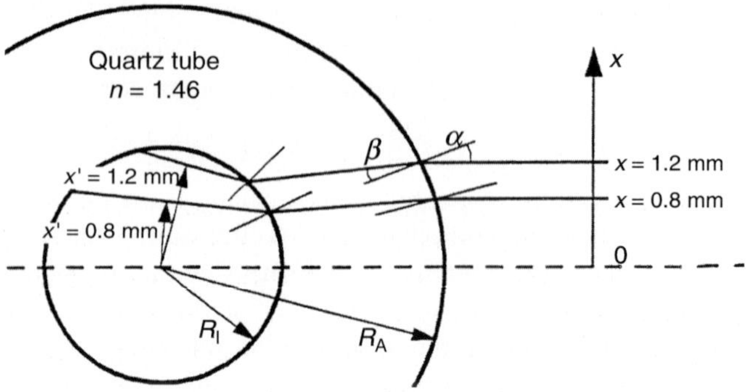

Fig. 3.38. Deflection of line-of-sight due to the refraction at the arc tube (here: the quartz vessel of a D2 automotive headlight lamp, $R_I = 1.35$ mm, $R_A = 3$ mm [132]). The electrodes are located above and below the drawing plane

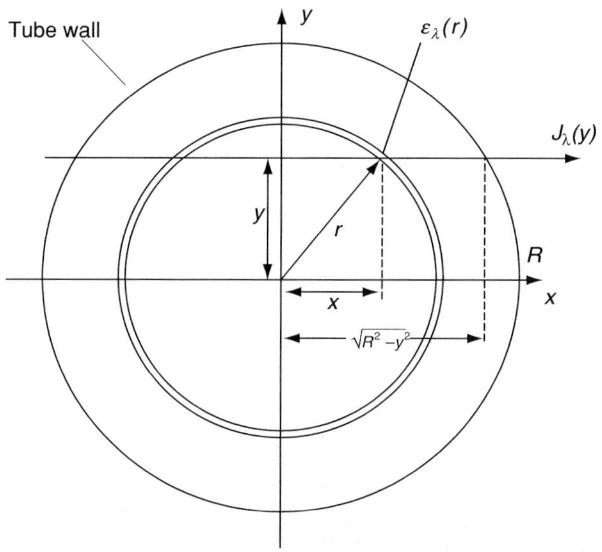

Fig. 3.39. Illustration of the inverse Abel transformation of the measured line-of-sight radiation intensity $J_\lambda(y)$ to get the wanted emission density $\varepsilon_\lambda(r)$ per steradian at wavelength λ. The electrodes are located above and below the drawing plane at $x = y = 0$

The inverse Abel transformation can be used if the radiation intensity of an optically thin line is measured, as, e.g., in [93] using the xenon atom line at 764 nm and the xenon ion line at 529.2 nm in a high-pressure xenon discharge (up to 4 MPa) or in [4, 83, 84, 101] using the mercury atom lines at 577 and 579 nm in different mercury discharge lamps (pressure between 0.5 and 3.7 MPa). Equation (2.23) can be used to compute the temperature profile $T(r)$ as soon as the emission density $\varepsilon_\lambda(r)$ per steradian is determined (cf. Sect. 2.2.3 and (2.23)[19]):

$$\varepsilon_{nm} = \frac{1}{4\pi}\frac{hc}{\lambda_{nm}} A_{nm}\, n \frac{g_n}{Z} \mathrm{e}^{-\frac{E_n}{kT}} L_{nm}(\lambda). \tag{3.33}$$

λ_{nm} is the wavelength of the emitted photon for a transition from energy level n to energy level m. A_{nm} is the transition probability for such a transition, n the atom density, g_n the statistical weight of energy level n, and Z the partition function. The factor $L_{nm}(\lambda)$ describes the line broadening of the transition as described in Sect. 2.2.3 Integrating $L_{nm}(\lambda)$ over all wavelengths yields 1. Nevertheless, there are two major problems using this method: The transition probability A_{nm} is often unknown and the line radiation in HID lamps is often not optically thin [132].

An alternative solution is the measurement of the continuous radiation due to recombination of free electrons with ions and electron-atom and electron-ion bremsstrahlung, as described in [93] in a high-pressure xenon discharge. However, the occurrence of line radiation in the wavelength range under consideration will often disturb this type of measurement.

Despite the frequent usage of the mercury lines at, e.g., 577 and 579 nm in [4, 83, 84, 101] for the determination of the plasma temperature in mercury high-pressure discharge lamps, it cannot always be assumed that these "diagnostic" lines are optically thin in HID lamps (cf. [132] citing [157]). An alternative method originating from Bartels (cf. [132] citing [7]) is the usage of the side-on measured emission spectra together with models for self-reversed lines. The Bartels method consists of the measurement of the radiation intensity of the two maxima of a self-reversed spectral line. Connecting this measured radiation intensity to the radiation of a black body radiator, a radial temperature distribution can be computed. This method was employed in [132] for the temperature profile determination of a mercury discharge lamp (6 MPa) and a metal halide lamp (3 MPa mercury + 6 hPa NaI). Moreover, the Bartels method was used in [100] to get the plasma temperature profiles in a D2 automotive headlight lamp. These results will be shown in Sect. 4.4.6 and compared to results from numerical simulations (Figs. 4.84 and 4.85). Another possibility to measure temperature profiles in mercury discharges is described in [6] using the self-reversed mercury lines at 404.7, 435.8, and 546.1 nm.

If translucent materials of the tube do not allow for optical or UV techniques because of refraction and scattering from the arc tube, X-ray techniques offer the possibility to measure the density/temperature of the plasma in the operating lamp with a spatial resolution that is not limited by the translucent materials (e.g., polycrystalline alumina (PCA), cf. Sect. 1.3.8). For that purpose, the arc is illuminated by X-rays and the resulting shadow is recorded by a two-dimensional array detector. Comparison of the shadow cast by an operating lamp with the shadow cast

[19] Instead of the radiation emission coefficient e_{nm} (radiation energy per unit time and volume) given in (2.23), the emission density ε_{nm} per steradian is given here.

by a cold lamp[20] yields the transmission of the Hg vapor alone. Spatially resolved column densities can be computed from the measured transmission of the Hg vapor, if the atomic absorption cross section is known. These column densities can be transformed into the absolute density distribution $\rho(r)$ depending on the radius r if rotational symmetry is assumed and inverse Abel transformation is applied. Finally, the gas or plasma temperature $T(r)$ can be calculated from the density distribution $\rho(r)$ if the pressure is known ($p = nkT$, cf. (2.13)). This method can notonly be applied to lamps with translucent arc tubes, but also to a switched off lamp to determine the time-dependent temperature of the gas during the cool-down after the arc has been extinguished. In this case, no radiation is generated by the gas, so that emission spectroscopy is not possible [42, 175].

3.2.2 Electrode Fall Voltage and Electric Field of Column

The total lamp voltage U_{tot} can easily be measured. U_{tot} is in general composed of the cathode fall voltage, the anode fall voltage, and the voltage drop across the plasma column. As indicated in Fig. 3.40, the voltage drop per unit length is constant across the plasma column, i.e., the electric field E_p is constant in the plasma column. Thus, the extension of the arc by increasing the electrode gap yields an increase of the total lamp voltage given by $E_p d$, where d is the additional extension of the arc. If the distance l_{ac} between the anode and cathode tip and the electric field in the plasma E_p are known, the equation

$$U_{tot} - E_p l_{ac} = U_c + U_a, \tag{3.34}$$

yields the sum of anode and cathode fall voltage. This sum is also known as the electrode sheath voltage (ESV).

The anode and cathode fall voltages are defined as the extrapolation of the linear part of the electric potential distribution on the lamp axis to the electrode tip positions (cf. Fig. 3.40). The difference between this extrapolation and the electric potential at the cathode surface ($U = 0\ V$) is the cathode fall voltage U_c, the difference between the extrapolation and the electric potential at the anode surface ($U = U_{tot}$) is the anode fall voltage U_a. This definition is similar to the experimental determination of cathode and anode fall voltages (see below and Sect. 3.2.3). The electrode fall voltages correspond to the additional power consumption in the near-electrode plasma, i.e., the *extra*[21] demand of electrical power input due to the losses to anode and cathode and due to the increased radiation output by the bright hot plasma spots.

These losses to anode and cathode are partly due to the large temperature gradient between hot plasma spots and electrode tip. The electrode tips in HID lamps have typical temperatures around 2,000–3,500 K, the typical plasma temperatures in the plasma column are around 4,000–7,000 K, at the hot plasma spots the

---

[20] In a cold lamp, the mercury vapor condenses on the wall or the electrodes.

[21] In contrast, the electrode fall voltages in sheath models (normally cathode fall voltage U_c) correspond to the total voltage drop over the sheaths under consideration without consideration of the "normal" power consumption per unit length in the arc and in general without consideration of constriction zones or hot plasma spots (cf. Sect. 4.1.1).

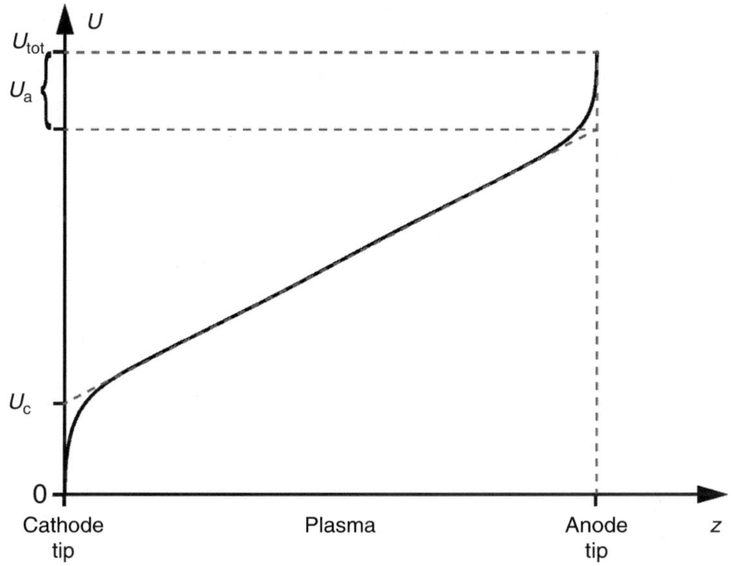

Fig. 3.40. Sketch of the electric potential in the plasma on the lamp axis of an HID lamp from cathode to anode tip

plasma temperature might be as high as 10,000 K (typical distance from hot plasma spot to electrode tip in HID lamps: 30–250 μm). Thus, the electrodes act as heat sinks for the plasma, i.e., a power flux from the plasma to the electrode tips maintains the electrode temperature. This heat is passed through the electrodes to the outside connections of the electrodes (electrode root) and power is radiated from the electrode surfaces according to the Stefan–Boltzmann law. This constant drain of power out of the plasma must be balanced by an increased electrical power input into the near-electrode plasma compared to the electrical power input into the plasma column. Additionally to the heat flux from the plasma to the electrodes, the cathode is heated by an ion current, leaving the ionization energy at the cathode during the neutralization of the ion. The generation of "new" ions in the near-cathode plasma consumes power which must also be balanced by an increased electrical power input into the near-cathode plasma. Moreover, the electron current at anode and cathode participates at the power balance of the electrodes[22]. This *additional* power demand is the reason for the electrode fall voltages (anode fall voltage and cathode fall voltage), i.e., the *additional* voltage drop in the near-electrode plasma to balance electrode losses. The *additional* power input into the near-electrode plasma can thus be described by U_cI and U_aI.

An alternative way of looking at the electrode fall voltages is to examine the electric field, the electrical conductivity, and the electric current density: The electric field is connected via the electric current density $j = \sigma E$ to the electrical

[22] More details on the power balance of cathode and anode can be found in Chap. 4.

conductivity. Due to the current continuity equation (cf. (4.6)), the electric field must be increased if the current must be conducted through a "sheath" with low electrical conductivity to preserve the total lamp current, i.e., one gets high electric fields in plasma areas where it is "tough" to get the imposed current through. The near-electrode plasma is relatively cold compared to the temperature in the plasma column or the hot plasma spots and, therefore, the electrical conductivity is extremely low (cf. Figs. 2.19 and 2.20). Thus, the electric field in the near electrode plasma is considerably higher compared to the electric field in the plasma column, resulting in the electrode fall voltages described earlier and visualized in Fig. 3.40. The corresponding high electrical power input ($\sigma E^2 = jE$) elevates the temperature of electrode and near-electrode plasma. An increased plasma temperature results in an increased electrical conductivity, a higher cathode temperature implies stronger electron emission, resulting as well in an enlarged electrical conductivity in the near-cathode plasma. Moreover, the temperature of the hot plasma spots is high for large electrical power inputs, resulting in high electron densities in this plasma region close to the electrodes (cf. Figs. 2.6 and 2.8). Due to diffusion of electrons from the hot plasma spots into the cold near-electrode plasma, the electrical conductivity is also increased (cf. Sect. 4.4). So the electric field can decrease to a value where power balance and current continuity are in balance. Moreover, the constriction of the arc in front of the electrode tips (constriction zones) is also responsible for an enlargement of the electric field due to the corresponding increase of the electric current density j (same current I through smaller area A).

The easiest way to measure electrode fall voltages (U_c, U_a) and electric field in the plasma column (E_p) is the measurement of the total lamp voltage in HID lamps with different electrode gaps but otherwise identical properties of the plasma column.[23] The measurements are easy, but ensuring identical properties of the plasma column in lamps with different electrode gaps is difficult. One possibility is described in [102], using high-pressure mercury discharge lamps with an electrode gap from 28 to 83 mm. The challenge in producing these lamps is the operating pressure of the mercury (in [102]: 0.6 MPa), which must be equal in all lamps. This is because the pressure has a great influence on the lamp voltage due to the strong dependence of the electrical conductivity on the pressure of the discharge (cf. Fig. 2.19). If this pressure effect is superimposed on the total lamp voltage change due to the different electrode gaps, the thus deduced electrode fall voltages (U_c,U_a) and electric field in the plasma column (E_p) will be wrong. Equal pressures require different mercury doses in the lamps with different electrode distances, which makes the production of appropriate test lamps very time consuming.

The result of the measurements described in [102] is shown in Fig. 3.41: The lamps were driven with sinusoidal lamp currents with frequencies in the range from 50 Hz to 5 kHz. The measured total lamp voltages for four different electrode gaps from 28 to 83 mm are shown as a function of time (two specimens of each lamp). In Fig. 3.42, these measured total lamp voltages are plotted vs. the electrode gap for

[23] This is a well-known method, described for example by Bauer and Schulz in 1954 in [10].

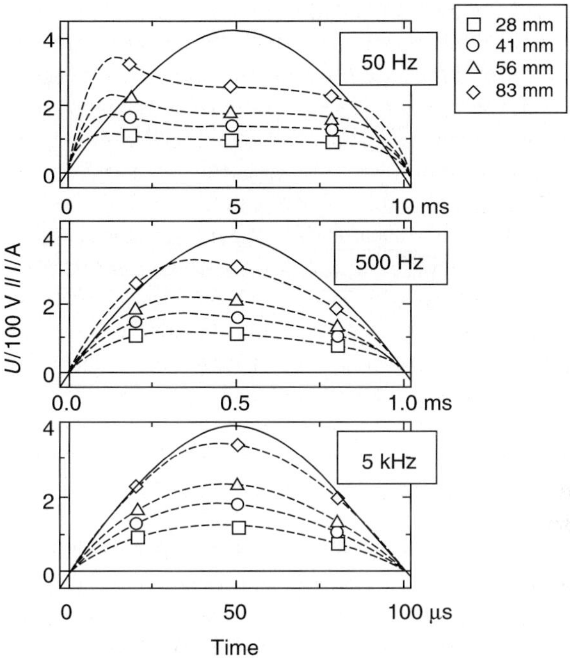

Fig. 3.41. Lamp current (*solid line*) and total lamp voltages for different electrode gaps (*dashed lines*, legend: top right) and different current frequencies vs. time (gas: mercury, pressure: 0.6 MPa, [102])

different points in time within one current cycle for the 50 Hz discharge. The slope of the lines[24] through the measured total lamp voltages vs. electrode gap is equal to the electric field in the plasma column E_p. The sum of anode and cathode fall voltage ($U_a + U_c$) is given by the extrapolation of these lines to zero electrode gap.

Figure 3.43 shows the evaluation of the slope of the lines shown exemplarily in Fig. 3.42 for the three different frequencies in the range from 50 Hz to 5 kHz vs. time, i.e., the electric field E_p in the plasma column vs. time. The dashed and dotted lines are obtained from numerical simulations for a 0.55 MPa (dashed line) to 0.65 MPa (dashed-doted line) mercury plasma as described in [102]. In the case of low frequencies (here: 50 Hz), the plasma cools down during the low current periods of the sinusoidal lamp currents. As soon as the lamp current increases, the plasma is heated up again, leading to a higher electric field (higher electrical power input) at the beginning of the cycle. The higher the frequency, the smaller this cooling down of the plasma column, i.e., the smaller the additional electrical power input necessary for heating up again the plasma.

Figure 3.44 shows the evaluation of the extrapolation of the lines shown exemplarily in Fig. 3.42 to zero electrode gap for the three different frequencies in

[24] Each line represents a different point in time.

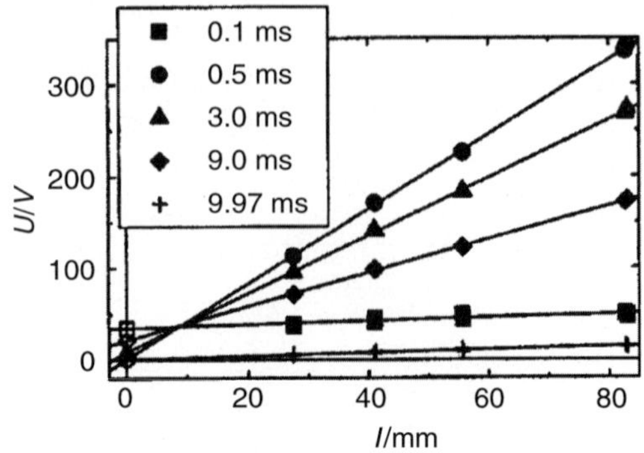

Fig. 3.42. Total lamp voltage vs. electrode gap *l* for 50 Hz at different points in time (gas: mercury, pressure: 0.6 MPa, [102])

the range from 50 Hz to 5 kHz versus time, i.e., the sum of anode and cathode fall voltage ($U_a + U_c$) vs. time. The dashed line corresponds to the total lamp voltage measured for the 28-mm lamp minus a calculated electric field in the plasma column according to a model described in [102] times the electrode gap, i.e., not the electric field in the plasma column obtained from the measurements described earlier and shown in Fig. 3.43 is taken for the determination of the dashed line, but a calculated electric field as indicated by the dashed and dotted lines in Fig. 3.43. The shaded band corresponds to the same for all other lamps with larger electrode gaps.

The maximum of the electrode fall voltages ($U_a + U_c$) at the beginning of the current cycle is due to the cooling down of the electrodes during the low current periods of the sinusoidal lamp currents. To sustain the increasing lamp current after current reversal, the cathode, for example, has to be heated up to emit sufficient electrons. Again, the higher the frequency, the smaller this cooling down of the electrodes, i.e., the smaller the additional electrical power input necessary for heating up the electrodes after current reversal. Due to the different time constants, the behavior of plasma and electrodes is different (compare Figs. 3.43 and 3.44).

Instead of using different lamps with different electrode gaps, one can use a lamp with movable electrodes, so that the electrode gap can be adjusted during operation. For this (and other) reasons, a model lamp was realized as described already in Sects. 3.1.4 and 3.1.9 (see also Figs. 3.5, 3.27, and 3.28). The lamp consists of a tube made of fused silica with an inner diameter of 9 mm. It is fixed at both ends in metallic end pieces. Electrode holders are inserted into the tube through axial cylindrical holes in the end pieces. They can be displaced within the tube by stepping motors. Tungsten electrodes of different length and diameter are soldered into the tips of the holders (Fig. 3.46). This model lamp is restricted to pressures up to 1 MPa (10 bar) and the usage of rare gases. A more detailed description of this model lamp can be found, e.g., in [43, 86, 114, 115, 121, 123].

The measured total lamp voltage for different lamp currents vs. electrode gap for pure tungsten electrodes with a diameter of 1 mm in an 2.6 MPa argon plasma is shown in Fig. 3.45 according to [114]. Again, the slope of the curves is equal

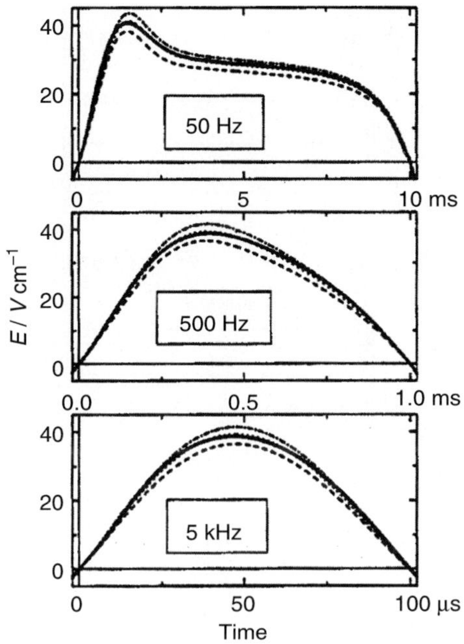

Fig. 3.43. Electric field E_p in the plasma column (*solid line*) vs. time obtained from the slope of the lines shown in Fig. 3.42. The *dashed and dotted lines* are obtained from numerical simulations for 0.55 MPa (*dashed* (lower) curve) to 0.65 MPa (*dashed-dotted* (upper) curve) according to [102]

to the electric field in the plasma column E_p, the sum of anode and cathode fall voltage ($U_a + U_c$) is given by the extrapolation of these curves to zero electrode gap. This model lamp can also be used for the probe measurements described in Sects. 3.2.3.

3.2.3 Plasma Potential

The model lamp described earlier and in Sects. 3.1.4 and 3.1.9 with movable electrodes can also be used for so-called probe measurements as described in [114, 115] using Langmuir probes.[25] For this reason, the arc tube of the model lamp is modified: Two tubes of smaller diameter are attached at the middle of the discharge tube so that a cross is formed (cf. Fig. 3.46). Two probes are inserted into the discharge tube through these side arms. They are made of thin tungsten wires shaped as half circles at their ends, so that both probes form a ring which surrounds the arc. One probe is operated at floating potential, the other is biased with a

[25] Langmuir probes have been widely used to study plasma parameters in low-pressure discharges in fluorescent lamps since the early 1950s [111, 130].

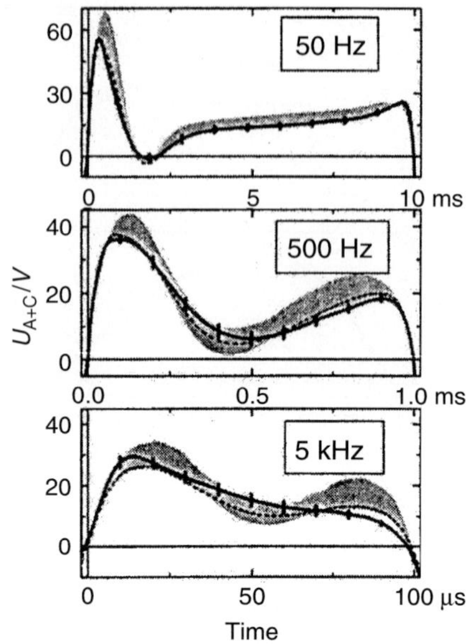

Fig. 3.44. Electrode fall voltages (*solid line*) vs. time obtained from the extrapolation of the lines shown in Fig. 3.42 to zero electrode gap. The *dashed line* corresponds to the total lamp voltage of the 28 mm lamp minus a calculated electric field in the plasma column according to a model described in [102] times the electrode gap, the *shaded band* corresponds to the same for all other lamps with larger electrode gaps

positive voltage ramp relative to the grounded cathode. The measured probe current as a function of the probe voltage (probe characteristic) can be used to deduce, for example, the plasma potential or the electron temperature at the position of the probe [114, 115]. Although the insertion of the probe in the plasma and the collection of charged particles by the probe disturbs the plasma, these disturbances are assumed to be localized in a thin layer around the probe, so that the plasma parameters are not altered by the probe measurements. The model lamp used for these probe measurements is restricted to pressures up to 1 MPa (10 bar) and the usage of rare gases, Langmuir probe measurements in real HID lamps are up to now out of reach.

The advantage of these probe measurements is the determination of the plasma potential as a function of the distance to the cathode tip (cf. Fig. 3.47). This plasma potential vs. distance to cathode tip can then be used to separately calculate anode and cathode fall voltages instead of the sum of anode and cathode fall voltages as described in Sect. 3.2.2 (cf. Figs. 3.40 and 3.47). Since the probes are fixed within the arc tube, the electrodes must be moved inside the discharge vessel (during lamp operation), keeping the electrode gap constant. Thus, it is possible to measure the plasma potential at different distances to the cathode tip.

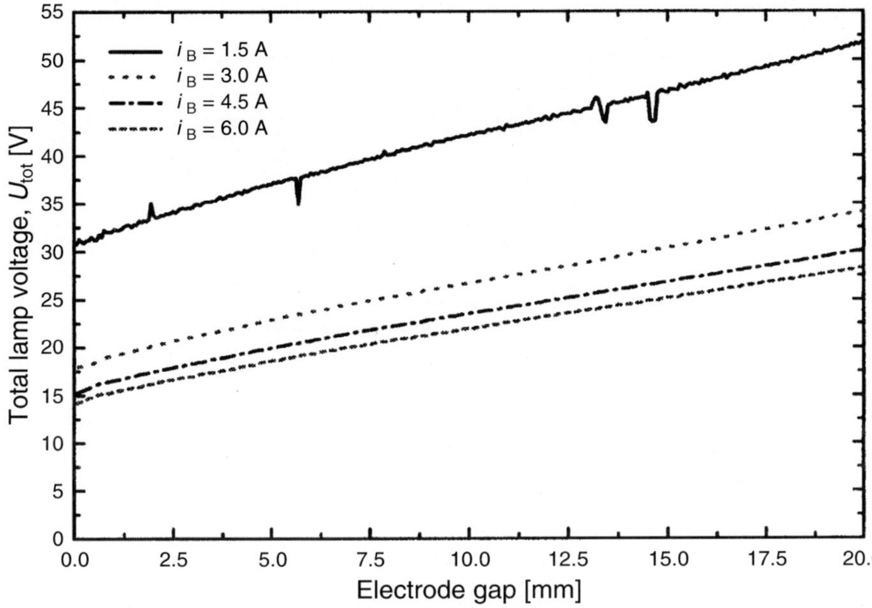

Fig. 3.45. Total lamp voltage versus electrode gap for different lamp currents i_B (electrode material: pure tungsten, electrode diameter: 1 mm, gas: argon, pressure: 0.26 MPa, [114])

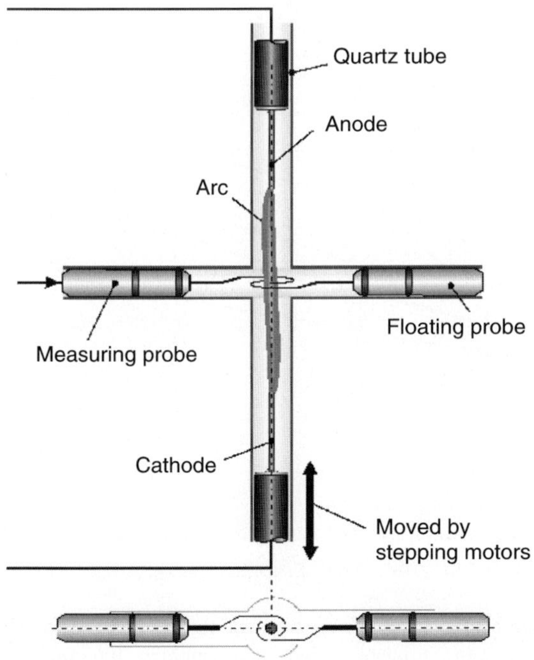

Fig. 3.46. Experimental setup for Langmuir probe measurements as described in [115]

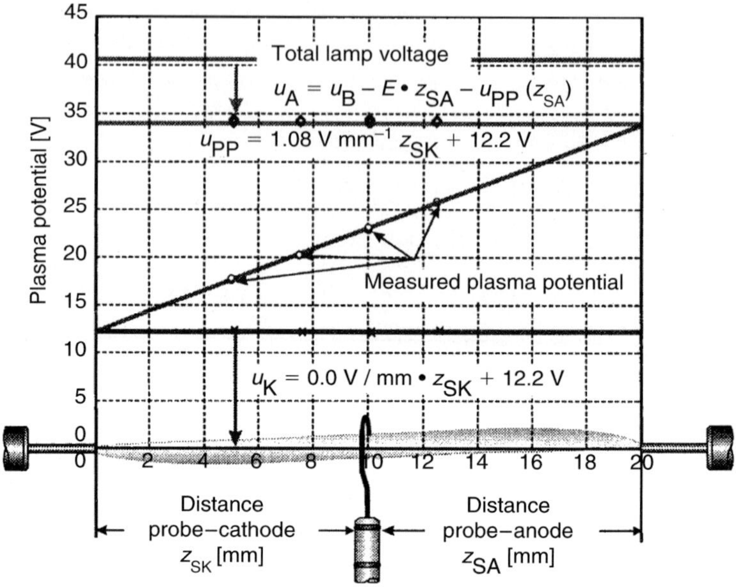

Fig. 3.47. Measured plasma potential vs. distance to cathode tip [114]

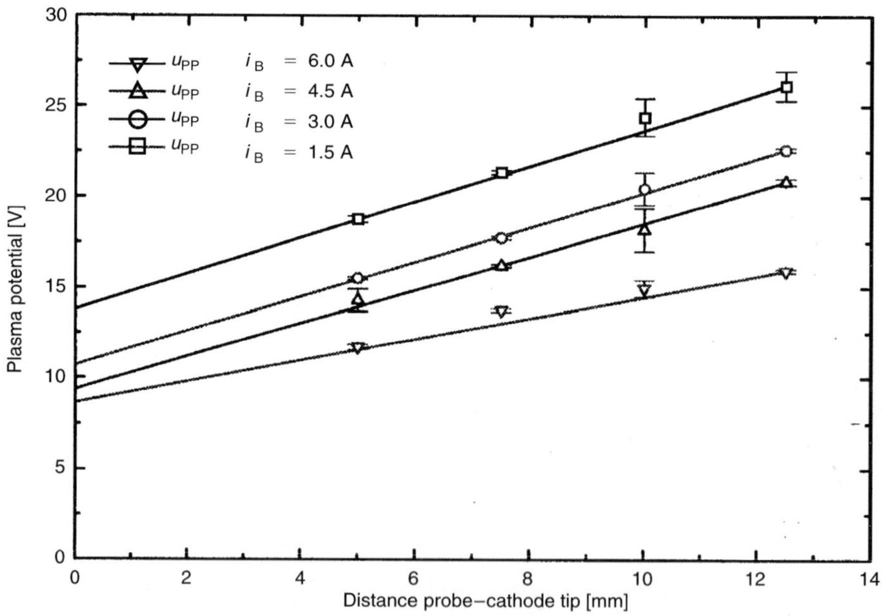

Fig. 3.48. Measured plasma potentials vs. distance to cathode tip for different lamp currents i_B (electrode material: pure tungsten, electrode diameter: 0.6 mm, gas: argon, pressure: 0.26 MPa, [114])

A visualization of the results obtained by such a measurement is given in Fig. 3.47. The line through the measured plasma potentials can be extrapolated to the position of anode and cathode tip. The difference between this extrapolation and the electric potential at the cathode surface (0 V) is the cathode fall voltage U_c, the difference between the extrapolation and the electric potential at the anode surface (U_{tot}) is the anode fall voltage U_a. Thus, the cathode fall voltage can be directly extracted from the measured plasma potential, the determination of the anode fall voltage requires additionally the measured total lamp voltage U_{tot}. The slope of the line through the measured plasma potentials corresponds to the electric field in the plasma column, E_p.

The measured plasma potentials vs. the distance to the cathode tip for different lamp currents from 1.5 to 6.0 A in a 0.26 MPa argon discharge with pure tungsten electrodes with 0.6 mm electrode diameter are given in Fig. 3.48. The cathode fall voltage (extrapolation of plasma potential at cathode tip position) decreases with increasing current from about 14 V at 1.5 A to about 9 V at 6 A. The power consumption in the near-cathode plasma corresponding to these cathode fall voltages, however, increases with increasing current from 21 W at 1.5 A to 54 W at 6 A ($U_c I$). The electric field in the plasma column (slope of the plasma potential curve) decreases with increasing current from 10 V cm^{-1} at 1.5 A to 6 V cm^{-1} at 6 A.

To be able to estimate the reproducibility of the results of the probe measurements, a series of ten independent measurements were performed using three different sets of identical electrodes (pure tungsten electrodes with a diameter of 1.5 mm and a length of 25 mm) in a 0.26 MPa argon discharge (cf. Fig. 3.49).

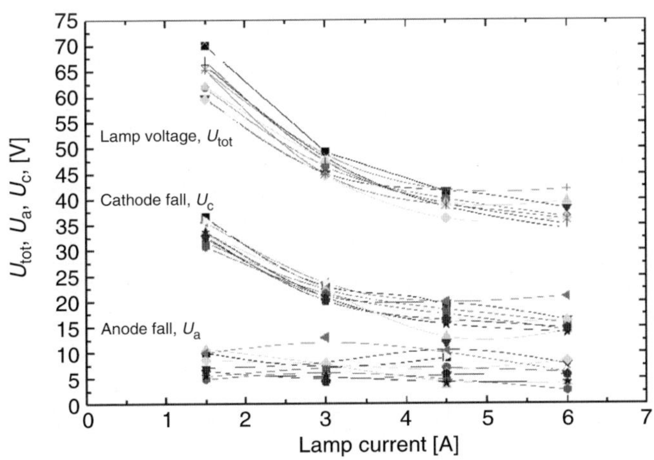

Fig. 3.49. Total lamp voltage, anode, and cathode fall voltages vs. lamp current from ten series of measurements using three different sets of identical electrodes (electrode material: pure tungsten, electrode diameter: 1.5 mm, electrode length: 25 mm, gas: argon, pressure: 0.26 MPa, [115])

The scattering of the total lamp voltage might be due to fluctuations in the arc column and at the electrodes and due to small tolerances between the three sets of electrodes used. The total lamp voltage for 1.5 A varies for example from 59 to 70 V, i.e., over quite a broad range. The cathode fall voltage, which is determined without the use of the measured total lamp voltage scatters for 3.0 A from 20 to 24 V and for 6 A from 14 to 21 V. The anode fall voltage seems to be independent of the lamp current *I*. The strong scattering of the anode fall voltage is due to the fact that the measured plasma potential curve and the measure total lamp voltage are necessary to determine the anode fall voltage. The anode fall voltage for the lamp under consideration in Fig. 3.49 varies from 2.5 to 12.5 V.

Again, if experimental results (e.g., electrode or plasma temperatures, electrode fall voltages, etc.) are compared with results from numerical simulations, it has to be kept in mind that the experimental results should not be overinterpreted, i.e., the measurements at similar lamps or at equal lamps at different times might yield different results (compare with Sects. 3.1.4, 3.1.7, and 3.1.13).

3.2.4 Determination of Lamp Pressure

An important parameter for the description of the plasma in HID lamps is the operating or working pressure. The gas and plasma properties and the transport coefficients of the plasma depend on the operating pressure of the lamp as described in Sects. 2.1, 2.2, and 2.3. Thus, many properties of the lamp, as for example the spectral power distribution (cf. e.g., Figs. 1.23 and 1.29), the power consumption, and the plasma temperature, strongly depend on the operating pressure. Nevertheless, the experimental determination of the working pressure in commercial HID lamps is quite difficult.

The operating or working pressure of a lamp should not be mixed up with the initial or cold fill pressure, i.e., the pressure within the cold lamp. The pressure of sodium, mercury, or metal halides in the cold lamp can normally be neglected. The inert, noble, or rare gas which is used as the starting gas, may have very different cold fill pressures, e.g., 100 hPa (0.1 bar) argon in a UHP lamp [88] or several bars of xenon in a D2 automotive headlight lamp. This pressure will of course rise during operation, reaching, e.g., 5 MPa (50 bar) of xenon in the D2 lamp.

A direct determination of the operating pressure with the help of pressure valves or a pressure gauge is only possible in pure noble gas discharges, as for example used in the model lamp described in Sects. 3.1.4, 3.1.9, 3.2.2, and 3.2.3 (cf. [43, 86, 114, 115, 121, 123] or the xenon lamp mentioned in Sect. 3.2.1 (cf. [93]). If metals like mercury or metal halides are present in the discharge, the pressure is determined by the cold spot temperature. This cold spot temperature, however, would be very low due to all sorts of extensions to the discharge tube for the determination of the lamp pressure by means of a pressure gauge. Thus, the use of pressure gauges in HID lamps is not possible apart from pure noble gas discharges.

The cold spot temperature can be used to determine the pressure in lamps containing metals as, e.g., low and high-pressure sodium lamps, fluorescent lamps, high-pressure mercury, or metal halide lamps. If the cold spot temperature is known, the vapor pressure of the gas can be calculate as outlined in Sect. 2.2.1 and shown in Figs. 1.17, 1.21, 1.25, and 1.32. However, since vapor pressure strongly depends on the cold spot temperature, this method is of limited precision. The method can be inverted to adjust the wanted operating pressure by using a suitably designed discharge tube yielding the appropriate cold spot temperature or by heating or cooling the cold spot temperature as described, e.g., in Sect. 3.1.12 or in [80].

The mercury or sodium vapor pressure can be determined by the self-reversed mercury 254 nm-line or sodium D-lines (589.0 and 589.6 nm) in HID lamps. The wavelength separation $\Delta\lambda$ between the maxima of the self-reversed resonance lines (sodium) or between the line-center and the long-wavelength self-reversal maximum (mercury) depends on the pressure, the radius of the discharge tube (R), and the temperature profile because all these parameters affect the possible absorption of the resonance lines especially in the outer region of the plasma. For typical temperature distributions in high-pressure sodium lamps, the relation between plasma temperature and sodium vapor pressure is rather weak, so that a rule-of-thumb can be given to determine the operating pressure in high-pressure sodium lamps [80]:

$$p_{\mathrm{Na}} \approx \frac{2.7 \pm 0.5}{\sqrt{\mathrm{R}}} \Delta\lambda. \tag{3.35}$$

Thus, the sodium pressure p_{Na} (in kPa) can be determined by the measured wavelength separation $\Delta\lambda$ (in nm) of the sodium D-lines and the radius R (in mm) with an inaccuracy of approximately 20% if a typical temperature profile is assumed.

Another method for the determination of the operating pressure can be used in high-pressure mercury lamps with unsaturated mercury vapor (no condensed mercury left at the inner tube wall) as, e.g., in the UHP lamp (cf. Sect. 1.3.6). If the temperature profile of the whole discharge volume or an average temperature is known, the pressure can be calculated using the ideal gas law (cf. Sect. 2.1.5):

$$p = nkT. \tag{3.36}$$

The integration of the particle density n over the whole discharge volume times the molar mass must yield the total amount of mercury placed in the lamp.

The pressure can also be determined if the dependence of the electrical conductivity on pressure and temperature is known. If the plasma temperature within the discharge has been measured, the pressure can be calculated using Ohm's law and the (known) lamp current as described in [84]. A similar method is described in [80] and has been used in [102].

Further discussions concerning the operating pressure of HID lamps can be found in Sects. 4.4.5 and 4.4.7.

4

Numerical Simulations

After the introduction of high-intensity discharge (HID) lamps in Chap. 1 and the detailed discussion of plasma and electrode properties in Chap. 2, Chap. 3 was concerned with experimental investigations of electrodes and plasma in HID lamps. The present chapter focuses on the numerical treatment of electrodes and plasma. The experimental and numerical investigations can greatly benefit from each other as will be shown in this chapter. The numerical simulation can explain many details of the experimental findings and give deep insight into the principles of HID lamps because many limitations of experimental methods do not exist for numerical simulations (e.g., accuracy in space and time). The description of numerical simulations of HID lamps treated in this chapter is based on the results and findings of Chaps. 1–3.

The chapter starts with electrode models, which are discussed in Sect. 4.1. The modeling of the plasma without the consideration of electrodes is the contents of Sect. 4.2. As described already in Chap. 3, there is a strong coupling between electrodes and plasma in HID lamps, so that numerical models should include both electrodes and plasma. Section 4.3 gives an introduction to such extended models. A central issue of Chap. 4 is such an extended model, namely the self-consistent electrode–plasma model which is the focus of the author's research work.[1] The model will be introduced and discussed in Sect. 4.4. Many interesting results obtained with the self-consistent electrode–plasma model will be presented, accompanied by numerous comparison between numerical and experimental results. As will be shown in Sect. 4.4, the numerical investigations are necessary to understand, improve, or develop new high-pressure discharge lamps and to improve the principal understanding of HID lamps.

[1] Part of Sect. 4.4 is published by the author in [53–65].

4.1 Modeling of Electrodes

The electrodes and the plasma are strongly coupled: Thermal conduction from the plasma to the electrode tips is heating anode and cathode. The cathode might be heated by an additional ion current generated in the near-cathode plasma,[2] the anode is heated through the electron current.[3] The electrons emitted by the cathode (cf. Sect. 2.4.4) and the electrons "extracted" from the cathode for the neutralization of ions cool the cathode. Moreover, anode and cathode are cooled due to radiation cooling, i.e., the black body radiation from their surface (cf. Sect. 2.4.3). Thus, the power balance of the electrode is strongly coupled to the plasma and its properties.

Moreover, the arc attachment to the electrode surfaces is not only determined by electrode properties like the electron emission current. The electric current must pass through the so-called constriction zones in front of the electrode tips, where the relatively broad electric current density profile of the plasma column is transformed to the relatively small electric current density profile in the near-electrode plasma. The extension of the arc attachment area on the electrode surfaces is influenced by the plasma and its ability to carry the electric current, i.e., a possible spot or diffuse arc attachment is not only determined by electrode properties (cf. e.g., Sects. 4.4.3, 4.4.4, 4.4.7, and 4.4.8).

Additionally, the work function, which determines the electron emission current of the cathode, can be influenced by constituents of the plasma as described in Sects. 3.1.11, 3.1.12, and 4.4.6. Tungsten, which is possibly evaporated from the electrode surfaces, might be transported through the bulk plasma and settle at other electrode positions, leading to the deformation of the electrode tips or to cold end attacks (cf. Sect. 2.4).

Despite the above, simple cathode models, neglecting to a large extent the influence of the plasma like heat conduction from the plasma to the cathode tip or the constriction zone in the near cathode plasma, are able to calculate as a first approximation the power balance of the cathode. Details like the ion to electron current ratio[4] or the arc attachment mode or area might be misjudged by simple cathode models. Details about these questions can be found in Sects. 4.1.1, 4.1.2, and Sect. 4.4 (especially Sects. 4.4.3, 4.4.4, 4.4.7, and 4.4.8).

---

[2] Due to the higher mobility of electrons compared to the mobility of ions, the lamp current in the bulk plasma is carried mainly by electrons (cf. Sect. 2.3.2). An ion current at the cathode surface, which is sort of balancing a possible deficiency of electron current due to insufficient electron emission at the cathode surface, must thus be "produced" in the near-cathode plasma.

[3] In the following "cathode" means "cathode" for dc currents and "electrode during cathode phase" for ac currents (likewise for the anode).

[4] If the heat conduction from the plasma to the cathode tip is neglected, the only heating mechanism for the cathode is the ion current, which will thus be overestimated (cf. Sect. 4.4.3).

Section 4.1.1 deals with a simple cathode model, some examples of electrode models are given in Sect. 4.1.2. In Sect. 4.1.3, differences between anode and cathode will be discussed.

4.1.1 Cathode Sheath Model

A simple and instructive cathode model is given by Neumann in [130] (based on [5], summarized in [31]). The temperature distribution within the cathode is calculated using the (time-independent) heat diffusion, heat conduction, or power balance equation:

$$\nabla \cdot (\lambda_E \nabla T_E) + \sigma_E E_E^2 = 0. \tag{4.1}$$

λ_E is the thermal conductivity of the electrode (normally tungsten, cf. Sect. 2.4.1), T_E the electrode temperature distribution within the cathode, σ_E the electrical conductivity of the electrode, and E_E the electric field within the electrode. The electric field within the cathode can be calculated using the current conservation within the cathode and the known electrical conductivity σ_E ($\nabla \cdot (\sigma_E E_E) = 0$, cf., e.g., [31]). The Joule heating term $\sigma_E \cdot (E_E)^2$ can be neglected in high-pressure discharge lamps [5, 18, 30, 130, 168].

The heat flow within the cathode is brought into the cathode via an ion current at the cathode tip. The model presented in [130] thus neglects the heat conduction from the plasma to the cathode tip. The reason for this approximation is that the model originates from low-pressure discharge lamp electrodes, where the heat conduction of the plasma to the electrode tips does not play a role. Instead, the ions are accelerated due to the voltage drop across the so-called space charge zone, which is assumed to be collision-free (which is true in low-pressure discharge lamps). This voltage drop is the cathode fall voltage, which is defined as the voltage drop across the space charge zone. This definition is different from that given in Sect. 3.2.2, where the cathode fall voltage (times the lamp current) was defined (similar to the experimental determination) as the *extra* amount of power necessary to balance the losses to the cathode and an increased radiation output by the bright hot plasma spots. Thus, the ions leave their ionization energy φ_i and their kinetic energy due to the collision-free acceleration in the space charge zone at the cathode surface. The electron, which is necessary to neutralize the ion at the cathode surface, cools the cathode (work function φ).

In high-pressure discharge lamps, however, the ions cannot be assumed to move collision-free through the space charge layer. Due to many collisions with other heavy particles, the ions leave their kinetic energy gained due to the acceleration in the cathode fall voltage within the near-cathode plasma. This energy is then transported to the cathode tip via heat conduction, so that the resulting power input into the cathode tip is comparable for both notions.

Apart from the heating due to an ion current, the cathode is cooled by electron emission (cf. Sect. 2.4.4) and radiation cooling (cf. Sect. 2.4.3), so that the resulting heat flux density q to the cathode surface is given by

$$q = \lambda_E \frac{\partial T_E}{\partial n} = j_{t,i}\,(eU_c + \varphi_i - \varphi) - (j_{t,\text{emission}} - j_{t,\text{back}})\varphi - \varepsilon\sigma_{SB}T_E^4. \tag{4.2}$$

$j_{t,i}$ is the particle flux density of ions, $j_{t,\text{emission}}$ the particle flux density of electrons emitted by the cathode, and $j_{t,\text{back}}$ the electron back diffusion particle flux density from the plasma to the cathode against the cathode fall voltage U_c. U_c is the cathode fall voltage, φ_i the ionization energy of the gas atoms, and φ the work function of the cathode. ε is the emissivity of the electrode material, σ_{SB} the Stefan–Boltzmann constant, and T_E the cathode temperature. The term $\partial T_E/\partial n$ denotes the component of the cathode temperature gradient normal to the cathode surface and λ_E the thermal conductivity of the cathode. The quantities $j_{t,i}$, $j_{t,\text{emission}}$, $j_{t,\text{back}}$, T_E, and $\partial T_E/\partial n$ depend on the position on the cathode surface, i.e., $j_{t,i}$, $j_{t,\text{emission}}$, and $j_{t,\text{back}}$ might be large at the very tip of the cathode but drop steeply for larger distances from the cathode tip. Thus, the heat flux density q to the cathode tip is large, positive, and dominated by electron and ion current near the cathode tip, whereas q is smaller, negative, and dominated by radiation cooling at the cathode surface below the cathode tip. The electrode properties λ_E and ε are temperature dependent, as described in Sects. 2.4.1 and 2.4.3, respectively. The cathode fall voltage U_c is independent of the surface position, i.e., U_c is assumed to be spatially constant.

The heating and cooling due to ion and electron current is localized at the cathode tip within the so-called arc attachment area. The radiation cooling is not localized to this arc attachment area, but due to its strong dependence on the cathode (surface) temperature, it is more important at the upper part of the cathode with higher temperatures.

To be able to calculate the cathode fall voltage U_c or the temperature distribution of the cathode T_E, some equations connecting the variables $j_{t,i}$, $j_{t,\text{emission}}$, $j_{t,\text{back}}$, U_c, and T_E are necessary. It would be beyond the scope of this section to go into all the details of this cathode model, so only two of these equations will be discussed here. The first equation is the Richardson–Schottky equation discussed in Sect. 2.4.4 (2.61). With a given electrode temperature T_E and electric field E_c in the plasma at the cathode surface, the electron emission current density $e{\cdot}j_{t,\text{emission}}$ can be calculated. The electric field E_c in the plasma at the cathode surface is calculated using the Poisson equation (4.10), the cathode fall voltage U_c, and the assumption that ions and electrons move collision-free through the space charge zone.

A further equation interlinks $j_{t,i}$, $j_{t,\text{emission}}$, and U_c using a power balance in the so-called ionization zone: The ionization zone is located between the space charge zone (adjacent to the cathode surface) and the plasma column. The ions which hit the cathode are generated within this ionization zone. The power necessary for this ion production in the ionization zone is provided by electrons emitted by the cathode and accelerated by the cathode fall voltage. The power flux from the electrons to the ionization zone is given by the particle flux density of electrons emitted by the cathode, $j_{t,emission}$ times the kinetic energy of each electron accelerated by the cathode fall voltage across the space charge layer, i.e., $e \cdot U_c$. Neglecting other terms of the power balance (like heat conduction, radiation, etc.), this energy is brought back to the cathode via the ion current. The power flux from the ionization zone to the cathode is given by the particle flux density of ions, $j_{t,i}$, times the ionization energy φ_i necessary to generate one ion. The power balance thus reads[5]:

$$eU_c\, j_{t,emission} = j_{t,i}\varphi_i. \tag{4.3}$$

So, the higher the cathode fall voltage U_c, the higher the ratio of ion to electron current. Since the ion current heats the cathode with the kinetic energy gained through the cathode fall voltage U_c (plus ionization energy and minus work function for the release of one electron), this will result in a strong heating of the cathode, resulting in an increase of the electron emission current, i.e., $j_{t,emission}$ rises. Thus, both the ion current and the cathode fall voltage will decrease. This interaction between electron emission current and ion impact heating the cathode is mainly responsible for the adjustment of the power balance of the cathode, the cathode temperature, and the ratio of ion to electron current.

In high-pressure discharge lamps, however, the ions and electrons cannot be assumed to move collision-free through the space charge zone. Electrons and ions together form the total current density j, so that the total electrical power input density into the plasma due to electrons and ions can be described by $j \cdot E$ (in low and high-pressure discharge lamps). The cathode fall voltage is "located" in the near-cathode plasma, i.e., the total electrical power input into the near-cathode plasma is (approximately[6]) given by $I \cdot U_c$ (again in low and high-pressure discharge lamps).

The simple cathode model described within this section splits this power into two parts: The electron current I_e leaves its power $I_e \cdot U_c$ in the ionization zone, where this power is "transformed" into ionization energy (4.3). This ionization energy is brought back to the cathode as described in the boundary condition given in (4.2). Moreover, the power gained by the ion current I_i in the cathode fall voltage (i.e., $I_i \cdot U_c$) is brought "directly" to the cathode surface (term $j_{t,i} \cdot e \cdot U_c$ in (4.2)). So, according to the model described within this section and together with the obvious relation $I = I_e + I_i$, it can be seen that the total electrical power input into the near-cathode plasma ($I \cdot U_c$) is heating the cathode. Additionally, the cathode is cooled due to electron emission ($I_e \cdot \varphi/e$) and due to the release of electrons to neutralize the ions ($I_i \cdot \varphi/e$). Again, with $I = I_e + I_i$ the cooling of the cathode can be described by $I \cdot \varphi/e$ according to the model used within this section. So, apart from the radiation cooling of the cathode, the cathode heating by the total lamp current I is given by $I \cdot (U_c - \varphi/e)$.

[5] In [130] the power balance is a little more elaborated, taking different cases like $j_{t,i} > j_{t,back}$ or $j_{t,i} < j_{t,back}$ and $j_{t,i} < j_{t,i,max}$ or $j_{t,i} = j_{t,i,max}$ into account, but (4.3) is the essential result.

[6] Depending on the definition of the cathode fall voltage, the total power input might be slightly different.

This result can be compared with the situation in high-pressure discharge lamps, where the assumption of a collision-free space charge zone is not valid. The cooling of the cathode is equal to the situation described earlier because the possible collisions within the plasma are irrelevant for the release of electrons from the cathode. Thus, the cathode is cooled by $I \cdot \varphi / e$. The electrons and ions do not move through the space charge zone without collisions, but leave their kinetic energy gained in the cathode fall voltage (i.e., $I \cdot U_c$) within the near-cathode plasma. Nevertheless, this electrical power input is (nearly completely) brought back to the cathode via heat conduction from the plasma to the cathode tip and via heating due to the ionization energy of the ions produced within the near-cathode plasma, so that the heating of the cathode due to the cathode fall voltage U_c can again be described by $I \cdot U_c$. Thus, apart from the radiation cooling of the cathode, the resulting cathode heating by the total lamp current I is again given by $I \cdot (U_c - \varphi / e)$.

So, the neglect of the heat conduction from the plasma to the cathode tip and the assumption of a collision-free space charge zone as used by the simple cathode model presented within this section does not alter the total power balance of the cathode. Since the cathode fall voltage and the cathode temperature are strongly coupled to the power balance, all cathode models will largely agree with respect to these quantities (as long as they allow for the interaction between ion and electron current). Nevertheless, the ratio of ion to electron current and the local variation of, e.g., T_E, $j_{t,i}$, or $j_{t,emission}$ might be quite different using different cathode models. The simple cathode model presented within this section will probably overestimate the ion current because the ion current must compensate the missing heat conduction from the plasma to the cathode tip. Moreover, the arc attachment area on the cathode tip is calculated independent of the constriction zone in the near-cathode plasma, thus it might be wrong. More details on these subjects and on different cathode models can be found in Sects. 4.4.3, 4.4.4, 4.4.7, and 4.4.8.

The calculation of, e.g., cathode fall voltage U_c or electrode temperature distribution T_E works as follows: First, a trial function for the electrode temperature distribution T_E and the cathode fall voltage U_c is assumed. The resulting electron emission current density, ion current density, and electron back diffusion current density can then be calculated. Integrating the resulting total current density over the cathode surface yields a calculated lamp current I_{calc}. This calculated lamp current must then be compared to the given lamp current I_{lamp}. If I_{calc} is larger or smaller than I_{lamp}, the assumed value for U_c must be reduced or increased,[7] respectively, until $I_{calc} = I_{lamp}$. Thus, the heat flux density q (4.2) is determined and can be used to calculate a new electrode temperature distribution T_E using the heat conduction (4.1) together with a fixed electrode root temperature or a

[7] An increase of the cathode fall voltage increases the power transfer to the ionization zone and thus the ion current. Moreover, the electric field at the cathode surface increases with increasing cathode fall voltage and thus the electron emission current increases. Concerning the calculation of the cathode temperature, a large cathode fall voltage leads to a large heat load of the cathode and thus a high cathode temperature, resulting in a large electron emission current.

fixed thermal heat flux through the electrode root. With this new electrode temperature distribution T_E, the iterative scheme described earlier is repeated until the convergence of T_E and U_c. The calculation of the electrode temperature distribution T_E using the heat conduction (4.1) as described earlier can be done, e.g., by using a finite difference method as described in [130] or by using a finite element method as done in [30–32], both allowing for variable electrode geometries.

4.1.2 Examples of Electrode Models

Similar models as the one described in Sect. 4.1.1 are used very frequently for cathode modeling. Bötticher et al. (e.g., [30–32]) take up the model described by Neumann in [130], extend the model to the transient case, and apply a commercial finite element analysis code (ANSYS) to solve the heat conduction equation in the cathode. Tielemans and Oostvogels [162] carried out electrode temperature measurements in high-pressure gas discharge lamps and successfully compared the measuring results with model calculations using an even more simplified model than that given in Sect. 4.1.1. More sophisticated cathode models are discussed by Benilov et al. (e.g., in [13–16, 18–20]), Wendelstorf ([167, 168], based on [20], including plasma), Waymouth (e.g., [165], including a thermal gradient sheath), and Schmitz and Riemann (e.g., [148–150]). Their focus is on the detailed description of the space charge and ionization layer in the near-cathode plasma. Morrow, Lowke et al. (e.g., [112, 113, 120, 144], including plasma) use a different approach, where the electron density is calculated using an electron continuity equation in a collision dominated plasma, i.e., the assumption of a collision-free space charge layer is not used. Moreover, they include the anode into their model. Some of the results obtained by Bötticher and Bötticher in [30] and by Tielemans and Oostvogels [162] will be presented within this section to illustrate the possibilities of electrode models. A comparison between four different cathode models will be presented in Sect. 4.4.3.

Tielemans and Oostvogels [162] carried out electrode temperature measurements in high-pressure gas discharge lamps and successfully compared the measuring results with model calculations using an even more simplified model than that given by Neumann [130] and described in Sect. 4.1.1. They neglect, for example, radial temperature gradients within the rod-shaped electrodes, so that they can use a one-dimensional heat conduction or heat transport equation (cf. (4.1)). Moreover, they use a constant thermal conductivity λ_E and emissivity ε (i.e., no dependence on electrode temperature T_E) and neglect Joule heating within the electrodes, so that the solution of (4.1) is further simplified. The power input into the cathode tip by the lamp current I is assumed to be uniform across the hemispherical electrode tip, i.e., the power $I\cdot(U_c - \varphi/e)$ is heating the cathode tip, whereas the radiation cooling at the electrode tip is neglected (radiation cooling is taken into account at the sides of the electrodes). The anode is heated by the electron current $(I\cdot\varphi/e)$ and an anode fall voltage U_a, which is set (in accordance with the temperature measurements) to $U_a = 1$ V. The resulting power input into the anode tip $(I\cdot(U_a + \varphi/e))$ is again assumed to be uniform across the hemispherical electrode tip. In the case of ac currents, the power input into the electrodes is assumed to be the arithmetical mean of anode and cathode heating, i.e., the average power input is given by[8] $(0.5 \cdot I \cdot (U_a + U_c))$. This average heating is used independent of the frequency or shape of the lamp current. Even more simplifications are used in [162], which will not be further discussed here.

Tielemans and Oostvogels carried out their measurements and calculations using argon (0.1 MPa) and mercury (0.3 MPa) high-pressure discharge lamps operated using dc and ac currents. The electrodes were made of tungsten or thoriated tungsten. The temperature was measured at the electrode tip, in the ac case during current reversal, i.e., between anode and cathode phases. The comparison of measured and calculated electrode temperatures gave a good agreement using $\varphi = 4.5$ eV (pure tungsten electrodes) and $\varphi = 3.3$ eV (thoriated tungsten electrodes) as the work function for electron emission. This confirms the above-mentioned important interaction between electron and ion current, adjusting the power balance of the cathode and the cathode temperature. The anode was described with the adjustable parameter U_a, so that agreement between measurements and calculations could be achieved by using the appropriate anode fall voltage (here: $U_a = 1$ V). Nevertheless, many of the assumptions used in [162] are quite crude and contradict to experimental findings. One example is the electrode fall voltages, which are assumed to be time-independent and frequency-independent, which contradicts to the findings in [102] (cf. Sect. 3.2.2). Thus, the model used in [162] is able to predict the power balance and electrode temperatures in a first approximation, but for further investigations a more sophisticated electrode model would be necessary.

Bötticher and Bötticher present in [30] a cathode model based on their former work [31], which again is based on the model by Neumann [130] described in Sect. 4.1.1. They extended the model given by Neumann to the transient case and apply a commercial finite element analysis code (ANSYS) to solve the heat conduction equation in the cathode. Moreover, the response time of the electrodes is determined by their heat capacity. Transient effects in the electric field and the cathode fall voltage in the plasma are neglected because they occur on a much smaller time scale than the thermal capacity effects.

Bötticher and Bötticher investigate the time-dependent cathode fall voltage and cathode temperature distribution in a 0.62 MPa high-pressure mercury discharge lamp, operated on a unipolar, sinusoidal lamp current with 50 Hz frequency.[9] The amplitude of the lamp current is varied between 3.3 and 6 A. Electrodes of different

---

[8] At first sight, the power input to the electrode tip and thus the resulting electrode temperature seems to be independent of the work function φ. But the cathode fall voltage strongly depends on the work function φ via the electron emission current, so that the work function φ is important for the power input and the temperature distribution of the electrodes.

[9] Because no anode model is applied, the investigation is restricted to the cathode phase even in the ac case. The sinusoidal lamp current is assumed to be rectified, so that the electrode stays in the cathode phase. Thus, the electrode runs through 100 cathode half-cycles per second.

geometry are investigated, varying the work function between 3.0 eV (thoriated tungsten) and 4.55 eV (pure tungsten).

One of the results presented in [30] is given in Fig. 4.1. A 9-mm long pure tungsten cathode with different coil radii was investigated with respect to the cathode fall voltage as a function of the sinusoidal lamp current. The diameter of the cathode tip was set to 0.4 mm, the radius of the coil (idealized by a thicker electrode radius as shown on the right-hand side of Fig. 4.1) was varied from 0.6 to 1.0 mm. The coil increases the radiation emitting surface, thus the losses due to thermal radiation increase with increasing coil radius. Even more important is the variation of the thermal inertia of the electrode for different coil radii. The larger the coil radius, the larger the thermal inertia of the electrode, i.e., the larger the response time of the electrode. In all cases ($R_{coil} = 0.6,\ 0.8,\ 1.0$ mm), the cathode fall voltage increases at the beginning of the current period, reaches a maximum, and decreases before the maximum current is reached. The higher the thermal inertia of the electrode, the larger the maximum and the steeper the increase and decrease of the cathode fall voltage during a current period. This is due to the power needed for the heating up of the cathode, which is higher for cathodes with larger thermal inertia. After the current maximum, the cathode fall voltage is lower compared to the time before the current maximum. This is due to the decreasing cathode temperature

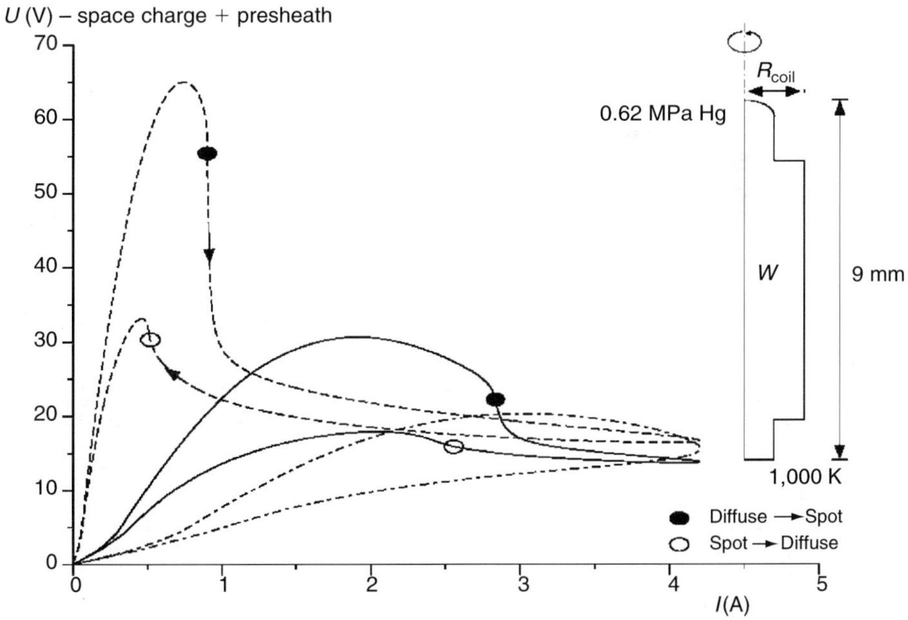

Fig. 4.1. Calculated current–voltage characteristics $U_c(I(t))$ for a rectified sinusoidal lamp current of 50 Hz for three different coil radii: *Dashed line*: $R_{coil} = 1$ mm, *solid line*: $R_{coil} = 0.8$ mm, *dash-dotted line*: $R_{coil} = 0.6$ mm (pure tungsten, radius of cathode tip: 0.4 mm, [30])

for the decreasing lamp current, so that no additional power input is needed for heating up the cathode.

In the case of the two large coil radii (R_{coil} = 0.8 (solid line) and 1.0 mm (dashed line)), a mode transition with respect to the arc attachment area can be observed: During the increase of the lamp current, the arc attachment area switches from a diffuse arc attachment to a spot-like arc attachment. This mode transition is accompanied by a steep decrease of the cathode fall voltage. The arc attachment mode changes back from the spot mode to the diffuse mode during the decreasing part of the current period.

The corresponding cathode temperature progression is shown in Fig. 4.2, though for a different coil radius (R_{coil} = 1.3 mm) and a different cathode material (pure tungsten for the results shown in Fig. 4.1, thoriated tungsten for the results shown in Fig. 4.2). However, the principal cathode temperature progression will be the same for both types of electrodes. The temperature at the cathode tip (T_{mid}) increases during the increasing part of the current cycle (after a short decrease of the cathode tip temperature after zero point). At the time of mode transition from spot to diffuse arc attachment, the cathode tip temperature steeply increases, whereas the temperature at the edge of the hemispherical cathode tip (T_{bor}) hardly changes. This steep increase of the cathode tip temperature with the simultaneous decrease of the

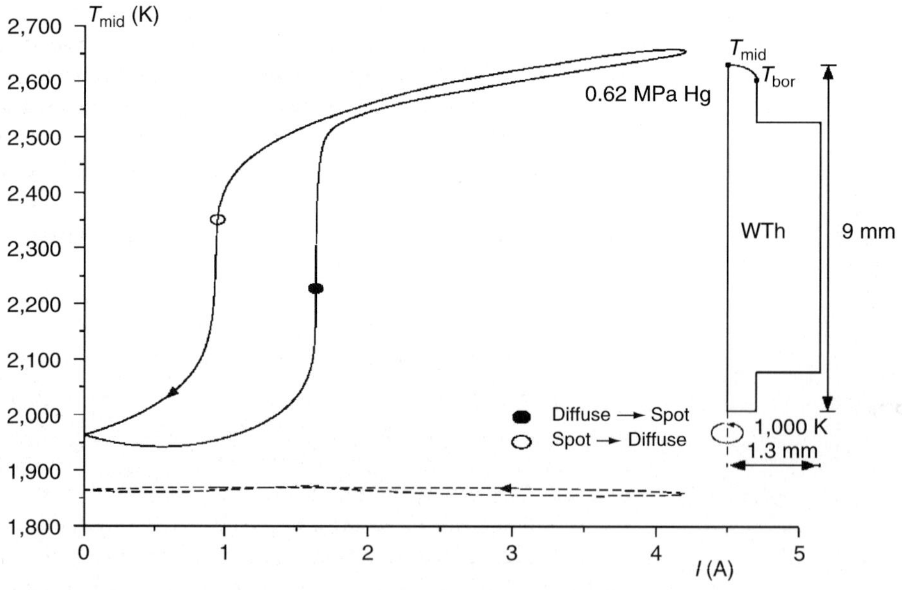

Fig. 4.2. Cathode tip temperature T_{mid} (*solid line*) and cathode temperature at the edge of the hemispherical cathode tip T_{bor} (*dashed line*) for a rectified sinusoidal lamp current of 50 Hz (thoriated tungsten, radius of cathode tip: 0.4 mm, coil radius: 1.3 mm, current amplitude: 4.2 A, [30])

cathode fall voltage suggests a more effective heating mechanism during the spot mode. The cathode tip temperature decreases after the maximum of the current is reached, the mode transition from spot to diffuse arc attachment is accompanied by a steep decrease of the cathode tip temperature. The spot–diffuse transition will also be discussed in Sect. 4.4.8.

The results of the numerical simulation are quite interesting because it is hardly possible to measure the very electrode tip temperature. The numerical results presented in [30] indicate in which range the electrode temperature might change during a current cycle and what parameters might be important for a possible mode transition. However, it has to be kept in mind that there is no anode phase taken into account for the investigations presented in [30]. Moreover, the calculated mode transitions do not take the constriction zone, the hot plasma spots, or the plasma column into account, which might have an important influence on the arc attachment area (cf. introduction to Sect. 4.1 and Sects. 4.4.3, 4.4.4, 4.4.7, and 4.4.8). The calculated mode transition presented in [30] is primarily induced by electrode properties. An advantage of the model from Bötticher and Bötticher is that operating conditions, cathode geometry, material properties, or the influence of different cooling and heating mechanism can easily be investigated. Examples of these variations are given in [31].

Thus, despite the assumption of a collision-free space charge zone and despite the neglect of the plasma (e.g., heat conduction of the plasma to the cathode tip, influence of the constriction zone on the arc attachment area on the cathode tip, radiation losses due to the hot plasma spots) apart from the assumed space charge and ionization zone directly adjacent to the cathode surface, electrode models are able to calculate or predict quite good results concerning the electrode power balance or the electrode temperature. For results concerning the local variation of, e.g., T_E, $j_{t,i}$, or $j_{t,emission}$ at the cathode or electrode tip, the arc attachment area on the electrodes tips, or the ratio of ion to electron current at the cathode, the plasma with plasma column, hot plasma spots, and constriction zones should be taken into account.

4.1.3 Cathode and Anode

HID lamps contain either a cathode and an anode (dc HID lamps) or electrodes which switch alternately from anode to cathode phase (ac HID lamps). The two examples concerning simple cathode models given in Sect. 4.1.2 already indicated that a missing anode model poses problems. Tielemans and Oostvogels used the adjustable anode fall voltage U_a [162], Bötticher and Bötticher used a rectified sinusoidal lamp current to get rid of the anode phase [30]. Within the numerical simulations, the only difference between anode and cathode[10] is found at the boundary between electrode and plasma, i.e., the boundary conditions at the electrode surfaces are different for anode and cathode.

[10] In the following "cathode" means "cathode" for dc currents and "electrode during cathode phase" for ac currents (likewise for the anode).

Lamp models including anode and cathode are given, for example, in Sect. 4.4 (cf. [53–65]), by Wendelstorf [167, 168], and by Morrow, Lowke et al. (e.g., [112, 113, 120, 144]). The differences between anode and cathode will be discussed in detail in Sect. 4.4, so that only a short overview will be given here.

In HID lamps, the heating and cooling of the cathode is primarily given by heating due to an ion current and due to heat conduction from the plasma to the cathode tip, and cooling due to an electron emission current and radiation cooling. Thus, the boundary condition at the cathode–plasma surface reads:

$$\lambda_E \frac{\partial T_E}{\partial n} = \lambda_P \frac{\partial T_P}{\partial n} + j_{t,i}(\varphi_i - \varphi) - j_{t,emission}\varphi - \varepsilon\sigma_{SB}T_E^4. \tag{4.4}$$

The terms $\partial T_E/\partial n$ and $\partial T_P/\partial n$ denote the component of the cathode and plasma temperature gradient normal to the cathode surface. λ_E and λ_P are the thermal conductivity of the cathode and the plasma, respectively, $j_{t,i}$ is the particle flux density of ions, $j_{t,emission}$ the particle flux density of electrons emitted by the cathode, φ_i the ionization energy of the gas atoms, and φ the work function of the cathode. ε is the emissivity of the electrode material, σ_{SB} the Stefan–Boltzmann constant, and T_E the cathode temperature. The quantities $j_{t,i}$, $j_{t,emission}$, T_E, $\partial T_E/\partial n$, and $\partial T_P/\partial n$ depend on the position on the cathode surface, i.e., $j_{t,i}$, $j_{t,emission}$, and $\partial T_P/\partial n$ might be large at the very tip of the cathode but drop steeply for larger distances to the cathode tip. Thus, the heat flux density to the cathode tip is large, positive, and dominated by electron and ion current and the heat flux from the plasma near the cathode tip, whereas heat flux density to the cathode is smaller, negative, and dominated by radiation cooling at the cathode surface below the cathode tip. λ_P, λ_E, and ε are functions of the plasma or cathode temperature. Mathematically, (4.4) couples T_E and T_P and accounts for the discontinuity of the heat flux on the cathode–plasma boundary.

In contrast to (4.2), in (4.4) the ion current is not assumed to bring the kinetic energy gained in the cathode fall voltage directly to the cathode surface. Instead, due to the many collisions of the ions in the near-cathode plasma, this energy is deposited in the near-cathode plasma. From there, it is mainly transported to the cathode tip via heat conduction, which is not included in (4.2) (cf. detailed comparison in Sect. 4.1.1).

The heating and cooling of the anode in HID lamps is primarily given by heating due to the electron current and due to heat conduction from the plasma to the anode tip, and radiation cooling. Thus, the boundary condition at the anode–plasma surface reads:

$$\lambda_E \frac{\partial T_E}{\partial n} = \lambda_P \frac{\partial T_P}{\partial n} + j_{t,e}\varphi - \varepsilon\sigma_{SB}T_E^4. \tag{4.5}$$

The anode is a more "passive" component compared to the cathode, lacking the self-regulating mechanisms of the cathode described in Sect. 4.1.1 (interaction between ion and electron emission current). Changes in the lamp current will thus lead to a larger impact on the anode temperature compared to the impact on the cathode temperature. This is because the lamp current is directly heating the anode via the term $j_{t,e}\varphi$ in (4.5), where $j_{t,e}$ is the particle flux density of electrons. Since the anode repels ions, there is no ion current at the anode surface.

The anode and the cathode fall voltage cannot be compared directly, i.e., if anode and cathode fall voltages are equal, this will not lead to equal anode and cathode temperatures. The power flux from the plasma to the electrode tips strongly correlates with the electrode fall voltages, but there is one important difference between anode and cathode: The electrons emitted by the cathode (as electron emission current and to neutralize incoming ions, cf. (4.4)) have a cooling effect on the cathode, whereas the electrons reaching the anode have a heating effect. Accordingly, the power we put in the cathode to continuously release electrons we get back at the anode. Therefore, equal anode and cathode fall voltages do not correspond to equal power fluxes to anode and cathode. If anode and cathode fall voltage are equal, the power flux to the anode would be higher than the power flux to the cathode.

The differences between anode and cathode or anode and cathode phase not only have an effect on the electrode temperature, but also the plasma near the electrodes is affected differently. Very different hot plasma spot temperatures can be measured, for example, in a 0.6 MPa high-pressure mercury discharge lamp as shown in Figs. 4.3–4.5 (cf. [99]). The arc constriction during the cathode phase is reported to be stronger compared to the arc constriction during the anode phase in this experimental setup, resulting in higher hot plasma spot temperatures during the cathode phase.[11] The influence of the electrodes and the different arc attachments are localized within the first millimeter of the plasma adjacent to the electrode tip, as can be seen in Fig. 4.5. For distances above 1 mm to the electrode tip, the smooth transition from the hot plasma spots to the plasma column are equal during anode and cathode phase. This knowledge is of particular importance for modern HID lamps with decreasing electrode gaps (D2 automotive headlight lamp: 4 mm, UHP lamp: 1 mm). The influence of the electrodes and the near-electrode plasma is more and more important in modern HID lamps. Besides the near-electrode plasma temperature, the electric field distribution within the near-electrode plasma is also different for anode and cathode, resulting in different anode and cathode fall voltages.

In Sect. 4.4.4, the results of a numerical model applied to a similar 0.6 MPa mercury HID lamp (operated on a dc current) will be presented, showing the same principal differences between the near-anode and near-cathode plasma. The hot plasma spot temperatures of the near-cathode plasma can reach considerably higher values compared to the hot plasma spot temperatures of the near-anode plasma. Moreover, it will be shown that the electrode geometry effects anode, cathode, near-anode plasma, and near-cathode plasma as well as anode and cathode fall voltage differently.

---

[11] The different hot plasma spot temperatures are not only due to the different arc attachments. They are also due to the principal differences in anode and cathode phase, as will be analyzed in Sect. 4.4.4.

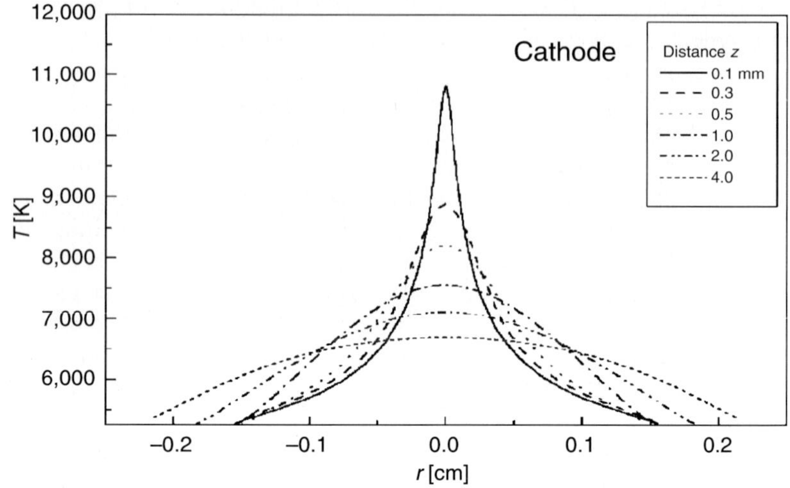

Fig. 4.3. Measured temperature profiles at different distances to the electrode tip during the cathode phase in a 0.6 MPa mercury HID lamp (150 W, 1.9 A, 50 Hz sinusoidal lamp current, [99])

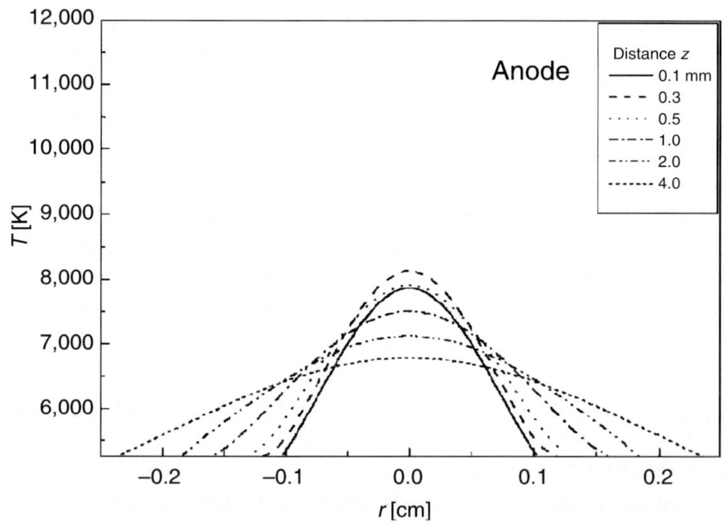

Fig. 4.4. Measured temperature profiles at different distances to the electrode tip during the anode phase in a 0.6 MPa mercury HID lamp (150 W, 1.9 A, 50 Hz sinusoidal lamp current, [99])

A good possibility to learn something about the differences of anode and cathode or anode and cathode phase is the external heating of the electrodes using a laser. The results concerning this type of electrode diagnostic are presented in Sects. 3.1.10 and 4.4.5.

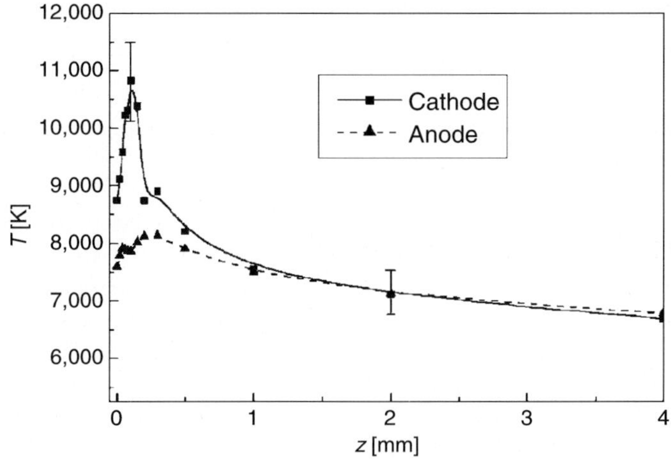

Fig. 4.5. Measured arc core temperature in a 0.6 MPa mercury HID lamp near the electrode tip (at $z = 0$) in anode and cathode phase (150 W, 1.9 A, 50 Hz sinusoidal lamp current, [99])

There are further heating and cooling terms which are not accounted for in (4.4) and (4.5) and which are not accounted for in most electrode models: The photons generated in the plasma might fall on the electrodes, resulting in an additional heating especially at the electrode tips. The photons will be partly reflected at the electrode surfaces, but due to an emissivity ε between 0.1 and 0.5, the absorption might be quite high.[12] This effect results in higher anode temperatures and lower cathode fall voltages, as can be deduced from the experiments and calculations concerning the external laser heating described in Sect. 4.4.5. Furthermore, the high energy photons of the plasma radiation might release electrons from the cathode, if the photon energy is higher than the work function for electron emission (photoemission, photoelectric effect). However, the latter effect should be sufficiently small with photoelectric yield values of the order of 10^{-4} according to [146]. Another heating or cooling term might be due to the convection cooling as described by Neumann [130] and Bötticher and Bötticher [31]. An estimate given in [31] shows that the influence might not be large, but this might differ from lamp to lamp.

Further interesting questions are connected with the material properties of hot or even molten tungsten. The most important parameter at the cathode is the work function for electron emission, which is already difficult to determine for solid tungsten (cf. Sects. 2.4.4 and 3.1.11) and might be different for molten tungsten. Moreover, the molten tungsten might react with different plasma constituents, changing thus the work function, as it is already observed for solid tungsten

[12] Investigations by Morrow and Lowke [120] concerned with a 200 A argon arc showed that the heating due to plasma radiation is very low, but the results might not be transferable directly to HID lamps.

(cf. Sects. 3.1.11, 3.1.12, and 4.4.6). Since the melting temperature of tungsten (3,680 K) is reached at least locally and temporarily at many electrode tips (cf. Sect. 3.1.13), this is an important but not yet well-understood subject.

4.2 Plasma Models

The plasma is responsible for the primary purpose of any lamp, the generation of light. The light output of a HID lamp is closely related to the temperature distribution within the plasma (cf. Sect. 2.2.3). The calculation of the plasma temperature necessitates the knowledge/calculation of the power loss due to radiation, the heat flux toward the discharge tube and toward the electrodes, and the electrical power input. Convection within the discharge will also alter the plasma temperature distribution because energy is redistributed within the plasma due to plasma flows. The plasma column is independent of the electrode properties or the arc attachment to the electrodes as long as the electrode gap is large enough. Thus, the properties of the plasma column can be treated without taking the influence of the electrodes into account. For short electrode gaps as they can be found in modern HID lamps (e.g., D2 automotive headlight lamp (4 mm), UHP lamp (1 mm)), this assumption is already problematic (cf. e.g., Sects. 4.4.4 and 4.4.7).

The governing equations for the plasma will be presented in Sect. 4.2.1, two examples of plasma models are given in Sects. 4.2.2 and 4.2.3.

4.2.1 Description of the Plasma

The plasma can be described by the electric potential within the plasma, the local plasma temperature, and the flow velocity. The calculation of these properties will be described in this section.

Electric Potential

The electrical power input into the plasma is closely related to the electric potential (V) and the electric current density (j). The electric current is conserved within the plasma, i.e., the current continuity equation can be used:

$$\nabla \cdot j = \nabla \cdot (\sigma E) = -\nabla \cdot (\sigma \nabla V) = 0. \tag{4.6}$$

E is the electric field and σ the electrical conductivity, which is proportional to the electron density as described in Sect. 2.3.2. If the electrical conductivity σ is known,[13] the electric potential V can be calculated. The electrical power input into the plasma is then determined by $\sigma \cdot E^2$ or $j \cdot E$.

[13] Usually, σ is calculated assuming LTE, so that σ depends only on the plasma composition, pressure, and the local plasma temperature (cf. Sect. 2.3.2). A possibility to account for deviations of the electrical conductivity from its LTE value in the near-electrode plasma due to diffusion of electrons from the hot plasma spots into the cold near-electrode plasma is described in Sect. 4.4.

Within the plasma column, the voltage drop per unit length is constant, which results in a constant electric field everywhere in the plasma column[14] (cf. Sect. 3.2.2). Since the electron density depends strongly on the plasma temperature, the electrical conductivity is very high in the hot center (core) of the plasma column and very low in the cooler outer parts of the plasma. Thus, the electrical power input ($\sigma{\cdot}E^2$) is high within the hot core of the plasma column and low in the outer plasma. Analogous, the electric current density is high in the hot center of the plasma column and very low in the cooler outer parts of the plasma. Thus, electrical power is released in the core of the plasma column, where it is partially transformed into radiation, which escapes from the plasma. The rest of the electrical power input is transported via heat conduction or convection to the discharge tube or the electrode tips (cf. e.g., Sect. 4.4.5, especially Figs. 4.67 and 4.68).

The transition from the plasma column, where the current density profile is relatively broad, to the arc attachment to the electrode tips, where the current density profile is much more constricted, is accompanied by an increase in the electric current density (same current I through smaller area A). This effect in the so-called constriction zone is partly balanced by an enlarged electrical conductivity due to the higher plasma temperature in the hot plasma spots, but the net effect is an increase in j and E in the constriction zone compared to the values within the cylindrical plasma column (cf. e.g., Sect. 4.4.5, especially Figs. 4.60–4.64).

Due to the low plasma temperature in the vicinity of the electrodes (between hot plasma spots and electrode tips), the electric field within this plasma region is even further enlarged compared to the electric field in the plasma column and the constriction zones. Due to the current continuity equation (4.6), the electric field must be high if the current must be conducted through a "sheath" with low electrical conductivity to preserve the total lamp current, i.e., one gets high electric fields in plasma areas where it is "tough" to get the given current through. Thus, the electric field in the near electrode plasma is considerably higher compared to the electric field in the plasma column, resulting in the electrode fall voltages described for example in Sects. 3.2.2, 3.2.3, and 4.1.1. The corresponding high electrical power input ($\sigma{\cdot}E^2$) in the near electrode plasma balances the losses to anode and cathode.

Due to the large temperature gradient between the hot plasma spots and the electrode tips and the resulting large gradient in electron densities, diffusion of electrons from the hot plasma spots into the cold near-electrode plasma increases the (LTE) electrical conductivity σ. This effect will be discussed in Sect. 4.4.

The determination of the electric potential V using (4.6) requires the adjustment of one more free parameter.[15] This is done by adjusting the given lamp current by setting the potential at the anode to the appropriate value of the total lamp voltage. The procedure is the following: First a "normalized" electric potential distribution

---

[14] The determination of this electric field in the plasma column in the case of a one-dimensional model describing only the cylindrical plasma column is described in Sect. 4.2.2.

[15] If a constant factor κ is introduced in (4.6) just in front of the electric potential V, it can be reduced, thus the electric potential is determined except for this constant factor κ.

V_n with the boundary conditions $V_n = 0$ at the cathode–plasma surface and $V_n = 1$ at the anode–plasma surface is calculated. This normalized potential V_n is then multiplied by a factor κ to get the "real" electric potential V, where κ is chosen in such a manner that the integral of

$$j = \sigma E = -\sigma \cdot \text{grad}(V) = -\kappa \cdot \sigma \cdot \text{grad}(V_n) \tag{4.7}$$

over a closed surface between anode and cathode (e.g., mid-plane, cathode surface, or anode surface) yields the given lamp current I. Alternatively, the factor κ can be used to fix the given total electrical power input P, which is determined by the integral of

$$jE = \sigma E^2 = -\sigma \cdot (\text{grad}(V))^2 = -\kappa \cdot \sigma \cdot (\text{grad}(V_n))^2 \tag{4.8}$$

over the discharge volume (cf. e.g., [74]).

At the inside wall of the discharge tube, it is assumed that the electric current density is parallel to the wall, i.e., there is no electric current within the tube material:

$$j_{\text{wall},\perp} = -\sigma \left(\frac{\partial V}{\partial n} \right)_{\text{wall}} = 0. \tag{4.9}$$

Charge Density

The electric field is related to the charge density ρ, which is the source of the electric potential or the electric field. The governing equation connecting electric field or electric potential with the charge density is Poisson's equation:

$$\nabla \cdot E = -\nabla \cdot \nabla V = -\Delta V = \frac{\rho}{\varepsilon_0} = \frac{n_i - n_e}{\varepsilon_0}. \tag{4.10}$$

The ∇ in front of the electric field E is the divergence operator, the ∇ in front of the electric potential V is the gradient operator, and Δ is the Laplace operator. ε_0 is the permittivity, n_i the ion density, and n_e the electron density.

If the electric field E is known (e.g., by using (4.6) with a known σ), the charge density ρ can be calculated. In the cylindrical plasma column, the electric field is constant, i.e., $\nabla \cdot E = 0$. Thus, the charge density is zero, i.e., $n_i = n_e$ (quasineutral plasma). The electric field in the near-cathode and the near-anode plasma is considerably higher than the electric field in the plasma column, i.e., there must be a space charge greater or lesser than zero. In front of the cathode, an excess of (positive) ions "shields" the cathode with respect to the plasma column, in front of the anode, an excess of electrons is responsible for the increase in the electric field.

If the charge density in the plasma is calculated at first, as for example in the space charge zone in certain cathode models, (4.10) can be used to calculate the electric field or the electric potential.

Power Balance

The power balance of the plasma is the core of every plasma model. Depending on the intention of the model under investigation, different forms are possible. If the pressure within the discharge is assumed to be constant, the power balance can be formulated with respect to the plasma enthalpy[16] h [73, 75, 168]:

$$\rho \frac{\mathrm{D}}{\mathrm{D}t} h = j \cdot E - u - \nabla \cdot q. \tag{4.11}$$

D/Dt is the total time derivative, ρ the mass density, j the electric current density, E the electric field, u the net radiation emission coefficient or radiation loss (cf. Sect. 2.2.3), and q the heat flux. The total time derivative D/Dt is defined as:

$$\frac{\mathrm{D}X}{\mathrm{D}t} = \frac{\partial X}{\partial t} + (v \cdot \nabla) X, \tag{4.12}$$

where v is the flow or gas velocity. Using (4.12), the relation $c_\mathrm{p} = (\mathrm{d}h/\mathrm{d}T)_{\mathrm{p=const.}}$ (specific heat at constant pressure), $q = -\lambda_\mathrm{P} \nabla T_\mathrm{P}$ (Sect. 2.3.3), and $j = \sigma \cdot E$, (4.11) transforms to

$$\rho c_p \frac{\partial T}{\partial t} + \rho c_p v \cdot \nabla T = \sigma \cdot E^2 - u + \nabla \cdot (\lambda_\mathrm{P} \nabla T_\mathrm{P}). \tag{4.13}$$

The plasma properties ρ, c_p, σ, u, and λ_P depend on the plasma composition, pressure, and the local plasma temperature (cf. Chap. 2). The electric field can be calculated using current continuity (4.6), the flow velocity v is determined by considering the momentum balance (see later). Thus, (4.13) serves for the calculation of the plasma temperature T.

The first term on the left-hand side of (4.13) represents the local temperature variation with time in the plasma, the second term the convective heat transfer. The first term on the right-hand side of (4.13) represents the electrical power input (Joule heating), the second term the radiation losses, and the third term the heat conduction. Equation (4.13) is also referred to as Elenbaas–Heller equation.

The boundary condition for the plasma temperature at the electrode–plasma surface is given by (4.4) (cathode) and (4.5) (anode). The plasma temperature at the inside wall of the discharge tube can either be set to a constant value or a prescribed function of the surface position, or it can be assumed that the heat flux to the wall must be balanced by the thermal radiation from the outer wall surface [111]:

$$q_{\mathrm{wall},\perp} = -\lambda_\mathrm{P} \left(\frac{\partial T_\mathrm{P}}{\partial n} \right)_{\mathrm{wall}} = \varepsilon_\mathrm{W} \sigma_\mathrm{SB} T_\mathrm{W}^4. \tag{4.14}$$

---

[16] The enthalpy H is defined as the sum of the internal energy U and the product of pressure and volume. The specific enthalpy h refers to the unit mass, i.e., $h = e + p/\rho$, where e is the internal energy per unit mass and ρ the mass density.

ε_W is the emissivity of the arc tube material and T_W the wall temperature. The latter method allows for the self-consistent calculation of the wall temperature as a function of surface position. An even better approach would be the calculation of the heat diffusion or heat conduction equation within the discharge tube material (cf. (4.1) without Joule heating term).

Momentum Balance

The momentum balance is considered to deliver the flow or gas velocity v [73, 75, 111, 168]:

$$\rho \frac{\mathrm{D}}{\mathrm{D}t} v = -\nabla p + \nabla \cdot (2\mu \underline{\underline{\varepsilon}}) + j \times B + \rho g. \tag{4.15}$$

ρ is the mass density, D/Dt the total time derivative, p the pressure, μ the dynamic viscosity, $\underline{\underline{\varepsilon}}$ the shear rate tensor, j the electric current density, B the magnetic field, and g the acceleration due to gravity.

The first term on the right-hand side of (4.15) is the force due to pressure gradients, the second term is the friction loss per volume, the third term the Lorentz force per volume due to the movement of charge carriers in a magnetic field, and the forth term is the gravity force per volume. The gravity force is responsible for the convective flows in HID lamps, which increase for increasing pressure (cf. Sect. 4.2.3). In horizontally operated lamps, the arc bows upwards due to convection, in vertically operated lamps, convective cells are induced by convective flows. The magnetic part of the Lorentz force ($j \times B$) is induced by the magnetic field caused by the lamp current. By applying a small external magnetic field (about 1 mT), the arc of a horizontally burning lamp can be straightened [10]. Equation (4.13) is also referred to as Navier–Stokes equation.

Mass Conservation

The set of equations is complete by the mass continuity equation:

$$\frac{\partial}{\partial t} \rho = -\nabla \cdot (\rho v). \tag{4.16}$$

ρ is the mass density and v the gas velocity.

Some Plasma Models

There are many plasma models taking the above equations either way into account. Thrum ([160], cf. Sect. 4.2.2) investigates the time-dependent properties of the plasma column in an HID lamp operated using a current superimposed by short current pulses (mercury and metal halide discharges). Similar investigation of the plasma column were carried out by Charrada and Zissis [37, 163] taking ac operation (50 Hz) in mercury and metal halide HID lamps into account. Giese ([73], cf. Sect. 4.2.3) is concerned with the convection in dc-operated mercury HID lamps operated in different burning positions. Charrada and Zissis [38, 39] also investigate the convection in vertically operated dc mercury HID lamps, whereas Palacin [134] investigates the convection in vertically operated rare gas HID lamps (also dc operation). Wendelstorf simulates dc-operated vertically burning HID lamps, taking convection and electrode–plasma interaction (especially at the cathode) into account [167, 168]. Galvez is also concerned with the electrode–plasma interaction in dc- operated HID lamps taking convection into account [69, 97]. Haverlag [87] presents a model of the cylindrical plasma column taking time-dependence into account but neglecting natural convection. He focuses on topics like radiation transport and demixing. Beks et al. [12] investigate dc-operated vertically operated lamps, taking natural convection into account. Fischer, Wiesmann, and the author of this book ([51, 53–65, 169] are concerned with plasma and electrode properties, focusing on the interaction between plasma and electrodes. Even ac currents and complex plasma compositions are taken into account (cf. Sect. 4.4).

In the following, two plasma models are described exemplarily: The model used by Thrum [160], investigating the time-dependent properties of the plasma column in HID lamps operated with a current superimposed by short current pulses (mercury and metal halide discharges) is described in Sect. 4.2.2. The results obtained by Giese [73], who is concerned with the convection in dc-operated mercury HID lamps operated in different burning positions, are presented in Sect. 4.2.3. Section 4.3 and 4.4 will be concerned with plasma models taking additionally to the plasma the discharge tube or the electrodes into account.

4.2.2 Example of a Plasma Column Model

The plasma model of Thrum [160] is concerned with the time-dependent properties of the plasma column in mercury and metal halide HID lamps. The focus of the model is the investigation of the effect of short current pulses, i.e., the dependence of efficiency or spectral power distribution on pulse width, repetition rate, or shape of the current pulses. The plasma is reduced to the cylindrical part of the plasma column, i.e., an infinite electrode gap is considered. Thus, the plasma column can be regarded as independent of the z-coordinate,[17] reducing the problem to a one-dimensional model with only radial dependences.

The current conservation (4.6) reduces to the integral of the electric current density $j = \sigma \cdot E$ over the cross section of the arc column:

$$I(t) = \int_{\text{cross section}} j \, dA = 2\pi E_z \int_{r=0}^{r=R} \sigma(T_P(r)) r \, dr. \tag{4.17}$$

[17] The z-axis is the symmetry axis, i.e., z measures, for example, the distance to the electrode tips in the arc core.

Because of symmetry reasons, the electric field is reduced to the z-component, independent of the radial position r. The electrical conductivity σ is the LTE electrical conductivity and thus depends only on the plasma temperature $T_P(r)$. The current continuity equation thus reduces to an integral from the arc core ($r = 0$) to the discharge tube wall ($r = R$) and connects the lamp current $I(t)$ with the electric field $E = E_z$.

The temperature distribution $T_P(r)$ is calculated according to (4.13), where the gas velocity v is not calculated using the momentum balance (4.15) but the mass conservation (4.16). Thus, the flow velocity v is not induced by convection, but it is due to the temperature and thus the particle density change induced by the current pulse: If the temperature T_P in the arc core rises due to a higher lamp current (current pulse), the resulting lower particle density n causes a gas flow into the colder parts of the plasma ($p = nkT_P = \text{const.}$). Mass inertia is not taken into account. Surplus mass disappears in the (not considered) cold plasma region at the electrode roots or at the cold spot, keeping the pressure constant. Furthermore, the temperature of the inside wall of the discharge vessel is assumed to be constant.

Thrum investigates Hg (3 and 6 MPa) and $HgSnI_2$ (3 and 6 MPa Hg and 0.4 MPa SnI_2) discharges, the basic load corresponds to a lamp current of 0.27–0.35 A, the current pulses are 5–100 μs long, the amplitude of the current during the pulse can be as high as 22 A. The average electrical power input into the plasma column corresponds to 8 W^{-1}mm.

The numerical investigations were accompanied by experimental investigations by Oettler [132]. The measured pulsed lamp current used for the experimental investigations can thus be used as an input parameter for the numerical simulations (cf. Fig. 4.6). The resulting plasma temperature in the arc core vs. time in the case of a 6 MPa mercury plasma is shown in Fig. 4.7. The agreement between experimental and numerical results is very good. As the current amplitude rises, the arc core temperature rises as well, reaching its maximum just before the end of the current pulse. The temperature increase amounts to more than 3,000 K. After the end of the

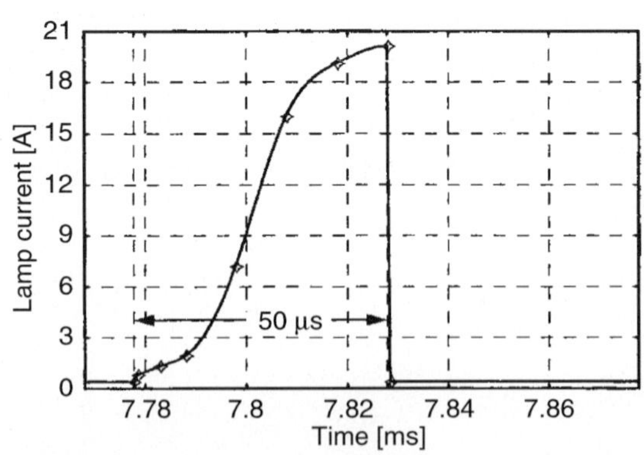

Fig. 4.6. Example of the imposed current pulse: The *solid line* represents the current pulse $I(t)$ used for the numerical simulation, the diamonds are measured values [160]

current pulse, the arc core temperature relaxes to its former value. The corresponding plasma column temperature vs. radius at different points in time during and after the current pulse is shown in Fig. 4.8. Even more data are shown in Fig. 4.9, illustrating the plasma column temperature vs. time and radius during and after the current pulse.

In Fig. 4.10, the electric field (independent of radius *r*) vs. time for the same 6 MPa mercury discharge is shown. The electrical power input per unit length into the plasma column is given by *E·I*, i.e., by the product of the two curves shown in Fig. 4.10. The electric field increases at the beginning of the pulse, i.e., the electrical power input is very high, heating up the plasma. The electric field reaches its maximum value well before the current maximum is reached. Nevertheless, the product *E·I* rises even after *E* has reached its maximum value due to the strongly increasing lamp current *I*. As soon as the lamp current is back to its starting value after the current pulse, the electric field drops below its initial (steady state) value,

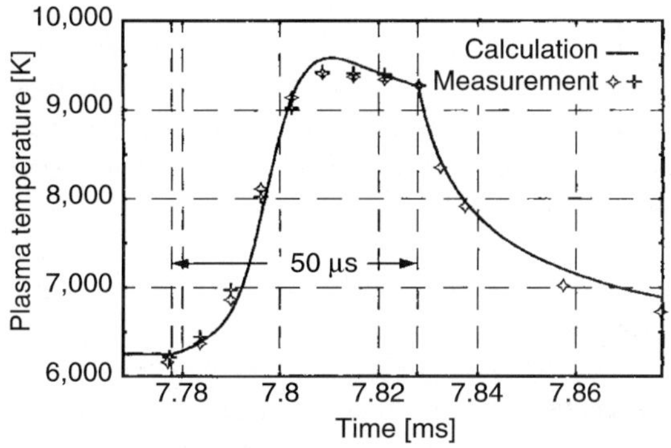

Fig. 4.7. Comparison of the calculated (*solid line*) and measured (*diamonds and crosses*) temperature in the arc core vs. time during and after the current pulse (Hg, 6 MPa, [160])

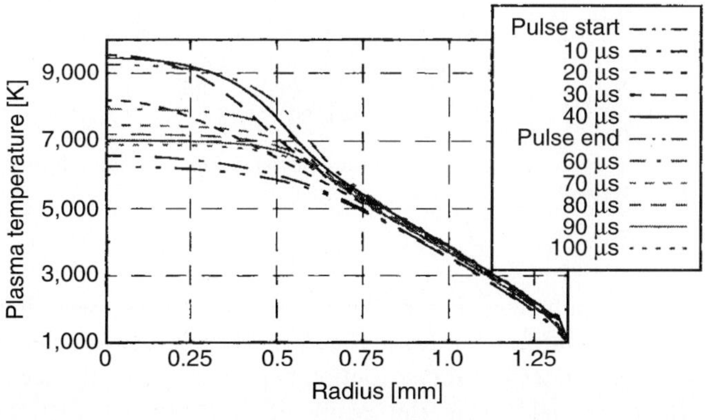

Fig. 4.8. Calculated plasma column temperature profiles vs. radius at different points in time during and after the current pulse (Hg, 6 MPa, [160])

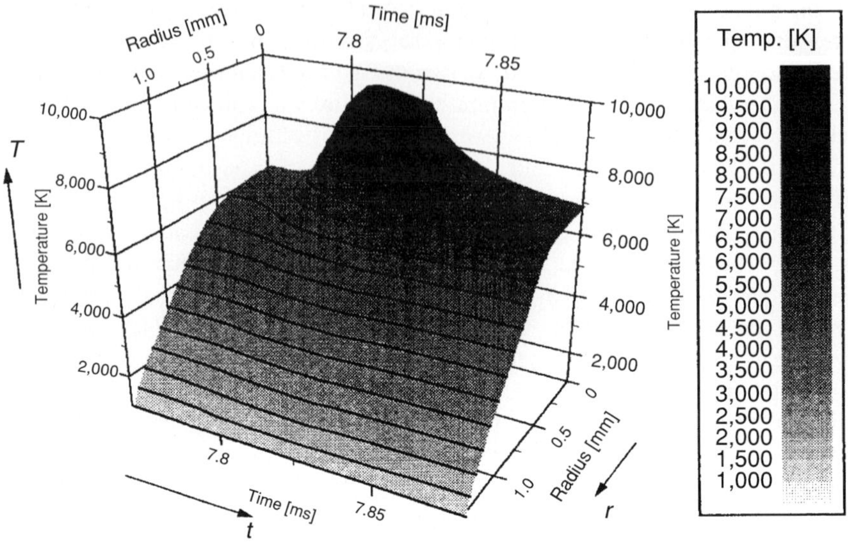

Fig. 4.9. Plasma temperature vs. time and radius during and after the current pulse (Hg, 6 MPa, [160])

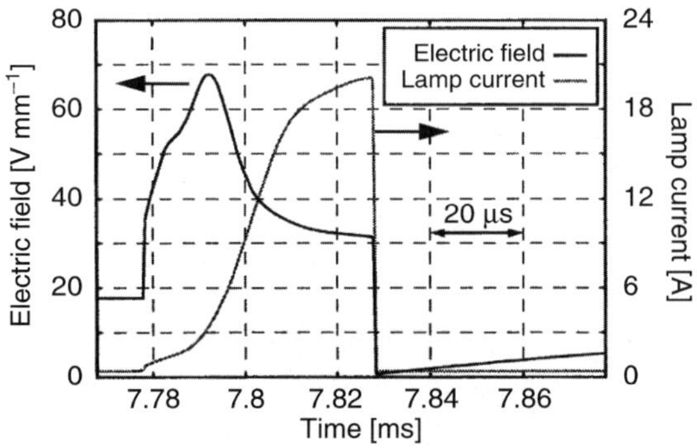

Fig. 4.10. Lamp current and electric field in the plasma column vs. time during and after the current pulse (Hg, 6 MPa, [160])

i.e., during the cooling down of the plasma, the electrical power input into the plasma is lower than during the steady-state operation. As soon as the plasma temperature has reached its initial distribution before the pulse, the electrical power input is "back to normal." This can easily be understood taking the electrical conductivity σ into account. The very high plasma column temperature after the pulse is responsible for a high electron density and thus a high electrical conductivity. Thus, a small electric field is sufficient to get the given lamp current through the plasma after the current pulse.

Finally, Fig. 4.11 shows the radial plasma velocity[18] vs. time and radius during and after the current pulse. Due to the steep temperature raise at the beginning of the current pulse, the plasma streams away from the core to the colder plasma regions (positive v_r). As soon as the current pulse is over, the plasma streams back to the cooling of arc core.

Similar investigation of the plasma column were carried out by Charrada and Zissis [37, 163] taking ac operation (50 Hz) in mercury and metal halide HID lamps into account.

4.2.3 Example of a Plasma Model Including Convection

Giese [73] is concerned with the convection in dc-operated mercury HID lamps operated in different burning positions. For the calculation of the electric potential he considers current continuity (4.6), using a fixed electrical power input instead of a given lamp current I (cf. (4.8)). The power balance (4.13) is solved for the time-independent case (dc current). To include natural convection caused by gravity, momentum balance (4.15) and mass continuity (4.16) are taken into account. Again, time dependence is not considered and the magnetic Lorentz force is neglected in the momentum balance equation (4.15).

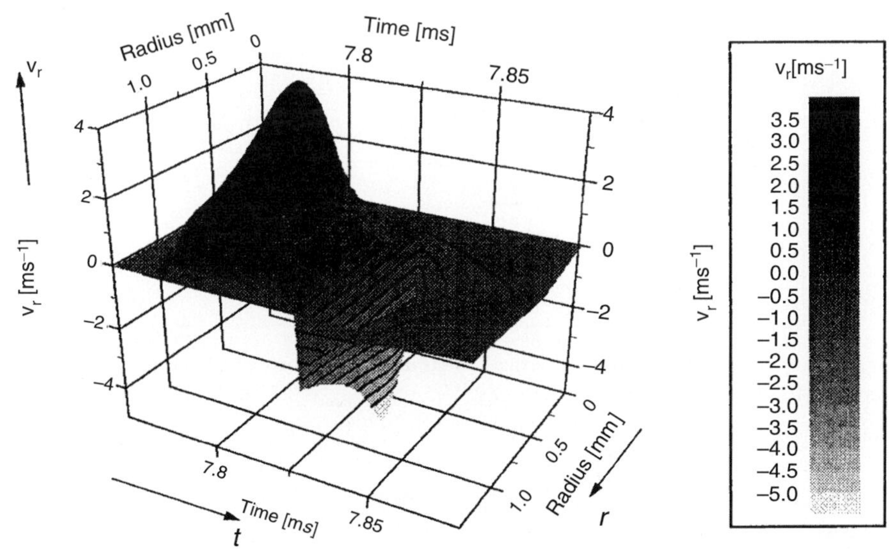

Fig. 4.11. Radial gas velocity v_r vs. time and radius during and after the current pulse (Hg, 6 MPa, [160])

[18] In accordance with the model assumptions, there is no velocity component in z-direction.

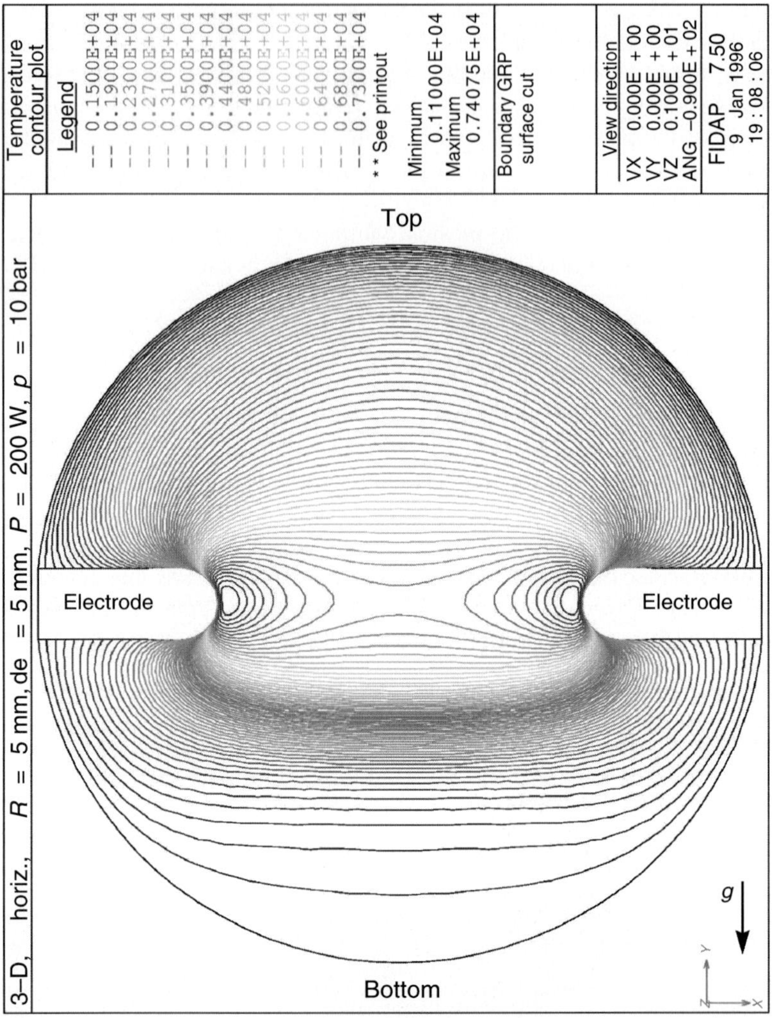

Fig. 4.12. Temperature profile (vertical longitudinal section) of a horizontally burning mercury HID lamp (Hg, 1 MPa, $R_{\text{lamp}} = d_{\text{elec}} = 5$ mm, $r_{\text{elec}} = 0.5$ mm, $P_{\text{el}} = 200$ W, [73])

Giese investigates mercury discharge lamps with pressures in the range from 0.1 to 10 MPa, 200 W electrical power input, and 5 mm electrode gap. The lamp can be operated vertically (rotational symmetry), horizontally, or with arbitrary orientation. The temperature at the inside wall of the discharge vessel and at the electrode roots is set to 1,100 K. The temperature at the electrode tips is set to 3,000 K, assuming a linear decrease of the electrode temperature from tip to root. A self-consistent calculation of the electrode temperature using (4.1) and appropriate boundary conditions (e.g., (4.4) and (4.5)) is not carried out, thus there is no difference

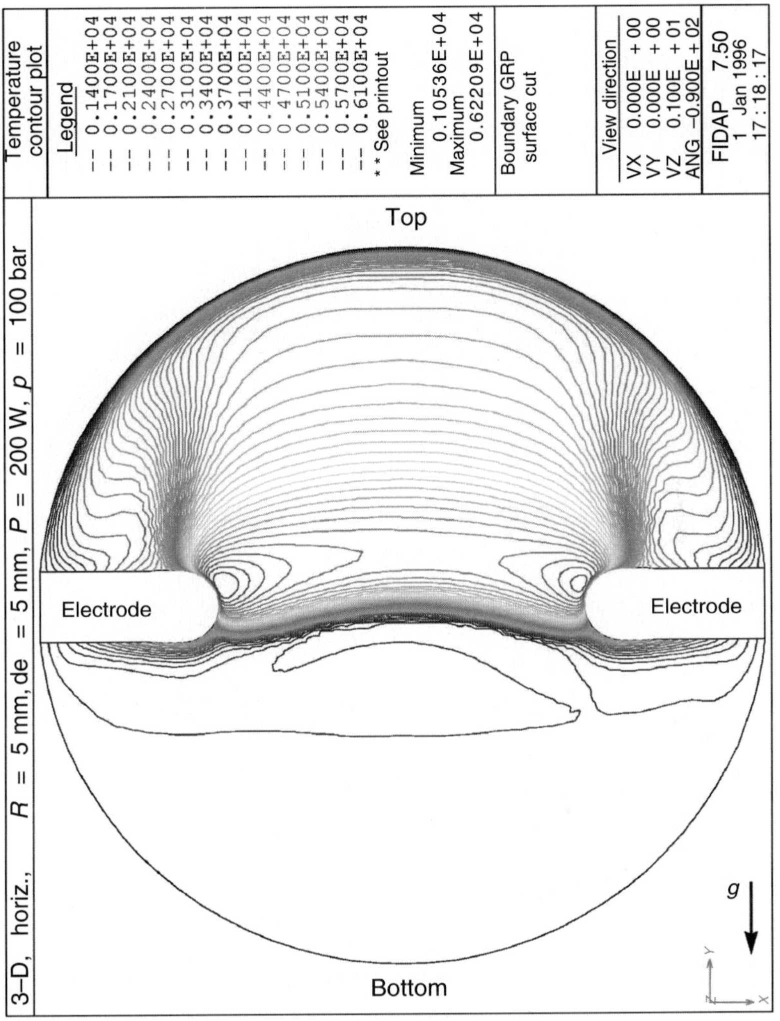

Fig. 4.13. Temperature profile (vertical longitudinal section) of a horizontally burning mercury HID lamp (Hg, 10 MPa, $R_{\mathrm{lamp}} = d_{\mathrm{elec}} = 5$ mm, $r_{\mathrm{elec}} = 0.5$ mm, $P_{\mathrm{el}} = 200$ W, [73])

between anode and cathode. As described in Sect. 4.1.3 (especially Figs. 4.3–4.5), this is not consistent with experimental findings (see also Sect. 4.4.4).

Some of the results obtained by this model are shown in Figs. 4.12–4.15. In Figs. 4.12 and 4.13, two horizontally burning lamps are shown, one with an operating pressure of 1 MPa (Fig. 4.12), the other with 10 MPa (Fig. 4.13). It is clearly visible that the bowing of the arc increases with increasing pressure. The temperature profile in the case of 1 MPa (Fig. 4.12) is largely independent of the convection, whereas convection alters the temperature distribution in the case of the 10 MPa discharge (Fig. 4.13) especially outside the core of the plasma column.

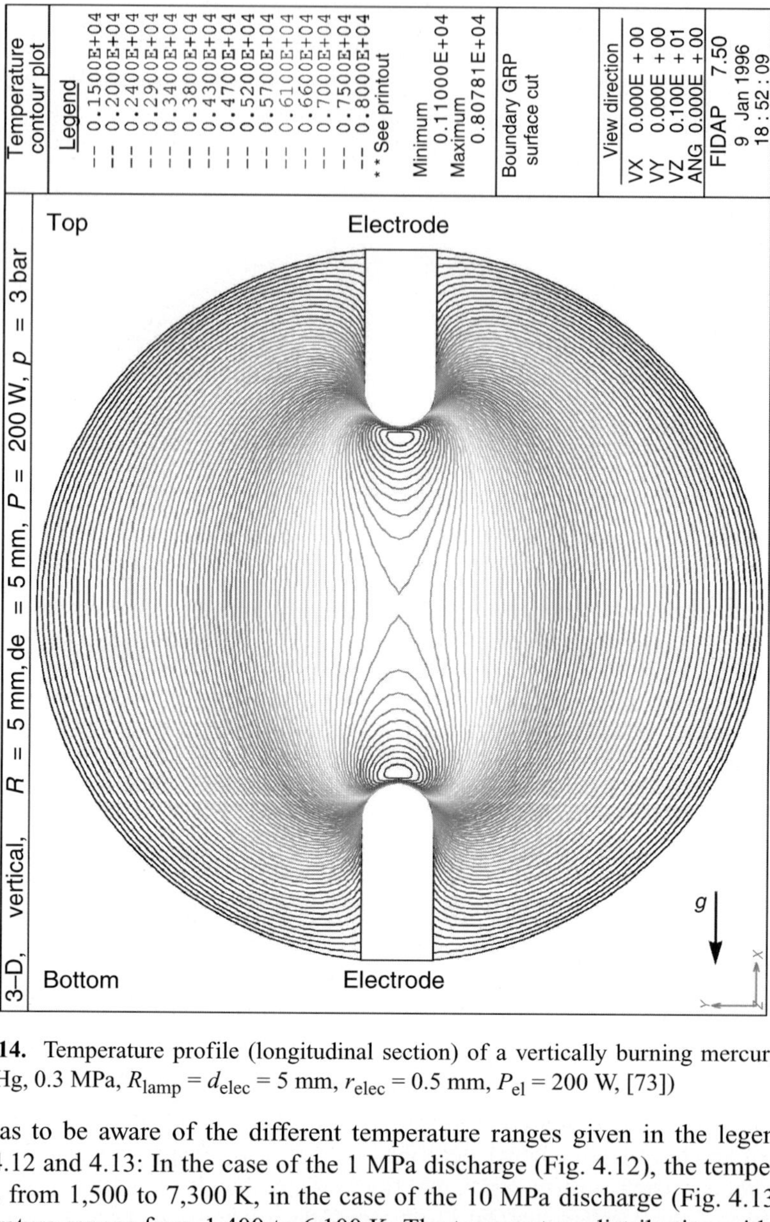

Fig. 4.14. Temperature profile (longitudinal section) of a vertically burning mercury HID lamp (Hg, 0.3 MPa, $R_{\text{lamp}} = d_{\text{elec}} = 5$ mm, $r_{\text{elec}} = 0.5$ mm, $P_{\text{el}} = 200$ W, [73])

One has to be aware of the different temperature ranges given in the legends of Figs. 4.12 and 4.13: In the case of the 1 MPa discharge (Fig. 4.12), the temperature ranges from 1,500 to 7,300 K, in the case of the 10 MPa discharge (Fig. 4.13), the temperature ranges from 1,400 to 6,100 K. The temperature distribution within the hot plasma spots in front of the electrodes is not realistic because there is no difference between anode and cathode (compare with Sect. 4.1.3 and Figs. 4.3–4.5). The same comparison is possible for two vertically burning lamps shown in Figs. 4.14 and 4.15. The vertically operated lamp with 0.3 MPa (Fig. 4.14) shows no influence of the convection, whereas the temperature distribution especially outside the core of the plasma column is altered by convection in the case of the

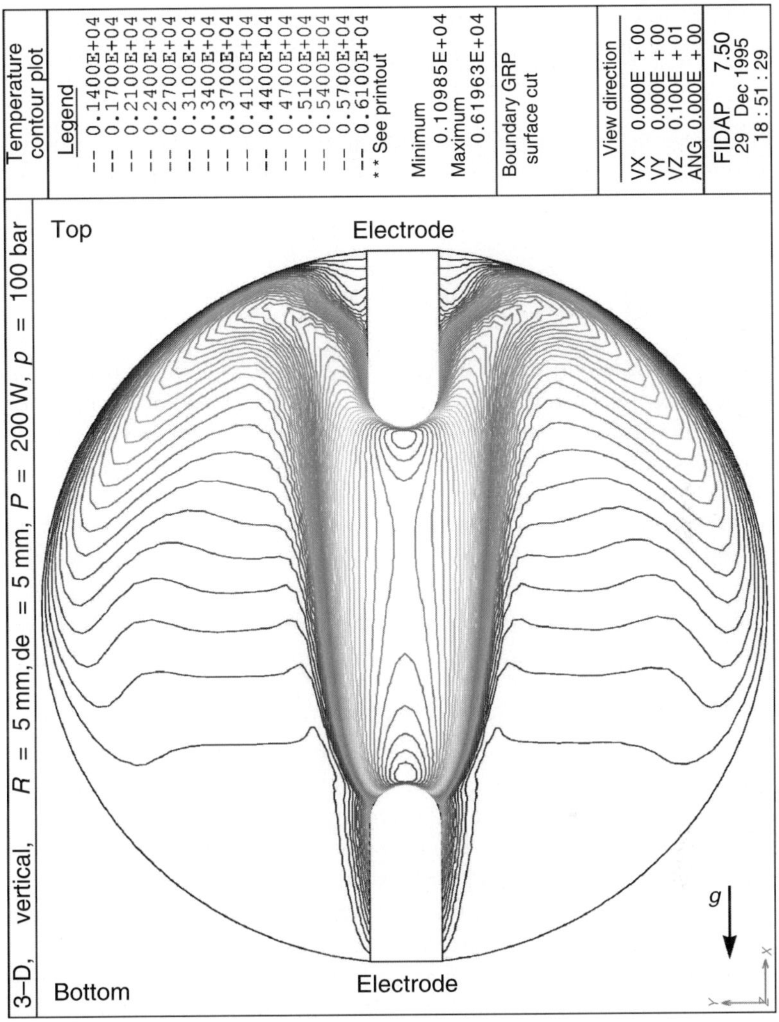

Fig. 4.15. Temperature profile (longitudinal section) of a vertically burning mercury HID lamp (Hg, 10 MPa, $R_{\mathrm{lamp}} = d_{\mathrm{elec}} = 5$ mm, $r_{\mathrm{elec}} = 0.5$ mm, $P_{\mathrm{el}} = 200$ W, [73])

10 MPa mercury HID lamp. Again, the temperature ranges given in the legends of Figs. 4.14 and 4.15 are different: In the case of the 0.3 MPa discharge (Fig. 4.14), the temperature ranges from 1,500 to 8,000 K, in the case of the 10 MPa discharge (Fig. 4.15), the temperature ranges from 1,400 to 6,100 K.

The convection in a lamp depends not only on the pressure, but also on the dimensions of the lamp, the lamp current and the resulting temperature profile, the electrode gap,[19] etc. The convection has an influence on electrode and discharge

---

[19] Due to the small electrode gap of about 1 mm, the arc bowing in an UHP lamp is very small, despite the very high pressure of 20 MPa (cf. Fig. 4.16).

vessel temperature. The latter will be discussed in Sect. 4.3, where further development (inclusion of the discharge vessel) of the model presented here is described. The inclusion of a radiative transfer calculation in the model of Giese [73] is given in [74].

Similar investigations of the convection in dc (vertically) operated mercury HID lamps were carried out by Charrada and Zissis [38, 39], whereas Palacin [134] investigates the convection in vertically operated rare gas HID lamps (also dc operation).

4.3 Extended models

The plasma has a large influence on electrode behavior: An ion current might heat the cathode, thermal heat flux from the hot plasma spots to the relatively cold electrodes is transporting energy from the plasma to the electrodes, the constriction zone of the plasma arc has an influence on the size of the arc attachment area on the electrode tips, and last but not least, plasma constituents might alter the work function for electron emission. On the other hand, electrode properties have an influence on the plasma,[20] e.g., the electrodes withdraw power from the plasma (resulting in an additional power consumption of the near-electrode plasma, leading to the so-called electrode fall voltages), the electrode geometry influences the constriction zone, the hot plasma spot temperature depends on the size of the arc attachment, etc. Thus, the modeling of electrodes should include the plasma and the modeling of the plasma should include the electrodes.

In Sect. 4.4, a model concerned with the coupling of electrodes and plasma will be introduced. It is based on an equation describing the enlargement of the electrical conductivity due to the diffusion of electrons from the hot plasma spots (high electron density) to the relatively cold near electrode plasma (low electron density). The temperature distribution in electrodes and plasma is thus calculated even for arbitrary time-dependent lamp currents. Other electrode–plasma models are given for example by Wendelstorf [167, 168] and Galvez [69, 97]. Both models are based on a plasma-sheath model comparable to that described in Sect. 4.1.1 and both models include convection. Wendelstorf considers vertically operated HID lamps (rotational symmetry), Galvez presents a horizontally operated lamp. Both models are used for time-independent lamp currents.

Besides the interaction between plasma and electrodes, the interplay between plasma and arc tube is important. The plasma heats the arc tube, so that the arc tube can be either too hot (resulting, e.g., in quartz recrystallization or softening of the quartz) or too cold (resulting, e.g., in condensation of mercury and thus a pressure drop or in the condensation of metal halides and thus a change in luminous efficacy and color rendering index). A model considering this interaction between plasma and arc tube has been developed at Philips, based on the model of Giese ([73], Sect.

[20] The influence of the different electrode phases (anode and cathode) on the plasma temperature can be seen in Figs. 4.3–4.5.

4.2.3). The modeling of the plasma includes convection, but there is no difference between anode and cathode. The heat conduction equation in the electrodes is taken into account (cf. (4.1)), but the power input into the electrode tip is not calculated self-consistent but supplied by the user to obtain reasonable electrode temperatures. The model is for time-independent lamp currents only (dc operation).

An example of calculated isotherms in the plasma, the arc tube, and the electrodes of a horizontally operated UHP lamp according to the model described here [88], is shown in Fig. 4.16. Because of the small electrode gap of about 1 mm, the arc bowing due to convection is very small (despite the very high pressure of 20 MPa). The temperature of the upper part of the arc tube should remain below 1,400 K to avoid softening or recrystallization of the quartz, the temperature of the lower part of the arc tube should remain above 1,200 K to avoid condensation of mercury. The mercury vapor in a UHP lamp is unsaturated, thus condensation of mercury at the cold spot of the arc tube would lead to a considerable pressure drop. This would alter the spectral power distribution, the luminous efficacy, and the lamp voltage.

An all-in-one device suitable for every purpose, the "eierlegende Wollmilchsau" as the German say (egg-laying wool-milk-sow, Jack of all trades, or fully integrated everything) has not yet been invented. It would include the modeling of the plasma, the electrodes, the discharge tube, and the electronics, taking the interactions among all these components into account. Arbitrary orientation (convection), time dependence (ac currents), exact radiation transport

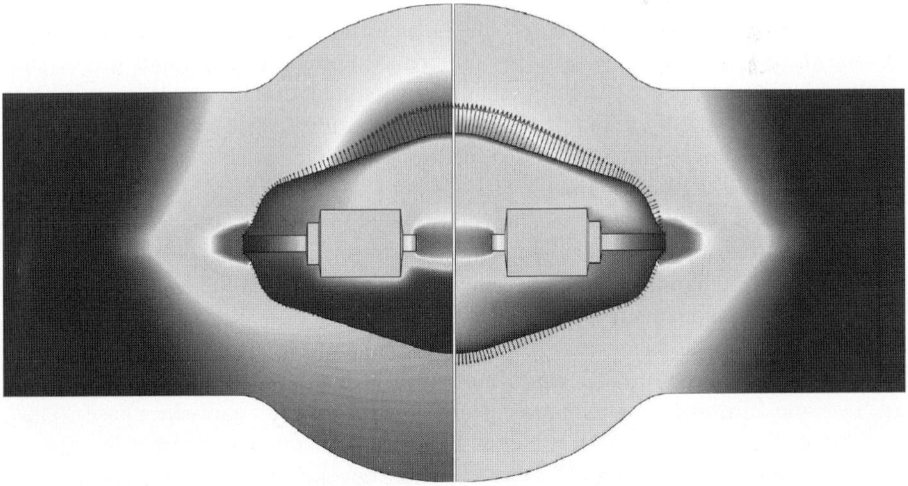

Fig. 4.16. Temperature of plasma, electrodes, and quartz vessel (vertical longitudinal section) of a horizontally burning UHP lamp including convection with (right-hand side) and without (left-hand side) inclusion of radiative transfer in the plasma (Hg, 20 MPa, courtesy of Philips, Aachen, similar image in [46])

calculations, and deviations from LTE in the near-electrode plasma should be part of this all-in-one device as well as the self-consistent calculation of the plasma composition depending on the cold spot temperature. Moreover, a change in electrode geometry during operation should be included and predicted. Up to now, this is a dream of the future. The next step would be an electrode–plasma–arc tube model, taking convection, time-dependence (ac operation), and electrode–plasma, plasma–arc tube, and electrode–arc–tube interaction into account. One of the problems is the computationally intensive nature of the problem, which requires massive computing time, another problem are uncertainties in the knowledge of material properties, e.g., the work function for electron emission. Nevertheless, it remains questionable if all lamp properties can be calculated on the basis of (macroscopic) lamp data like geometry, material, lamp current, etc. Experimental investigations reveal limitations of numerical simulations, for example if macroscopically identical lamps show a different behavior in spite of identical operation conditions (cf. for example Sects. 3.1.7 and 3.1.13).

4.4 A Self-Consistent Electrode–Plasma Model

Even though HID lamps are known for several decades now, the important plasma–electrode interactions are still not well understood. Because HID lamps are usually operated on ac currents (electrodes switch alternately from anode to cathode phase), time-dependent simulations including realistic and verified anode and cathode models are essential. This is the research area of the author and in the following of this section it is reported on the progress in this field of research. Large parts of this sections are published already in [53–65].

Because of the high pressure and the high electrical power input in HID lamps, the plasma is very hot. Typical temperatures in the plasma column are around 4,000–7,000 K, at the hot plasma spots the plasma temperature might be as high as 10,000 K. In contrast to low-pressure discharge lamps with much lower heavy particle temperatures, this results in a strong interaction between plasma and electrodes. An example of this interaction is the heat conduction from the hot plasma spots to the electrode tips (temperature difference of about 4,000–6,000 K over a distance of about 30–250 μm). Furthermore, there is cooling/heating by electrons, heating by ions, and cooling by thermal radiation at the electrode surface. Unfortunately, this important and complex interaction between plasma and electrodes is a very localized and fast process, which is very difficult or impossible to measure in an experimental setup. Moreover, HID lamps are usually operated on ac, i.e., the electrodes switch alternately from anode to cathode phase. Thus, numerical simulations of HID lamps including plasma and electrodes are an important tool to understand, improve, and develop new HID lamps. Time dependence and realistic anode and cathode models are essential for the numerical treatment of ac-operated HID lamps.

The electrodes are not only responsible for delivering the electric current into the plasma, but may also limit the lifetime of the lamp if reignition of the lamp is no longer possible due to electrode failures, or by blackening the inner walls of the discharge by tungsten evaporation. Furthermore, arc stability is strongly influenced by the shape of the electrode tips (e.g., [119]) and the efficacy can drop drastically if electrode fall voltages rise (cf. Sect. 4.4.4). Thus, a detailed knowledge of the time-dependent properties of plasma and electrodes and their mutual interactions is necessary. Important features to be determined are, for example, temperature distributions within plasma and electrodes, heat fluxes to electrodes, electrode losses, and type of arc attachment to the electrodes.

In this section, a model concerned with the coupling of electrodes and plasma will be introduced. The model is able to treat dc- and ac-operated HID lamps with pressures above 0.5 MPa. It is based on an equation describing the enlargement of the electrical conductivity due to the diffusion of electrons from the hot plasma spots (high electron density) to the relatively cold near-electrode plasma (low electron density). This is a different approach compared to the common plasma-sheath models described for example in Sect. 4.1.1. The plasma-sheath models start their description at the electrode surface and the plasma adjacent to the electrode surface, proceeding from there to the so-called Saha-plasma, where the electron and ion density can be calculated using the Saha equation (cf. Sect. 2.2.2). In contrast to these plasma-sheath models, the description of the model described in this section starts in the LTE plasma, where all plasma properties can be calculated using the LTE assumption (cf. Sect. 2.1.8). Due to the diffusion of electrons from the hot plasma spots to the relatively cold near-electrode plasma, the electrical conductivity in the near-electrode plasma is considerably higher than the LTE value of the electrical conductivity.[21]

This approach corresponds to the observations in HID lamps of the pressures discussed here. There is a smooth transition from plasma column to hot plasma spots and constriction zones to the near-electrode plasma, without visible sheaths or layers as in low-pressure discharge lamps. A further advantage of the model is the necessity of only one additional differential equation (for the electrical conductivity σ) to describe the whole discharge area. The absence of different plasma sheaths not only evades the necessity of many equations describing the plasma properties in the different sheaths, but also one gets rid of boundary conditions between these plasma sheaths. The disadvantage is that this model is accompanied by a large numerical effort: It is necessary to have a very fine grid in front of the electrodes to include the progression of the electric field and the electrical conductivity in the near-electrode plasma (cf. Sect. 4.4.2).

The model, which includes plasma column, hot plasma spots in front of the electrodes, constriction zones, and near-electrode non-LTE plasma as well as anode and cathode, delivers, for example, the time-dependent power balances of plasma

---

[21] The LTE electrical conductivity is very low in the near-electrode plasma due to the relatively low plasma temperature (cf. Sect. 2.3.2). In the following, the LTE electrical conductivity will be referred to by σ_{LTE}, the "real" electrical conductivity including the enlargement in the near-electrode plasma due to diffusion processes is referred to by σ.

and electrodes, plasma and electrode temperatures, electric potential and electric field in the plasma, time-dependent total lamp voltage and electrode fall voltages, etc. In Sects. 4.4.3–4.4.8 the model is applied to very different HID lamps. It is compared to experimental results and other numerical models to validate the model and to give a deeper insight into regions, where measurements are not possible. Furthermore, principal studies are presented concerning for example the consequences of different electrode tip geometries or the differences between anode and cathode. In this way, the relevant physical processes concerning plasma, electrodes, and interactions between them can be identified. These results are important for the understanding and the further development of HID lamps which, due to their small dimensions, are often experimentally inaccessible. Thus, modeling gets more and more important.

4.4.1 Model Equations and Boundary Conditions

The model presented here for the numerical simulation of high-pressure discharge lamps was first proposed by Fischer [51] and further improved by the author of this book (cf. e.g., [53, 56, 57]). It uses a complete power balance for plasma and electrodes (see. (4.18) and (4.19)), electric current conservation (4.22), and, as a special feature, an additional differential equation accounting for deviations from local thermal equilibrium conditions (LTE conditions) due to strong diffusion of electrons and ions in close proximity to the electrodes (see. (4.42a) or (4.42b)). The diffusion of electrons from the hot plasma spots close to the electrodes (with high plasma temperatures resulting in high electron densities) is directed into the relatively cold plasma regions immediately in front of the electrodes (with low plasma temperatures resulting in extremely low (LTE) electron densities). This creates a strongly enhanced non-LTE electrical conductivity σ in regions close to both cathode and anode compared to the LTE electrical conductivity σ_{LTE}.

The enlarged non-LTE electrical conductivity σ leads to realistic values of the electric field in near-electrode regions (resulting in realistic anode and cathode fall voltages) and offers the possibility to connect the numerical simulation of the plasma to the electrodes (plasma–electrode interaction). This is important for HID lamps which have often small electrode gaps of some millimeters, leading to a smooth transition from non-LTE near-cathode plasma to hot plasma spot, constriction zone, and plasma column to the anodic plasma with constriction zone, hot plasma spot, and non-LTE near-anode plasma. The model under consideration here includes the entire discharge plasma (plasma column, hot plasma spots in front of electrodes, near-electrode non-LTE plasma) as well as anode and cathode, without dividing the plasma into different layers or sheaths. Thus, a special focus of the model is the investigation of the lamp as a whole and the results cover many details of the lamp. The many possibilities of the model will be demonstrated in Sects. 4.4.3–4.4.8.

Plasma Temperature Distribution

The time-dependent plasma temperature distribution $T_P(\vec{r}, t)$ is determined by the Elenbaas–Heller equation[22]

$$\nabla\cdot(\lambda_P \nabla T_P)+\sigma(\nabla V)^2-u=\rho c_p \frac{\partial T_P}{\partial t}. \tag{4.18}$$

$\lambda_P(p,T_P)$ is the thermal conductivity, $\sigma(\vec{r}, t)$ the non-LTE electrical conductivity (cf. (4.42a) and (4.42b)), $V(\vec{r}, t)$ the electric potential (cf. (4.22)), $u(p,T_P)$ the net radiation emission coefficient, $\rho(p,T_P)$ the mass density, and $c_P(p,T_P)$ the specific heat at constant pressure.[23] $u(p,T_P)$ is calculated taking line radiation and radiation due to recombination and bremsstrahlung into account. However, in plasmas containing mercury, the Hg resonance lines at 185 and 254 nm are not included in $u(p,T_P)$; instead they are treated separately as radiative energy transport, using the approximation of radiation diffusion. For spectral lines which are partially absorbed within the plasma, a radiation transport calculation yields spectral line weakening factors χ (cf. [160] and Sect. 2.2.3): For Hg, χ is 0.4 for the lines at 365.02, 365.48, and 546.08 nm, and 0.5 for the line at 435.8 nm. In plasmas containing xenon, χ is 0.7 for the xenon lines at 820.86, 828.24, 823.39, and 834.91 nm (cf. Sect. 2.2.3, especially Figs. 2.13 to 2.16). The thermal conductivity $\lambda_P(p,T_P)$ consists of three parts: transport of kinetic energy, transport of reaction energy (in general, ionization, recombination, dissociation, and formation of molecules), and radiation diffusion. The latter is important for high temperatures (>4,000–5,000 K) and is treated by radiation transport calculation for the mercury resonance lines at 185 and 254 nm (cf. Sects. 2.2.3 and 2.3.4, especially Figs. 2.25 and 2.26). At the inner walls of the discharge tube, the temperature is set to a constant value depending on the lamp under consideration. The boundary conditions at the electrode surface are given below (cf. (4.20) and (4.21)).

Electrode Temperature Distribution

The computation of the time-dependent electrode temperature distribution $T_E(\vec{r}, t)$ is based on the heat flux equation (cf. Sect. 4.1.1) using the thermal conductivity $\lambda_E(T_E)$ (cf. Sect. 2.4.1), the mass density $\rho_E(T_E)$ [31], and the specific heat $c_E(T_E)$ [31] of the electrode material:

$$\nabla\cdot(\lambda_E \nabla T_E)=\rho_E c_E \frac{\partial T_E}{\partial t}. \tag{4.19}$$

At the outside connections to the electrodes (electrode root), the temperature is set to a constant value depending on the lamp under consideration.

---

[22] See also Sect. 4.2.1 and (4.13). A similar equation is, for example, used in [87].

[23] The specific heat at constant pressure for a 0.6 MPa mercury discharge is shown in Fig. 4.42 for a D2-lamp plasma in Fig. 4.76.

At the boundary between cathode and plasma, T_P and T_E are coupled (cf. Sects. 4.1.1 and 4.1.3). Cooling by electron emission, heating by ion current, thermal heat flux from the plasma toward the cathode, and radiation cooling are considered:

$$\lambda_E \frac{\partial T_E}{\partial n} = \lambda_P \frac{\partial T_P}{\partial n} + j_{t,i}(\varphi_i - \varphi_{cool}) - j_{t,emission}\varphi_{cool} - \varepsilon\sigma_{SB}T_E^4. \tag{4.20}$$

φ_i is the ionization energy of the gas (e.g., 10.434 eV for mercury and 12.13 eV for xenon), φ_{cool} the work function for cathode cooling due to the release of an electron (cf. Sect. 4.4.6), $j_{t,i}$ the particle flux of ions, $j_{t,emission}$ the particle flux of emitted electrons, ε is the emissivity of the electrode material (cf. Sect. 2.4.3), and σ_{SB} the Stefan–Boltzmann constant. The terms $\partial T_E/\partial n$ and $\partial T_P/\partial n$ denote the component of the temperature gradient normal to the electrode surface. The electron emission $j_{t,emission}$ at the cathode is determined by equations given by Dyke and Dolan [47]. These equations include thermal electron emission, field-enhanced thermal electron emission, and field emission (cf. Sect. 2.4.4). The ion current density $j_i = ej_{t,i}$ at the cathode is set to $|j| - |j_e|$, if $|j| > |j_e|$ and zero otherwise (cf. [144, 167, 168]).

At the boundary between plasma and anode, thermal heat flux from the plasma toward the anode and radiation cooling are taken into account. The ion flux is zero at the anode and the electron flux $j_{t,e}$ is heating the anode, where φ_{heat} is the work function for anode heating due to the electron current (cf. Sects. 4.1.3 and 4.4.6):

$$\lambda_E \frac{\partial T_E}{\partial n} = \lambda_P \frac{\partial T_P}{\partial n} + j_{t,e}\varphi_{heat} - \varepsilon\sigma_{SB}T_E^4. \tag{4.21}$$

The electrode properties like thermal conductivity or work function will be discussed in detail in Sect. 4.4.6.

Electric Potential in the Plasma

The electric potential $V(\vec{r}, t)$ is calculated using the current continuity equation (cf. Sect. 4.2.1):

$$\nabla \cdot j = \nabla \cdot (\sigma E) = -\nabla \cdot (\sigma \nabla V) = 0. \tag{4.22}$$

j is the electric current density, $\sigma(\vec{r}, t)$ the non-LTE electrical conductivity (cf. (4.42a) and (4.42b)), and E the electric field.

The explicit dependence of the electric potential $V(\vec{r}, t)$ on time t is neglected, since transient effects in the electric field occur on a much smaller time scale than the thermal capacity effects [30]. The power needed for the generation of the ion current I_i at the cathode (I_i is the integral of $ej_{t,i}$ ($j_{t,i}$: ion particle flux) over the cathode surface, cf. (4.20)) is not included in the power balance (4.18). This power is given by $I_i\varphi_i/e$, where φ_i is the ionization energy in eV. The electric potential at the cathode, V_0, can therefore be determined from the total power balance of the lamp:

$$V_0 = \frac{I_i}{I(t)} \frac{\varphi_i}{e}. \tag{4.23}$$

Thus, $V_0 I(t)$ equals the power needed for the generation of the ion current I_i heating the cathode. This is a common technique also used in a similar manner in plasma–sheath models (cf. Sect. 4.1.1). This leads to an exact balancing of all terms in the power balance (electrical power input, radiation, conductive losses). The voltage V_0 is part of the total cathode fall voltage U_c, which is determined by linear extrapolation of the electric potential distribution towards the cathode.[24] In the original form of the model as proposed by Fischer [51], the electric potential at the cathode was set to zero, neglecting the power needed for the generation of an ion current component at the cathode.

The determination of the electric potential V using (4.22) requires the adjustment of one more free parameter. This is done by adjusting the given time-dependent lamp current $I(t)$ by setting the potential at the anode to the appropriate value of the total lamp voltage. The procedure is the following: First a "normalized" electric potential distribution V_n with constant values $V_n = 0$ at the cathode and $V_n = 1$ at the anode surface is calculated. This normalized potential is then multiplied by a factor κ to get the "real" electric potential V, where κ is chosen in such a manner that the integral of $j = \sigma \cdot E = -\sigma \cdot \text{grad}(V) = -\kappa \cdot \sigma \cdot \text{grad}(V_n)$ over a closed surface between anode and cathode (e.g., midplane, cathode surface, or anode surface) yields the given lamp current $I(t)$ (cf. Sect. 4.2.1).

The anode and the cathode fall voltages U_a and U_c are computed by extrapolation of the linear part of the electric potential distribution on the lamp axis toward anode or cathode (cf., e.g., Figs. 3.40, 4.44, 4.59, and 4.80). At first sight, these electrode fall voltages are determined by the computed electrical conductivity σ (cf. (4.42a) and (4.42b)) and the resulting electric potential computed with (4.22). However, the physics behind this is more complex: The anode and cathode fall voltages correspond to an additional electrical power input into the plasma close to the electrodes ($I \cdot U_a$ and $I \cdot U_c$) compensating additional losses in the near-electrode plasma (cf. Sect. 3.2.2). These additional losses are radiation emitted by the hot plasma spots, heat conduction from the hot plasma spots to the comparatively cold electrodes, and ion current production in the near-cathode plasma (the latter only near the cathode surface). The latter two "losses" elevate the electrode temperature, leading to a higher electrical conductivity (strong temperature dependence) in the near-electrode plasma. This results in lower electrode fall voltages. Thus, we have a self-regulating mechanism, in this case responsible for the proper adjustment of electrode fall voltages and electrode temperatures. The electrode fall voltages are therefore related in the same manner to the power balance (4.18) and (4.19) as to the current continuity equation (4.22) and the electrical conductivity equation (4.42a) or (4.42b). This self-adjustment must be allowed for in a given model (like the model discussed here) without being perturbed by unnecessary restrictions.

---

[24] V_0 represents the part of the cathode fall voltage U_c that is necessary to "produce" a sufficient number of ions near the cathode surface (power consumption $I \cdot V_0$). The remainder of the cathode fall voltage times the lamp current I (i.e., $I \cdot (U_c - V_0)$) balances the additional power consumption due to heat conduction to the cathode and the additional radiation emitted by the hot plasma spot near the cathode.

Electrical Conductivity in the Plasma

Using only (4.18) and (4.19) to determine the plasma temperature distribution $T_P(\vec{r}, t)$ and the electrode temperature distribution $T_E(\vec{r}, t)$ together with (4.22) for the electric potential $V(\vec{r}, t)$, the numerical simulation of high-pressure discharge lamps (including the plasma close to the electrodes) would result in severe problems. In the "cold" plasma immediately in front of the electrodes, the LTE electrical conductivity σ_{LTE} would be extremely low. To transport the lamp current $I(t)$ through this "cold" region, the electric field would have to be very high indeed, leading to unrealistic high electrode fall voltages much above the measured total lamp voltage. To include deviations from the LTE electrical conductivity σ_{LTE} in the plasma very close to the electrodes caused by diffusion of electrons from the hot plasma spots (high temperature and thus high electron density) toward the plasma near cathode and anode tip (low temperature and thus low electron density), an additional differential equation to determine a non-LTE electrical conductivity σ has to be derived (cf. Fig. 3.40). Near the cathode surface, the electric field drives away the electrons from the near-cathode plasma into the direction of the plasma column. Nevertheless, diffusion of electrons from the hot plasma spots and electron emission at the cathode surface increase the electron density and thus the electrical conductivity (cf. Fig. 4.17). Near the anode surface, the electric field "supports" the electron diffusion, but the electrons are absorbed by the anode surface.

We start from fundamental diffusion equations for the mass flux [90]:

$$\sum_{\substack{k=1\\k\neq j}}^{v} \frac{n_j n_k}{n^2 D_{jk}} (V_k - V_j) = d_j - \mathrm{grad}(\ln T_P) \sum_{\substack{k=1\\k\neq j}}^{v} \frac{n_j n_k}{n^2 D_{jk}} \left(\frac{D_k^T}{n_k m_k} - \frac{D_j^T}{n_j m_j} \right) \tag{4.24}$$

and

$$d_j = \mathrm{grad}\left(\frac{n_j}{n}\right) + \left(\frac{n_j}{n} - \frac{n_j m_j}{\rho}\right) \mathrm{grad}(\ln p) - \frac{n_j m_j}{p\rho}\left(\frac{\rho}{m_j} X_j - \sum_{s=1}^{v} n_s X_s\right). \tag{4.25}$$

The different quantities in (4.24) and (4.25) are the following: v number of plasma components, n_j particle density of component j, n overall particle density, ρ overall mass density, D_{jk} binary diffusion coefficients, V_k mean velocity of component k, T_P plasma temperature; D_k^T multicomponent thermal diffusion coefficient, m_k mass of particle k, p total pressure, X_j external forces (e.g., forces due to an electric field in the case of discharge lamps).

Using all numerical results obtained with the final form of the model (4.18)–(4.23) and (4.42a), (4.43), (4.44), and (4.45) for Hg- and Xe-discharges (Hg: 4 and 8 MPa, 0.2–1.0 A; Xe: 1 MPa, 1.5–6.0 A), we can evaluate the importance of the single terms in (4.24) and (4.25). Thus, maximum local and global errors in (4.24) and (4.25) caused by the neglect of single terms can be obtained. The given errors in the following were obtained using this procedure. High pressures, high lamp currents, and small work functions for electron emission result in small errors. The

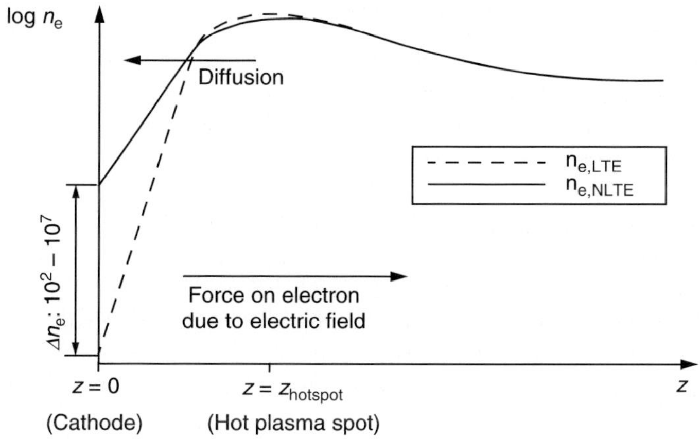

Fig. 4.17. Sketch of the deviation of the electron density (and thus the electrical conductivity) from its LTE value due to diffusion of electrons from the hot plasma spots ($z = z_{hotspot}$) to the near-electrode plasma (here: cathode at $z = 0$) on the lamp axis of a HID lamp

pressure is the most important factor. For pressures above 0.5 MPa the model is applicable, for lower pressures the assumptions might not be fulfilled in a sufficient manner. Nevertheless, due to the importance of other factors, like boundary conditions for the heat flux at anode and cathode surface (4.20) and (4.21) or the work function for electron emission, the results would probably be quite good as well for lower pressures.

Neglecting thermal diffusion (proportional to grad($\ln T_P$) in (4.24) produces a total error of 10% or less. Increasing lamp current and gas pressure decreases the error due to the neglect of thermal diffusion. Because the total pressure p is assumed to be constant, pressure diffusion is absent (proportional to grad($\ln p$) in (4.25)).

If we have the three components electrons (j or k = "e"), ions (j or k = "i"), and neutral atoms (j or k = "n") in the plasma and if we take the electric field E as the external force, we can derive the following two equations from (4.24) and (4.25):

$$\frac{n_e n_i}{n^2 D_{e,i}}(V_i - V_e) + \frac{n_e n_n}{n^2 D_{e,n}}(V_n - V_e) = \mathrm{grad}\left(\frac{n_e}{n}\right) + \frac{n_e eE}{p}\left(1 + \frac{m_e}{\rho}(n_i - n_e)\right) \tag{4.26}$$

and

$$\frac{n_i n_e}{n^2 D_{e,i}}(V_e - V_i) + \frac{n_i n_n}{n^2 D_{i,n}}(V_n - V_i) = \mathrm{grad}\left(\frac{n_i}{n}\right) - \frac{n_i eE}{p}\left(1 - \frac{m_i}{\rho}(n_i - n_e)\right) \tag{4.27}$$

Assuming small ionization degrees less than 1% (weakly ionized plasma, cf. Sect. 2.2.2), using $j_{t,n} = -j_{t,i}$ (particle conservation), $D_{e,i} \ll D_{e,n}$ (maximum local error less than 6%, overall error much smaller), and $n_e \cdot D_{i,n} \ll n_n \cdot D_{e,i}$ (maximum local error less than 0.05%) yields the following two equations:

$$-j_{t,e}\frac{1}{n}\left(\frac{n_i}{nD_{e,i}}+\frac{1}{D_{e,n}}\right)+j_{t,i}\frac{1}{n}\frac{n_e}{nD_{e,i}}=\text{grad}\left(\frac{n_e}{n}\right)+\frac{n_e eE}{p}\left(1+\frac{m_e}{\rho}(n_i-n_e)\right) \quad (4.28)$$

and

$$j_{t,e}\frac{1}{n}\frac{n_i}{nD_{e,i}}-j_{t,i}\frac{1}{n}\frac{1}{D_{i,n}}=\text{grad}\left(\frac{n_i}{n}\right)-\frac{n_i eE}{p}\left(1-\frac{m_i}{\rho}(n_i-n_e)\right). \quad (4.29)$$

$j_{t,e}=n_e{\cdot}V_e$ is the particle flux density of electrons and $j_{t,i}=n_i{\cdot}V_i$ the particle flux density of ions. For the weakly ionized plasmas considered here, further simplifications can be made, yielding the following set of equations:

$$-j_{t,e}\frac{1}{n}\left(\frac{n_i}{nD_{e,i}}+\frac{1}{D_{e,n}}\right)+j_{t,i}\frac{1}{n}\frac{n_e}{nD_{e,i}}=\text{grad}\left(\frac{n_e}{n}\right)+\left(\frac{n_e eE}{p}\right) \quad (4.30)$$

and

$$j_{t,e}\frac{1}{n}\frac{n_i}{nD_{e,i}}-j_{t,i}\frac{1}{n}\frac{1}{D_{i,n}}=\text{grad}\left(\frac{n_i}{n}\right)-\frac{n_i eE}{p}. \quad (4.31)$$

These two equations differ from those given by Fischer [51]. Combining (4.30) and (4.34), the electric field can be eliminated, yielding:

$$j_{t,e}\frac{1}{n}\left(\frac{n_e-n_i}{nD_{e,i}}-\frac{1}{D_{e,n}}\right)+j_{t,i}\frac{1}{n}\left(\frac{n_e}{nD_{e,i}}-\frac{n_e}{n_i D_{i,n}}\right)=\text{grad}\left(\frac{n_e}{n}\right)+\frac{n_e}{n_i}\text{grad}\left(\frac{n_i}{n}\right). \quad (4.32)$$

Within the two pairs of parenthesis on the left-hand side of (4.32), the first terms are small compared to the second terms, yielding

$$-j_{t,e}\frac{1}{nD_{e,n}}-j_{t,i}\frac{1}{nD_{i,n}}\frac{n_e}{n_i}=\text{grad}\left(\frac{n_e}{n}\right)+\frac{n_e}{n_i}\text{grad}\left(\frac{n_i}{n}\right). \quad (4.33)$$

The right-hand side of (4.33) can be transformed using

$$\text{grad}\left(\frac{n_e}{n}\right)=\text{grad}\left(\frac{n_e}{n_i}\frac{n_i}{n}\right)=\frac{n_e}{n_i}\text{grad}\left(\frac{n_i}{n}\right)+\frac{n_i}{n}\text{grad}\left(\frac{n_e}{n_i}\right) \quad (4.34)$$

or

$$\text{grad}\left(\frac{n_e}{n}\right)+\frac{n_e}{n_i}\text{grad}\left(\frac{n_i}{n}\right)=2\,\text{grad}\left(\frac{n_e}{n}\right)-\frac{n_i}{n}\text{grad}\left(\frac{n_e}{n_i}\right). \quad (4.35)$$

Because $(n_i/n){\cdot}\text{grad}(n_e/n_i) \ll \text{grad}(n_e/n)$ (total error less than 1%) we can combine (4.33) and (4.35) to give:

$$j_{t,e}=-j_{t,i}\frac{n_e D_{e,n}}{n_i D_{i,n}}-2nD_{e,n}\,\text{grad}\left(\frac{n_e}{n}\right). \quad (4.36)$$

Using the equation

$$\operatorname{div} j = e \operatorname{div}(j_{t,i} - j_{t,e}) = 0 \tag{4.37}$$

and the approximation $n_i \cdot D_{i,n} \ll n_e \cdot D_{e,n}$ (overall error less than 1%) we get

$$\operatorname{div} \operatorname{grad}\left(\frac{n_e}{n}\right) = -\frac{1}{2nD_{e,n}} \frac{n_e D_{e,n}}{n_i D_{i,n}} \operatorname{div} j_{t,i}. \tag{4.38}$$

The next step is to transform the expression $\operatorname{div} j_{t,i}$ in an appropriate manner. In, e.g., [51, 53, 55–65, 129, 158, 169], the local production and annihilation terms of ions taken into account are the following:

$$\operatorname{div} j_{t,i} = R^+ - R^- = q^I n_e n_n - q^R n_e n_i - q^C n_e n_i n_n, \tag{4.39a}$$

where q^I, q^R, and q^C are (temperature dependent) rate coefficients for ionization, radiative recombination, and three-body collisional recombination. This means, it is assumed that the collisional ionization ($q^I \cdot n_e \cdot n_n$) predominates the production of ions (and electrons) and the loss of charged particles is mainly by radiative recombination ($q^R \cdot n_e \cdot n_i$) and by three-body recombination with a neutral atom being the third particle ($q^C \cdot n_e \cdot n_i \cdot n_n$). A detailed balance, i.e., each process in equilibrium with its inverse process, is not assumed. The inverse process of radiative recombination ($q^R \cdot n_e \cdot n_i$) would be the ionization of a neutral by a photon (emitted, e.g., during a previous recombination). In total thermal equilibrium this detailed balance is realized, resulting in the well-known Planck radiation formula (black body radiation, cf. Sect. 2.4.3). In a HID lamp, however, the radiation is not in equilibrium, leading to the characteristic spectral power distribution different for each type of lamp (cf. Sect. 1.3 and 2.2.3). The radiation coming from radiative recombination will most probable leave the discharge without ionizing a neutral particle[25] leading to the recombination continuum (cf. Sect. 2.2.3). Nevertheless, the high pressure of the plasma in HID lamps leads to a collision-dominated plasma where radiation processes do not disturb the local thermal equilibrium noticeably, which is established by the countless collisions between the plasma constituents. In a similar manner, the three-body recombination with an electron as the third particle (inverse reaction of collisional ionization ($q^I \cdot n_e \cdot n_n$)) is assumed to be rather unlikely in the weakly ionized plasma of the HID lamps under consideration here (ionization degree less than 1%, cf. Sect. 2.2.2). The three-body recombination with a neutral as the third particle ($q^C \cdot n_e \cdot n_i \cdot n_n$) is assumed to be much more probable. This reasonable procedure leading to (4.39a) yields finally (4.42a) and very good results

[25] A more serious problem for this radiation will be the passage of the tube walls. Furthermore, this high energy radiation might heat the electrodes or release an electron from the cathode. However, the latter effect should be sufficiently small with photoelectric yield values of the order of 10^{-4} according to [146].

compared to experimental findings (cf. Sects. 4.4.3–4.4.8). Nevertheless, there is a widespread alternative emphasizing the detailed balance notion (e.g., page 71 in [168], page 4228 in [17], or page 258 in [16]):

$$\operatorname{div} j_{\mathrm{t,i}} = R^{+} - R^{-} = q^{\mathrm{I}} n_{\mathrm{e}} n_{\mathrm{n}} - q^{\mathrm{I'}} n_{\mathrm{e}}^{2} n_{\mathrm{i}}. \tag{4.39b}$$

In (4.39b), one reaction and its inverse reaction are considered, but the remaining two reactions and their corresponding inverse reactions are neglected ($q^{\mathrm{R}} \cdot n_{\mathrm{e}} \cdot n_{\mathrm{i}}$, $q^{\mathrm{C}} \cdot n_{\mathrm{e}} \cdot n_{\mathrm{i}} \cdot n_{\mathrm{n}}$, and corresponding inverse reactions). Without judging the quality of these two approaches (4.39a) and (4.39b), we will continue the derivation of the necessary differential equation for a non-LTE electrical conductivity, taking both approaches parallel into account.[26] We will use the resulting equations for the non-LTE electrical conductivity (4.42b) for the numerical simulation of an external laser heating experiment (Sect. 4.4.5).

As described in Sect. 4.1 and as it will be shown in Sects. 4.4.3–4.4.8, there should not be a large difference between the results obtained by using these two notions because the important fundamental interactions between plasma and electrodes are included in the boundary conditions (e.g., (4.20), (4.21), (4.43), (4.44), and (4.45)). The self-regulation of the cathode–plasma interaction and the power balances of electrodes and plasma are more crucial than certain differences in the description of the near-electrode plasma. Moreover, the plasma and electrode properties are more important for the electrode–plasma interaction than the details of the electron and ion production and annihilation terms, for example, the very important work function for electron emission $\varphi_{\mathrm{emission}}$ (cf. Sect. 4.4.6) or the thermal conductivity λ_{P} affecting the thermal heat flux from the hot plasma spots to the electrodes.

Assuming $n_{\mathrm{n,LTE}} = n_{\mathrm{n}}$ (weakly ionized plasma, see earlier), we can express all production and annihilation terms with the help of q^{I} in LTE using $\operatorname{div} j_{\mathrm{t,i}} = 0$:

$$q^{\mathrm{R}} + q^{\mathrm{C}} n_{\mathrm{n}} = q^{\mathrm{I}} \frac{n_{\mathrm{n}}}{n_{\mathrm{i,LTE}}} \tag{4.40a}$$

and correspondingly

$$q^{\mathrm{I'}} = q^{\mathrm{I}} \frac{n_{\mathrm{n}}}{n_{\mathrm{i,LTE}}^{2}} \tag{4.40b}$$

Combining (4.38), (4.39a), and (4.40a) and (4.38), (4.39b), and (4.40b), respectively, we get

$$\operatorname{div} \operatorname{grad}\left(\frac{n_{\mathrm{e}}}{n}\right) = -\frac{q^{\mathrm{I}}}{2D_{\mathrm{i,n}}} \frac{n_{\mathrm{e}}^{2}}{n_{\mathrm{i}}} \left(1 - \frac{n_{\mathrm{i}}}{n_{\mathrm{i,LTE}}}\right) \tag{4.41a}$$

[26] All subsequent equations will be labeled additionally with an *a* or *b* according to the corresponding approach of (4.39a) or (4.39b).

and

$$\operatorname{div}\operatorname{grad}\left(\frac{n_e}{n}\right) = -\frac{q^I}{2D_{i,n}}\frac{n_e^2}{n_i}\left(1-\frac{n_e n_i}{n_{i,LTE}^2}\right). \tag{4.41b}$$

Assuming n_e/n proportional to the electrical conductivity σ (cf. Figs. 4.18–4.20, this assumption is fulfilled, for example, for mercury, xenon, and complex plasma compositions like the one in the D2 lamp[27]), setting $\alpha \equiv n_i/n_e$, and using the relations $n_e/n_{e,LTE} = \sigma/\sigma_{LTE}$ and $n_{i,LTE} = n_{e,LTE}$ we finally arrive at (4.42a) and (4.42b):

$$\Delta\sigma = -\frac{q^I n}{2D_{i,n}\alpha}\left(1-\alpha\frac{\sigma}{\sigma_{LTE}}\right)\sigma \tag{4.42a}$$

and

$$\Delta\sigma = -\frac{q^I n}{2D_{i,n}\alpha}\left(1-\alpha\left(\frac{\sigma}{\sigma_{LTE}}\right)^2\right)\sigma, \tag{4.42b}$$

where Δ is the Laplace operator, q^I the rate coefficient for ionization, n the overall particle density, $D_{i,n}$ the binary diffusion coefficient between ions and atoms, and α the ratio n_i/n_e. Fischer [51] arrives at the same differential equation (4.42a), despite his derivation is different from the one given here. However, the boundary conditions at anode and cathode surface for the non-LTE electrical conductivity σ derived using (4.30) are significantly different from those given by Fischer (see later). The non-LTE electrical conductivity σ replaces σ_{LTE} in (4.18) and (4.22) and provides realistic results for the electric field, electric potential, plasma temperature, and electrode temperature. Using only this one additional differential equation (which is valid in the whole plasma, LTE, and non-LTE regions) closes the gap between LTE plasma and electrode surfaces without the "low-pressure approach" of dividing the plasma into different layers or sheaths. In the LTE regions (i.e., plasma column and a good deal of the hot plasma spots) (4.42a) or (4.42b) computes σ equal to σ_{LTE}, only in the small areas of cold plasma immediately in front of the electrodes, there are significant deviations between σ and σ_{LTE} (cf. Figs. 4.65 and 4.66). The ratio $\alpha = n_i/n_e$ is calculated using Poisson's equation ((4.10)). It turns out that α has little or no effect on (4.42a) and (4.42b) since $\alpha = 1$ in the entire bulk of the plasma. Only in the plasma immediately in front of the electrodes, deviations from $\alpha = 1$ occur, but within this regions $\sigma \gg \sigma_{LTE}$, so that the two α in (4.42a) and (4.42b) cancel each other.

[27] The constant of proportionality for a 0.6 MPa mercury plasma is identical to the constant of proportionality for a 4 MPa mercury plasma, i.e., this constant appears to be valid for a wide range of pressures.

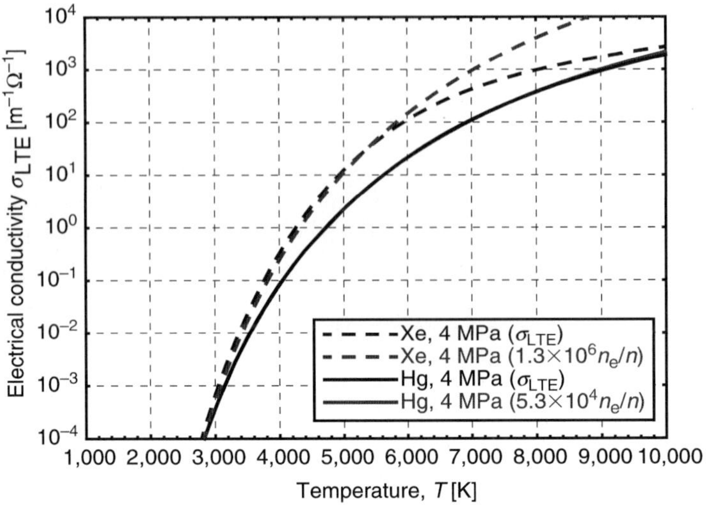

Fig. 4.18. LTE electrical conductivity σ_{LTE} compared to the ratio $n_{\mathrm{e,LTE}}/n$ for a 4 MPa xenon or mercury plasma vs. plasma temperature

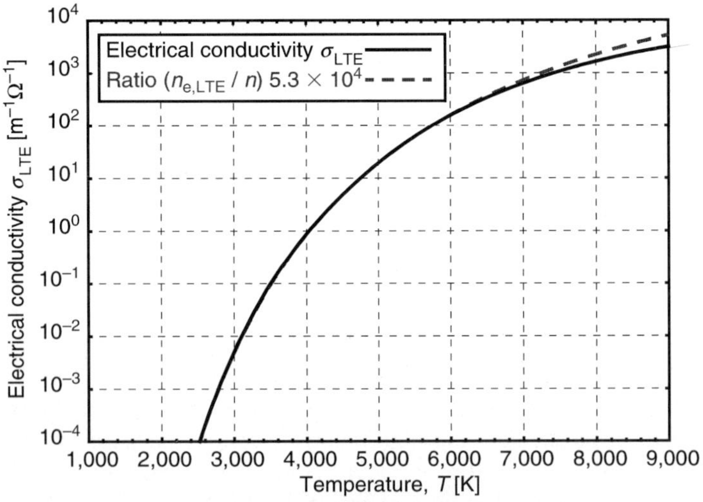

Fig. 4.19. LTE electrical conductivity σ_{LTE} compared to the ratio $(n_{\mathrm{e,LTE}}/n)\times 5.3\times 10^4\ \mathrm{m}^{-1}\,\Omega^{-1}$ for a 0.6 MPa mercury plasma vs. plasma temperature [63]

The only difference between (4.42a) and (4.42b) is the square of the $\sigma/\sigma_{\mathrm{LTE}}$ term.[28] The consequences following from this difference have been analyzed by

[28] Taking into account all terms of the detailed balance, we get the same type of equation as (4.42b), but with a $q^{\mathrm{I}}_{\mathrm{eff}}$ depending on n_{e}, q^{I}, q^{R}, and q^{C} [91]

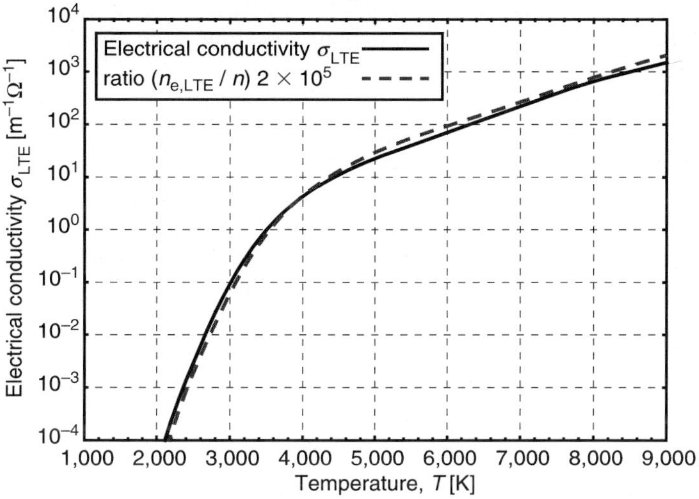

Fig. 4.20. LTE electrical conductivity σ_{LTE} compared to the ratio $n_{e,LTE}/n{\cdot}2{\times}10^5\ m^{-1}\Omega^{-1}$ for a D2 lamp plasma (Xe, Hg, NaI, ScI_3) vs. plasma temperature [59]

modeling the external laser heating of anode and cathode as described in Sect. 4.4.5 with both equations (4.42a) and (4.42b) independently. The differences between the results are very small. For the lamp described in Sect. 4.4.5, the cathode fall voltage U_c is 11.2 V (4.42a) and 11.4 V (4.42b), respectively, the anode fall voltage U_a is 3.3 V (4.42a) and 3.6 V (4.42b), respectively, the electrode and plasma temperatures deviate less than 3% from each other. This confirms the findings of other investigations (Sects. 4.4.3–4.4.8), showing that the details of the numerical model do not play a decisive role as long as the fundamental interactions between plasma and electrodes are properly included in the boundary conditions (in the model described here for example (4.20), (4.21), (4.43)–(4.45) and as long as the numerical model does not interfere with these boundary conditions. At the cathode, for example, the interplay between ion current and electron emission current "adjusts" cathode fall voltage, cathode temperature, and the ratio of electron to ion current: A cold cathode does not emit enough electrons to sustain the given lamp current I, with the result that ion current provides the difference between lamp current and electron emission current. The generation of this ion current requires electrical power input into the near-cathode plasma, which is reflected in a large cathode fall voltage. The ions deposit their ionization energy at the cathode surface, heating up the cathode. This boosts the electron emission current, which entails a reduced ion current. Thus, the cathode temperature is adjusted to a value, where the ratio of ion to electron emission current corresponds to a stable power balance of plasma and cathode.

A further example of the self-regulation or self-adjustment of plasma and electrode properties in HID lamps is the temperature of the arc column. A "cold" plasma arc implies low electrical conductivity (cf. Figs. 4.18–4.20), leading to high electric fields and, therefore, a high electrical power (heating) density (see (4.18)). This increases the temperature of the arc, leading to much higher electrical conductivities. Due to the strong dependence of the electrical (LTE) conductivity on the plasma temperature, this mechanism smoothes out small errors and uncertainties in the total value of σ_{LTE}.

The details of the numerical model under consideration in the near-electrode plasma have not (and should not have) much impact on these processes. However, without consideration of (for example) heating of the electrodes by thermal heat conduction from the plasma to the electrodes in the boundary conditions at the electrode surfaces (see (4.20) and (4.21)), problems may arise.[29] But even then, the interplay between ion current and electron emission "will do the job." The ion current will be higher, compensating the missing power due to heat conduction from the plasma to the cathode. This leads to a smaller electron emission current and a somewhat lower cathode temperature. However, since electron emission current depends strongly on electrode temperature, these differences are normally not large (cf., Sect. 2.4.4). An important factor is the proper choice of the work function for electron emission, $\varphi_{\text{emission}}$ (strongly affecting electron emission), which has more influence on the numerical results than details of the numerical model under consideration (cf., e.g., Sects. 4.4.3 and 4.4.6). Further influencing factors are the exact geometry of the electrode tips and the ability of the model to deal with more than one dimension: In Sect. 4.4.4 it will be shown in detail how different electrode tip geometries affect the electrode and plasma temperatures as well as the electrode fall voltages. A one-dimensional numerical model is not able to account for these effects due to different electrode tip geometries (independent of the physical details of the model). More examples of the self-regulating mechanisms in HID lamps will be discussed in the following sections.

The results presented in Sects. 4.4.3–4.4.8 are obtained using (4.42a) in combination with (4.18) and (4.19) (plasma temperature distribution $T_P(\vec{r}, t)$ and electrode temperature distribution $T_E(\vec{r}, t)$) and (4.22) (electric potential distribution $V(\vec{r}, t)$). As stated earlier, the differences to the results obtained with (4.42b) are negligible. Using only this additional equation ((4.42a) or (4.42b)) closes the gap between LTE plasma and electrode surfaces without the "low-pressure approach" of dividing the plasma into different layers or sheaths. Realistic results for the electric field, the electric potential, and the plasma temperature in the whole discharge region (LTE and non-LTE regions) and the electrode temperature distributions are thus provided by the numerical simulations.

Equation (4.42a) or (4.42b) is the central point of the model used within this and the following sections; it is the major difference compared to other (plasma or electrode) models. Nevertheless, this equation should not be overestimated, as described earlier and in the following sections. The advantage of the model used

---

[29] In the lamp described in Sect. 4.4.5 used for the comparison of (4.42a) and (4.42b), the thermal heat flow from the plasma to one electrode is between 3.7 and 11.1 W, compared to up to 4 W ion heating during the cathode phase, 8.2 W electron heating during the anode phase, radiation cooling between 7.9 and 13.7 W, and external laser heating up to 4.5 W.

within this and the following sections is that plasma and electrodes are regarded as an integrated whole, allowing for interactions between them. The self-regulating mechanisms are not "concealed" by too many assumptions but can do their "job" to give realistic results for the complete lamp system of electrodes and plasma. The given boundary conditions are reduced to a minimum, like temperature at the inside wall of the discharge vessel (not very important for the results) or temperature at the electrode roots. Geometrical effects induced, for example, by different electrode tip geometries are allowed for, the plasma itself establishes its way from cathode tip over constriction zone and hot plasma spot to the arc column and further on to the anode with the related constriction zone and hot plasma spot. Still, there are many general uncertainties in plasma properties (λ_P, u, σ_{LTE}, ...), work function for electron emission $\varphi_{emission}$, etc. which will be addressed in the following sections. The results presented within the following sections encourage further studies with the model leading to an improved understanding of complex HID lamps.

At the inside wall of the lamp bulb, σ is set to its LTE value at wall temperature. At the cathode surface we derive the following boundary condition for σ: Starting from (4.30), using $p = nkT$ (cf. Sect. 2.1.5), $n_e/n \propto \sigma$, $n_e/n_{e,LTE} = \sigma/\sigma_{LTE}$, and $j_{t,i} \ll neED_{e,i}/kT$, we get

$$\sigma_{cathode} = \frac{kT}{eE}\left[-j_{t,e}\frac{\sigma_{LTE}}{n_{e,LTE}}\left(\frac{n_i}{nD_{e,i}} + \frac{1}{D_{e,n}} \right) - \operatorname{grad}\sigma \right]. \tag{4.43}$$

This equation is consistent with the diffusion equations, but different from that used by Fischer [51]. Wiesmann [169] uses the same model but a much simpler boundary condition (cf. (4.44)), which converges faster numerically, but which is not strictly compatible with the fundamental diffusion equations from which (4.42a) and (4.42b) are derived. Nevertheless, the numerical results obtained with boundary condition (4.44) are not very different from those obtained with boundary condition (4.43), thus (4.44) will be used for the investigations presented in the following sections. This simpler boundary condition is similar to cathode boundary conditions given by Benilov et al. (e.g., [21]) or Lowke et al. (e.g., [120] or [144]). Starting point is the assumption that the electrons emitted by the cathode have a thermal temperature distribution according to the cathode temperature, i.e., $v_{th} = (8k_B T_{cathode}/\pi m_e)^{1/2}$ (see also [151]). If we denote the electron emission current density at the cathode with $j_{e,cathode}$ and use $j_{e,cathode} = ev_{th}n_{e,cathode}/4$, we get

$$\sigma_{cathode} = n_{e,cathode}\frac{\sigma_{LTE}}{n_{e,LTE}} = \frac{4j_{e,cathode}\sigma_{LTE}}{ev_{th}n_{e,LTE}}. \tag{4.44}$$

Starting again from (4.30), using $j_{t,i} = 0$ and $j_{t,e} = -\sigma E/e$ at the anode surface, $n_e/n \propto \sigma$, and $n_e/n_{e,LTE} = \sigma/\sigma_{LTE}$, we derive the following boundary condition for σ at the anode[30]:

[30] The factor 16.49 can be obtained using expressions for $D_{e,n}$ and σ_{LTE} from simplified kinetic theory [53].

$$\sigma_{\mathrm{anode}} = -\frac{16.49kT}{e}\frac{\partial\sigma/\partial n}{\partial V/\partial n}. \tag{4.45}$$

Again, this boundary condition is also different from that used by Fischer [51].

A summary of the self-consistent electrode–plasma model including all differential equations and boundary conditions is given in Fig. 4.21. The numerical procedure to solve the model equations and boundary conditions is described in Sect. 4.4.2.

Rate Coefficient for Ionization

One of the plasma properties used for the self-consistent electrode–plasma model described earlier is the rate coefficient for ionization. It is needed for the calculation of the non-LTE electrical conductivity σ (4.42a) or (4.42b). If we define the "adjustment length" λ_σ as

$$\lambda_\sigma \equiv \sqrt{\frac{2D_{\mathrm{i,n}}}{q^{\mathrm{I}}n}} \tag{4.46}$$

we can rewrite the non-LTE electrical conductivity equation (4.42a) to give

$$\Delta\sigma = -\frac{1}{\lambda_\sigma^2\alpha}\left(1-\alpha\frac{\sigma}{\sigma_{\mathrm{LTE}}}\right)\sigma \tag{4.47}$$

Thus, the adjustment length λ_σ is a measure for the extend of the non-LTE region in the near-electrode plasma. The larger λ_σ, the larger the region where $\sigma \gg \sigma_{\mathrm{LTE}}$ (cf. Figs. 4.65 and 4.66).

The adjustment length λ_σ is proportional to the square root of the binary diffusion coefficient of atoms and ions $D_{\mathrm{i,n}}$, i.e., the larger the diffusion, the larger the region where $\sigma \neq \sigma_{\mathrm{LTE}}$, which complies with our expectations and the model assumptions.[31] On the other hand, the larger the rate coefficient for ionization q^{I} and the larger the total particle density n (i.e., the larger the probability for ionization or recombination), the smaller the adjustment length λ_σ and thus the region where $\sigma \neq \sigma_{\mathrm{LTE}}$. This also complies with our expectations and the model assumptions because ionizations or recombinations counteract the diffusion.

All plasma and electrode properties necessary for the self-consistent electrode–plasma model have been described in Chap. 2, except for the rate coefficient for ionization q^{I}. The rate coefficient for ionization q^{I} is defined as the number of ionizations per unit volume and time for a given electron and neutral particle density: $R^+ = q^{\mathrm{I}}n_{\mathrm{e}}n_{\mathrm{n}}$ (cf. (4.39a) or (4.39b)). To determine concrete values for q^{I} for different plasmas, two possible ways have been used, which are described in the following.

One way is to use the probability P_{i} for an ionization per cm at the reference temperature $T_0 = 0°\mathrm{C}$ and the reference pressure of 1 mmHg by electrons as a

[31] As it can be seen in (4.38), λ_σ is actually proportional to $(D_{\mathrm{i,n}}/D_{\mathrm{e,n}})D_{\mathrm{e,n}} = D_{\mathrm{i,n}}$.

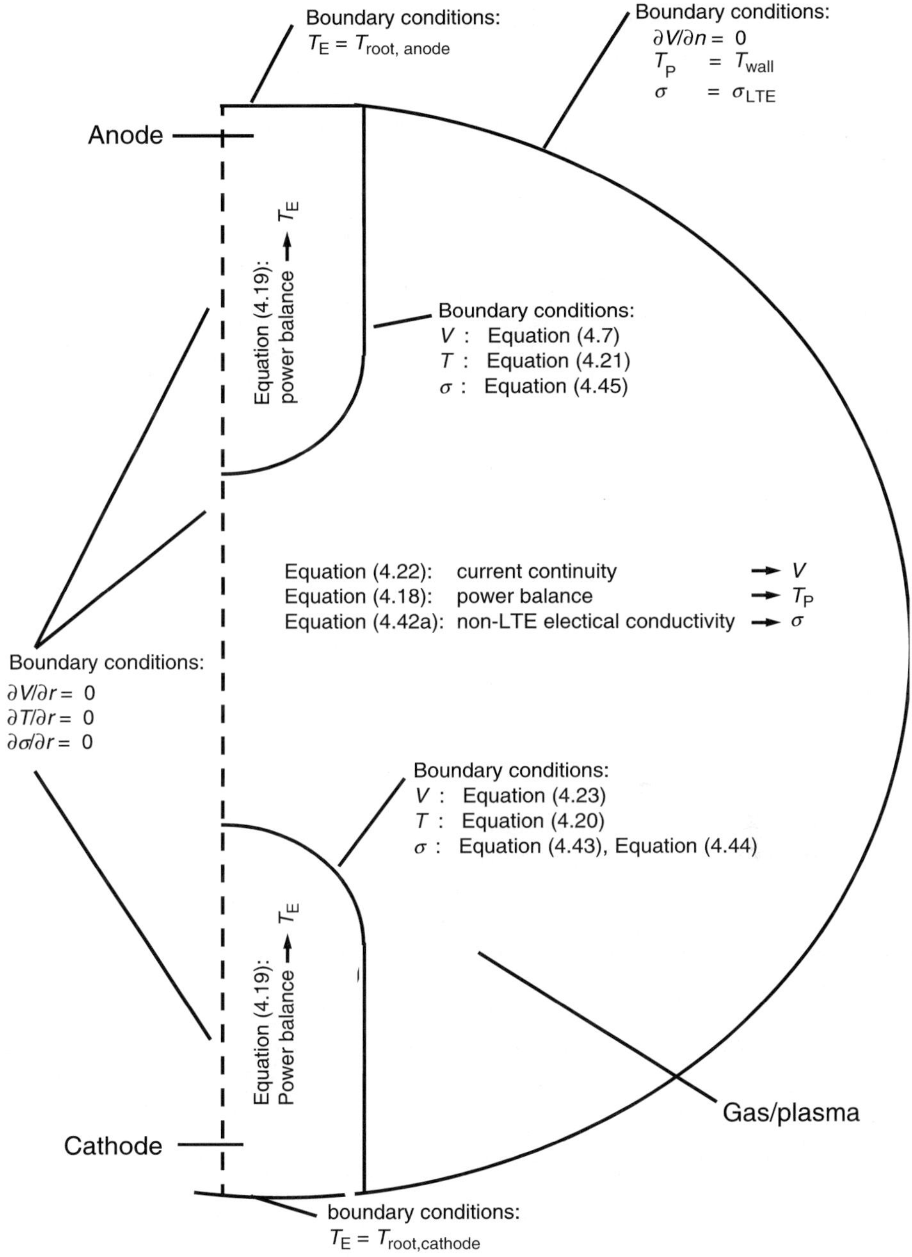

Fig. 4.21. Summary of the self-consistent electrode–plasma model including all differential equations and boundary conditions. The electrodes might change their "function" (anode/cathode) with time in the case of ac operation

function of electron energy u_e (cf. [24, 35]). Using $P_i(u_e)$, the ionization frequency ν_i per electron of the energy u_e at the pressure p can be calculated [35]:

$$\nu_i = pP_i(u_e)v_e(u_e) = pP_i(u_e)\sqrt{\frac{2u_e}{m_e}}. \tag{4.48}$$

Assuming a Maxwellian energy distribution $f(u_e,T)$ for the electrons as described in Sect. 2.1.3, the mean ionization frequency $\bar{\nu}_i$ per electron at temperature T is given by

$$\bar{\nu}_i = nkT_0 \int_{u_{ion}}^{\infty} v_e(u_e)P_i(u_e)f(u_e,T)\,du_e, \tag{4.49}$$

where u_{ion} is the ionization energy of the atoms in the gas/plasma. To convert the units given in [35] to SI units, we have to multiply P_i with the factor $0.75025 \cdot (\mathrm{cm} \cdot \mathrm{mmHg})/(\mathrm{m} \cdot \mathrm{Pa})$. Because we have the relation (cf., e.g., [49])

$$R^+ = \bar{\nu}_i n_e = q^I n_e n_n, \tag{4.50}$$

we can calculate q^I using the approximation $n_n = n$ (weakly ionized plasma):

$$q^I = 0.75025 \frac{\mathrm{cm} \cdot \mathrm{mmHg}}{\mathrm{m} \cdot \mathrm{Pa}} kT_0 \int_{u_{ion}}^{\infty} v_e(u_e)P_i(u_e)f(u_e,T)\,\mathrm{d}u_e, \tag{4.51}$$

The measured ionization probabilities P_i as a function of electron energy for mercury are taken from [131], for argon the values of P_i are taken from [24].

An alternative way to determine concrete values for q^I for different plasmas is to use the cross section for ionization Σ_i as a function of electron energy u_e. According to Sect. 2.1.6, the mean free path for electrons between two ionizations in a gas/plasma is given by

$$\ell_e = \frac{1}{n\Sigma_i(u_e)}, \tag{4.52}$$

so that the ionization frequency per electron of the energy u_e is given by

$$\nu_i = \frac{v_e(u_e)}{\ell_e(u_e)} = \sqrt{\frac{2u_e}{m_e}}\, n\Sigma_i(u_e). \tag{4.53}$$

According to the procedure described earlier, the rate coefficient for ionization q^I is given by:

$$q^{\mathrm{I}} = \int_{u_{\mathrm{ion}}}^{\infty} v_{\mathrm{e}}(u_{\mathrm{e}}) \Sigma_{\mathrm{i}}(u_{\mathrm{e}}) f(u_{\mathrm{e}}, T) \, \mathrm{d}u_{\mathrm{e}}. \tag{4.54}$$

The measured cross sections for ionization Σ_{i} as a function of electron energy for xenon and argon are taken from [106].

Figure 4.22 shows the calculated rate coefficient for ionization q^{I} for mercury, xenon, and argon plasmas vs. plasma temperature. They are independent of pressure p. The higher the plasma temperature, the higher the average kinetic energy of the electrons, and thus the higher the rate coefficient for ionization. The lower the ionization energy of the atoms, the easier it is for the electrons to ionize the atoms. Thus, the rate coefficient for ionization is the highest in the case of mercury ($u_{\mathrm{ion}} = 10.434$ eV), the next one is xenon ($u_{\mathrm{ion}} = 12.13$ eV), then argon ($u_{\mathrm{ion}} = 15.76$ eV). The q^{I} values for argon have been calculated using both methods described earlier with different experimental data. They agree very well. In the case of a complex plasma composition as for example in the D2 lamp (Fig. 4.23), the rate coefficient for ionization is defined as the weighted average value[32] of partial ionization rate coefficients $q^{\mathrm{I}}_{\mathrm{Xe}}$, $q^{\mathrm{I}}_{\mathrm{Hg}}$, $q^{\mathrm{I}}_{\mathrm{Na}}$, and $q^{\mathrm{I}}_{\mathrm{Sc}}$ of most frequently occurring neutral atoms (cf. [129, 158]).

Because the non-LTE region is small for large values of q^{I}, and because a small non-LTE region complies well with the model assumptions, mercury or metal halide HID lamps are ideal objects for the model under consideration in this and the

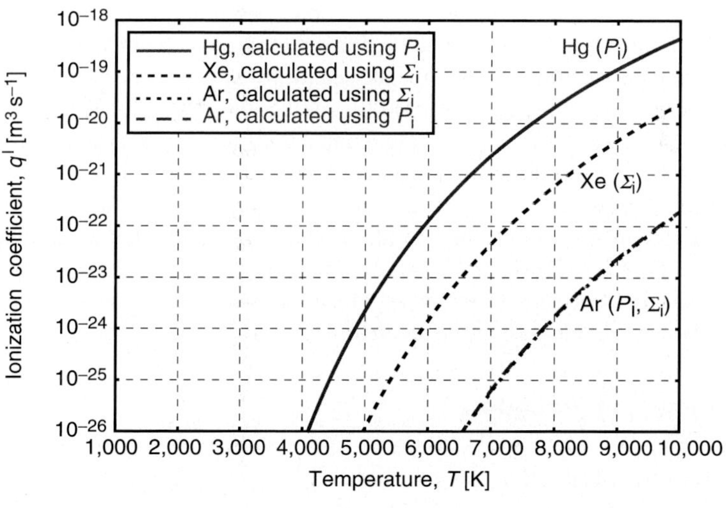

Fig. 4.22. Rate coefficient for ionization q^{I} for different plasmas vs. plasma temperature

[32] A similar approach is used by Benilov et al. [22].

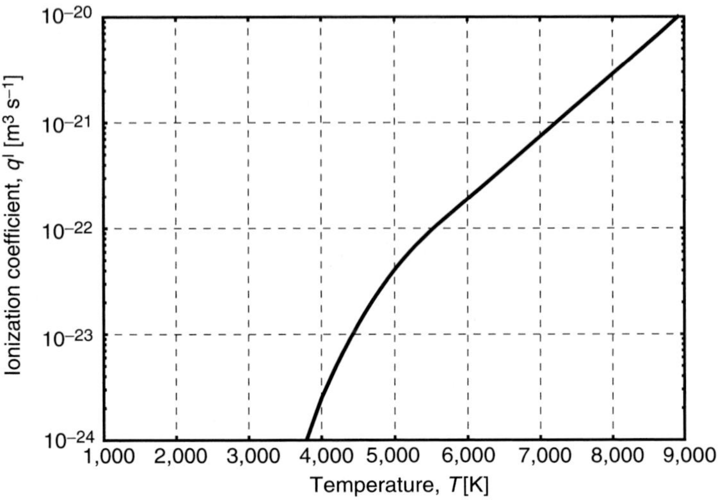

Fig. 4.23. Rate coefficient for ionization q^I for a D2 lamp plasma (Xe, Hg, NaI, ScI_3) vs. plasma temperature

following sections. Numerical simulations taking xenon HID lamps into account are also possible, but the pressure has to be higher than in mercury or metal halide discharges to comply with the model assumptions (above 0.5–1.0 MPa, depending on lamp current, work function for electron emission, etc.). The pressure in argon HID lamp must be even higher, if all model assumptions shall be fulfilled. Nevertheless, due to the importance of other factors, like boundary conditions for the heat flux at anode and cathode surface see (4.20) and (4.21) or the work function for electron emission, the results should be quite good as well for lower pressures.

Fischer [51] uses a constant rate coefficient for ionization q^I assuming a Maxwellian energy distribution corresponding to $T = 4{,}000$ K, i.e., there is no temperature dependence of q^I in the calculations of Fischer. The quantities n and $D_{i,n}$ are also set to their values at $T = 4{,}000$ K in [51], so that the adjustment length λ_σ is constant in his calculations. Lowke and Quartel [113], who are using a similar approach taking ambipolar diffusion of electrons into account, also use a constant value for their recombination coefficient γ.

4.4.2 Numerical Procedure

Finite Element Method

The model equations described earlier are solved using the finite element method on an unstructured grid assuming rotational symmetry. One important advantage of this method is that regions with large gradients (temperature, electric field, electrical conductivity) can be spatially resolved with a much finer mesh than other regions, leading to a very fine mesh in the vicinity of the electrode tips (very high spatial gradients of, e.g., temperature and electric field) and a rather low density of mesh points elsewhere. Furthermore, it is possible to consider arbitrary rotationally symmetric electrode and discharge vessel shapes. Unfortunately, the mesh immediately in front of the cathode has to be extremely fine to resolve spatial changes of the non-LTE electrical conductivity σ and the electric field E. This results in a mesh with up to a million unknowns, which have to be calculated using an iterative scheme. The computing time can be up to a couple of weeks for a single dc discharge. To reduce the computing time and to be able to carry out even more time consuming time-dependent calculations as presented in the following sections, an improved numerical approach for the description of the plasma immediately in front of the cathode surface has to be used. This is done by using improved boundary conditions for σ, V, and T at the cathode–plasma surface. Based on these new boundary conditions, the extremely small mesh size at the cathode tip can be increased, reducing the computing time.[33] Nevertheless, the grid has still a very small mesh near the electrode tips. As an example of the finite element mesh, the temperature distribution in the plasma and one electrode together with the finite element mesh is shown in Fig. 4.24. Clearly visible is the refinement of the mesh near the electrode tip.

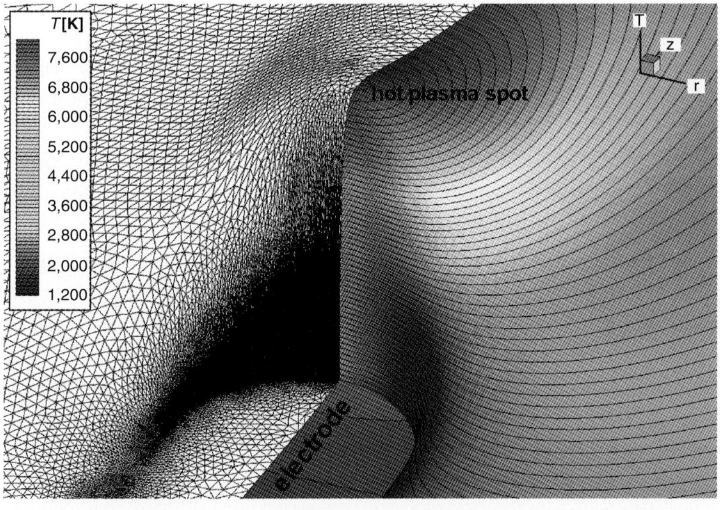

Fig. 4.24. Calculated electrode and plasma temperature distribution during anode phase near electrode tip with hot plasma spot (right-hand side) and finite element mesh (left-hand side) of a D2 automotive headlight lamp (Xe/Hg/NaI/ScI_3, $p = 7$ MPa, $I = 0.4$ A, 400 Hz rectangular-wave lamp current, $\varphi_{\text{emission}} = 2.6$ eV, $\varphi_{\text{cool/heat}} = 4.5$ eV)

[33] At the anode surface, the gradients of the electrical conductivity and the electric field are smaller, so that the problem is reduced to the cathode.

Iteration Loop

The differential equations necessary for the model described in the previous section are: current continuity to calculate the electric potential distribution $V(\vec{r}, t)$ (4.22), power balance of plasma and electrodes to calculate the plasma temperature distribution $T_P(\vec{r}, t)$ (4.18) and the electrode temperature distribution $T_E(\vec{r}, t)$ (4.19), and the non-LTE electrical conductivity equation (4.42a). These differential equations are coupled: The determination of the electric potential requires the non-LTE electrical conductivity distribution $\sigma(\vec{r}, t)$. To calculate σ, the temperature-dependent plasma properties and thus the plasma temperature distribution $T_P(\vec{r}, t)$ must be known. And the calculation of the plasma temperature distribution $T_P(\vec{r}, t)$ necessitates the electric potential distribution $V(\vec{r}, t)$ and the non-LTE electrical conductivity distribution $\sigma(\vec{r}, t)$. Thus, the differential equations are solved using an iteration loop. An overview of the differential equations and the iteration loop is given in Fig. 4.25.

The calculation starts with a trial temperature distribution and the initial value $\sigma = \sigma_{LTE}(T_P)$. Then, the first electric potential distribution can be calculated, followed by the determination of the plasma and electrode temperature distribution. After this, the iterative loop shown in Fig. 4.25 is passed through until convergence of temperature, electric potential, and electrical conductivity. All plasma and electrode properties are temperature dependent, i.e., they depend on the calculated temperature distribution. Because the differential equation for the determination of the non-LTE electrical conductivity σ is strongly nonlinear, it has proved beneficial to solve this equation in several loops until convergence of the electrical conductivity is reached. This reduces the total number of outer iteration loops, so that the total computing time is reduced.

The time dependence of the results can either be caused by a time-dependent lamp current $I(t)$ (boundary condition for electric potential, see Sect. 4.4.1) or by external, time-dependent disturbances, for example external laser heating as described in Sect. 4.4.5. The numerical time steps are automatically adapted according to the changes in the results from the preceding time step (adaptive step size). If the results do not vary much within one time step, the step size for the next time step is increased. If the results within one time step do vary too much, this time step is discarded and the calculation is repeated with a smaller step size. In the case of the external laser heating, the time-dependent lamp current $I(t)$ (1.8 A, 0.05 Hz rectangular-wave current) is superimposed by a local, time dependent, external laser heating of one electrode. The resulting time steps vary between 5×10^{-5} s near current reversal and 3×10^{-3} s near laser on/off, and up to 0.8 s during quasi stationary operation.

Improved Numerical Approach

Time dependent calculations (time-dependent lamp current or external, time-dependent disturbances) are only possible because calculation time has been reduced by increasing the extremely small mesh size at the electrode tips and thus

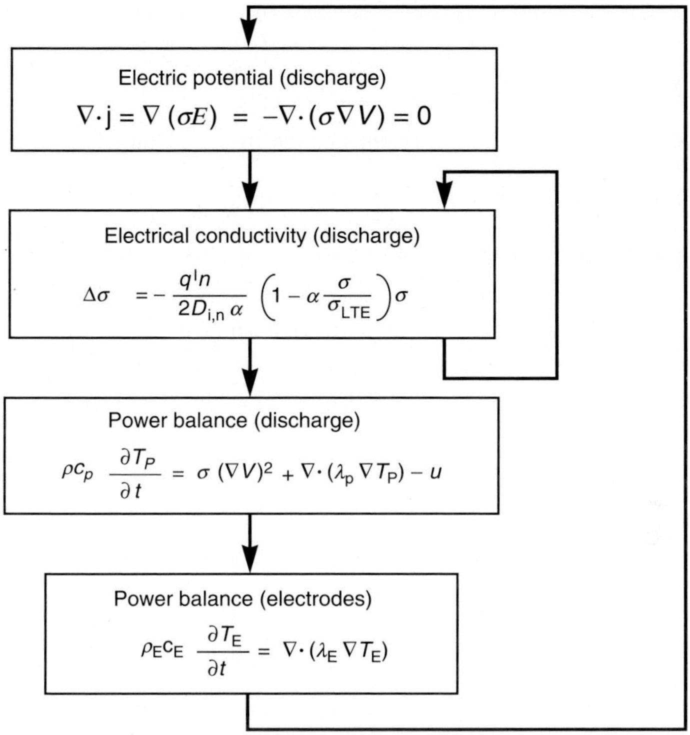

c_{E}	specific heat of electrodes [J kg^{-1} K^{-1}]
c_p	specific heat of plasma at constant pressure [J kg^{-1} K^{-1}]
$D_{\mathrm{i,n}}$	binary diffusion coefficient ion–atom [m^{2}s^{-1}]
E	electric field [V m^{-1}]
j	electric current density [A m^{-2}]
n	particle density [m^{-3}]
q^{I}	rate coefficient for ionization [m^3 s^{-1}]
T_{E}	electrode temperature [K]
T_{P}	plasma temperature [K]
u	net radiation emission coefficient [W m^{-3}]
V	electric potential [V]
α	ratio of ion to electron density
λ_{E}	thermal conductivity of electrodes [W K^{-1} m^{-1}]
λ_{P}	thermal conductivity of plasma [W K^{-1} m^{-1}]
ρ	mass density of plasma [kg m^{-3}]
ρ_{E}	mass density of electrodes [kg m^{-3}]
σ_{LTE}	LTE electrical conductivity [Ω^{-1} m^{-1}]
σ	non-LTE electrical conductivity [Ω^{-1} m^{-1}]

Fig. 4.25. Iteration loop of the self-consistent electrode–plasma model

reducing the number of unknowns. This has been done by using an improved numerical approach for the description of the plasma immediately in front of the cathode surface. At the anode surface, the gradients of the electrical conductivity and the electric field are smaller, so that the problem is reduced to the cathode.

Using the experiences and the detailed analysis of a multitude of numerical results obtained with the model for the dc case, improved boundary conditions for σ, V, and T at the cathode–plasma surface can be found. These new boundary conditions allow for a coarser grid in the vicinity of the cathode surface and reduce computing times dramatically without a significant change in the obtained results (deviation less than 1%).

A detailed analysis of all numerical results obtained with the time-independent, dc version of the model revealed that the gradient of the electrical conductivity σ normal to the cathode surface ($\partial\sigma/\partial n = \mathrm{grad}_n\sigma$) is constant over a certain distance perpendicular to the cathode surface. That means that starting from the cathode surface, σ increases linearly with increasing distance from the cathode surface into the plasma. Furthermore, the electric current density j is also constant over a certain distance to the cathode surface. This is due to the fact that the cathode surface lies at constant potential, so that the electric field E is perpendicular to the cathode surface. The electric current density $j = \sigma E$ has the same direction as the electric field E. Together with the current continuity equation (4.22) it follows that j is constant over a certain distance on a path perpendicular to the cathode surface.

The "sheath" over the cathode surface, where $\mathrm{grad}_n\,\sigma$ and j are constant on a path perpendicular to the cathode surface, has a thickness which is much larger than the typical grid spacing of the finite element grid[34] but much smaller than the typical electrode dimension (typical electrode diameter between 0.25 and 2 mm). Therefore, one can reduce the grid spacing by including this thin "sheath" with constant thickness δ in the corresponding boundary conditions. The thickness δ of the "sheath" used for the numerical simulations is chosen much smaller than the thickness of the "sheath" where $\mathrm{grad}_n\,\sigma$ and j are constant. However, it has to be emphasized that this is not a "physical sheath" with certain assumptions concerning physical processes (like the ionization layer or the space charge sheath in layer or sheath models for low-pressure discharges), but only a simplification in the numerical treatment of the problem which leaves the results unchanged (deviation < 1%). In the following, this "numerical sheath" will be called "skin" to distinguish it from "sheaths" in traditional layer approaches.

The modification of the boundary condition for σ at the interface between plasma and cathode skin ($\sigma_{\mathrm{SkinCathode}}$) is simple: At the plasma border of the skin, σ equals the value at the cathode surface given by (4.44) plus the linear increase with the constant slope $\mathrm{grad}_n\,\sigma$ over the skin thickness δ (with $\mathrm{grad}_n\,\sigma$ depending on the position on the cathode surface):

---

[34] Example: In a 0.6 MPa mercury discharge with a cathode diameter from 0.5 to 0.7 mm and a lamp current from 0.4 to 3.0 A (cf. Sect. 4.4.4), the thickness of the "sheath" where $\mathrm{grad}_n\,\sigma$ and j are constant is approximately 10 μm, the chosen "sheath" thickness δ for the numerical simulations is only 800 nm, so that the assumption of constant $\mathrm{grad}_n\,\sigma$ and j is assured. The grid spacing at the cathode surface for the calculations including the "sheath" can be chosen to be approximately 500 nm, the necessary grid spacing at the cathode surface if no "sheath" is used must be as small as 20–50 nm.

$$\sigma_{\text{SkinCathode}} = \sigma_{\text{cathode}} + \delta \operatorname{grad}_{\text{n}} \sigma = \frac{4 j_{\text{e,cathode}} \sigma_{\text{LTE}}}{e v_{\text{th}} n_{\text{e,LTE}}} + \delta \operatorname{grad}_{\text{n}} \sigma. \tag{4.55}$$

The electric current density j (which is constant on a path perpendicular to the cathode surface within the skin) can be expressed as the product of the electrical conductivity σ and the electric field E either at the cathode surface or at the plasma border of the skin:

$$j = \sigma_{\text{cathode}} E_{\text{cathode}} = \sigma_{\text{SkinCathode}} E_{\text{SkinCathode}}. \tag{4.56}$$

Combining (4.55) and (4.56) we get

$$E_{\text{cathode}} = E_{\text{SkinCathode}} \left(1 + \frac{e v_{\text{th}} n_{\text{e,LTE}}}{4 j_{\text{e,cathode}} \sigma_{\text{LTE}}} \delta \operatorname{grad}_{\text{n}} \sigma \right). \tag{4.57}$$

This equation has to be solved iteratively at every point at the cathode surface as the electron emission current density $j_{\text{e,cathode}}$ depends on the electric field strength E_{cathode}.

The potential drop over the cathode skin can be computed by integrating the electric field E over the skin thickness δ:

$$U_{\text{SkinCathode}} = \int_0^{\delta} E \mathrm{d}z = \int_0^{\delta} \frac{j}{\sigma_{\text{cathode}} + z \operatorname{grad}_{\text{n}} \sigma} \mathrm{d}z. \tag{4.58}$$

Considering that j and $\operatorname{grad}_{\text{n}} \sigma$ are constant on a path perpendicular to the cathode surface within the skin, one gets[35]:

$$U_{\text{SkinCathode}} = \frac{j(r=0)}{\operatorname{grad}_{\text{n}} \sigma(r=0)} \ln(\sigma_{\text{cathode}}(r=0) + z \operatorname{grad}_{\text{n}} \sigma(r=0)) \Big|_{z=0}^{z=\delta}. \tag{4.59}$$

This potential drop over the cathode skin implies an electrical power input into the plasma within the cathode skin and must therefore be considered in the boundary condition connecting the temperatures of cathode and plasma at the cathode surface (4.20):

$$\lambda_E \frac{\partial T_{\text{E}}}{\partial n} - \lambda_{\text{P}} \frac{\partial T_{\text{P}}}{\partial n} = j U_{\text{SkinCathode}} + j_{\text{t,i}} (\varphi_{\text{i}} - \varphi_{\text{cool}}) - j_{\text{t,emission}} \varphi_{\text{cool}} - \varepsilon \sigma_{\text{SB}} T_{\text{E}}^4. \tag{4.60}$$

Again, it has to be emphasized that these new boundary conditions do not include new physical effects. They are (only) based on a detailed analysis of the many numerical results obtained with the model described in this section, which showed that j and $\operatorname{grad}_{\text{n}} \sigma$ are constant on a path perpendicular to the cathode surface within a small skin over the cathode surface. This knowledge has been used

[35] The cathode lies at constant potential, so that in the plasma immediately adjacent to the cathode surface the potential lines are parallel to the cathode surface. Due to the thinness of the skin, the skin boundary lies on a constant potential line, so that the integral in (4.58) must be evaluated at a single path only (here: lamp axis).

to reduce the numerical "burden" caused by the extremely fine mesh in the vicinity of the cathode surface by extending the boundary conditions on the cathode–plasma surface and allowing for a coarser mesh near the cathode. This improvement of the numerical approach does not change the obtained results significantly (deviation less than 1%), but reduces the computing times by a factor of more than 10.

In the following sections (Sects. 4.4.3–4.4.8), the self-consistent electrode–plasma model described in Sects. 4.4.1 and 4.4.2 is used for the numerical simulation of different dc- and ac-operated HID lamps with different plasma compositions (mercury, xenon, or metal halide plasmas). Each of the following sections focuses on one or several different aspects of HID lamp modeling.

4.4.3 Comparing Different Cathode Models

Following up the description of cathode sheath models in Sect. 4.1.1, we now compare the numerical results for the cathode obtained with the self-consistent electrode–plasma model described in Sects. 4.4.1 and 4.4.2 with results obtained by three other numerical models concerned with cathode behavior [65]. The three other models are based on cathode sheath models as described in Sect. 4.1.1. A so-called model lamp was agreed upon for this comparison (called LS8-lamp[36]) to produce comparable results using these four different numerical models. This LS8-lamp is cylindrical with an inner diameter of 9 mm, an electrode gap of 20 mm, electrode lengths of 14 mm, and electrode diameters of 0.6 mm for both identical electrodes. The electrodes are cylindrical with a hemispherical tip (cf. Fig. 4.26). The lamp fill is xenon with an operating pressure of 1 MPa. Moreover, the following other parameters and data were used by all four participants in this comparison of cathode models to allow for maximum comparability. Work function for electron emission and cooling/heating of the cathode/anode surface[37]: $\varphi = \varphi_{\text{emission}} = \varphi_{\text{cool}} = \varphi_{\text{heat}} = 4.55$ or 3.5 eV, $A_R = 1.2\times10^6\ \text{A}\cdot\text{m}^{-2}\cdot\text{K}^{-2}$ (cf. Sect. 2.4.4), thermal conductivity $\lambda_E(T_E)$ of the electrodes according to [94] (cf. Fig. 2.27), emissivity $\varepsilon(T_E)$ of the electrode material according to [174] (cf. Fig. 2.29), lamp current I between 1.5 and 6.0 A, inner wall temperature of the cylindrical lamp bulb 1,000 K, and outer electrode ends kept at 300 K. The plasma properties of a 1 MPa xenon plasma are given in Chap. 2.

[36] Details of this lamp are the result of a discussion between experimentally and theoretically working research groups in the context of the LS8-conference focusing on the comparison of modeling results and experimental data. Hence, the lamp is called LS8-lamp. The LS8-conference (8th International Symposium on the Science and Technology of Light Sources) took place at Greifswald from August 30th–September 3rd 1998.

[37] The same work functions (4.55 and 3.3 eV) have been used, for example, by Tielemans and Oostvogels [162].

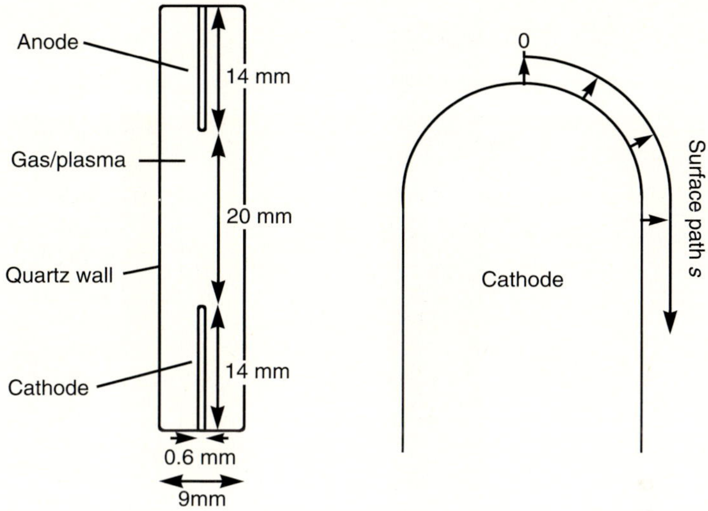

Fig. 4.26. Geometry of LS8-lamp and definition of surface path s at cathode tip [56]

First Impression

To give a first impression of the conditions in the LS8-lamp, Figs. 4.27 and 4.28 show an example of the temperature distribution within plasma and electrodes and the electric potential distribution in the plasma, respectively. These results are obtained using the self-consistent electrode–plasma model described in Sects. 4.4.1 and 4.4.2. Due to the rotational symmetry, only one-half of the lamp is shown. The characteristic hot plasma spots located very close to the electrode tips (distance to electrode tips: 30–250 μm) with temperatures between 7,500 and 9,000 K are clearly visible. Because of the large electrode gap of the LS8-lamp (20 mm) compared, for example, to the D2 lamp (4 mm, cf. Sect. 4.4.6, Figs. 1.35, 4.24, 4.79, 4.84, and 4.85), a cylindrical plasma column can be observed.

Participating Models

In addition to the self-consistent electrode–plasma model described in Sects. 4.4.1 and 4.4.2 (labeled "Flesch" in Figs. 4.29–4.38), three other models participated in the comparison. Different from the high-pressure model with non-LTE electrical conductivity described in Sects. 4.4.1 and 4.4.2, the other three models are based on a traditional layer or sheath approach for modeling the plasma close to the cathode surface (cf. Sect. 4.1.1, sheath thickness some 100 nm (space charge layer) and some 10 μm (ionization layer), respectively). The model used by Bötticher (cf. Sect. 4.1.2 and [31]) consists of a space charge layer and an ionization layer only, with no connection to an LTE plasma with the characteristic hot plasma spots and the cylindrical plasma column. The rotationally symmetric cathode body is treated in two spatial dimensions. The model used by Schmitz [148–150] includes a space

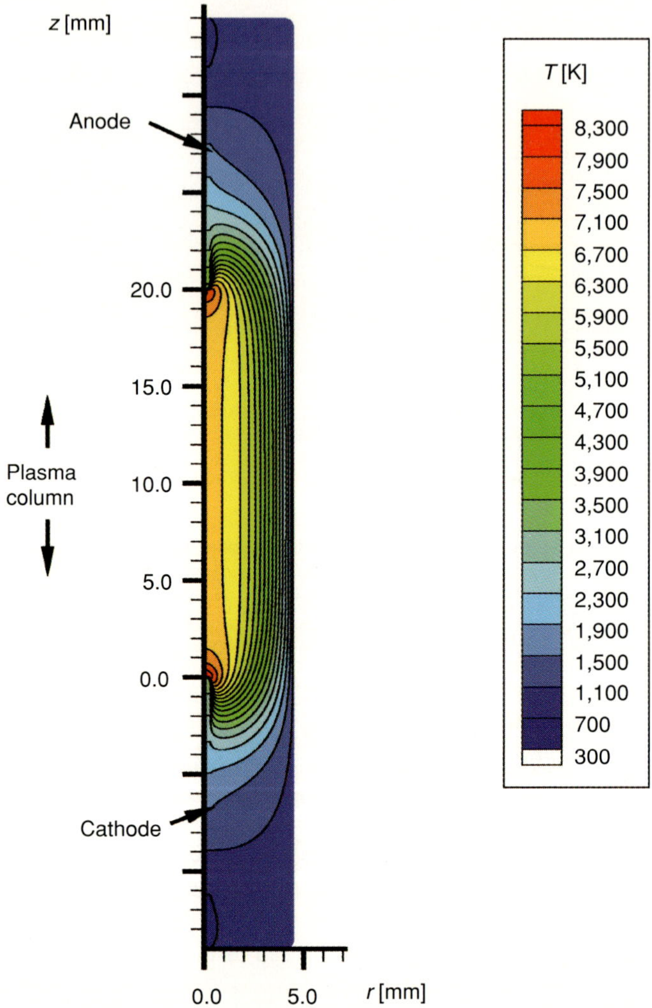

Fig. 4.27. Calculated temperature distribution in the plasma and the electrodes (LS8-lamp, Xe, $p = 1$ MPa, $I = 4.5$ A (dc), $\varphi = 4.55$ eV, [56])

charge layer, plus a so-called Knudsen-layer, and a presheath (ionization layer), also with no connection to the LTE plasma with the characteristic hot plasma spots and the cylindrical plasma column. For simplification, Schmitz divides the cathode into an inhomogeneous and a homogeneous domain of temperature distribution, resulting in an average cathode temperature over the arc attachment on the cathode tip, an average electric current density of the arc attachment, and so forth[38] instead of distributions depending on the surface path s (see Figs. 4.29–4.34). Since both

[38] To include the data of Schmitz in Figs. 4.29–4.34 over the surface path s, the area of the arc attachment is transformed into a corresponding radius.

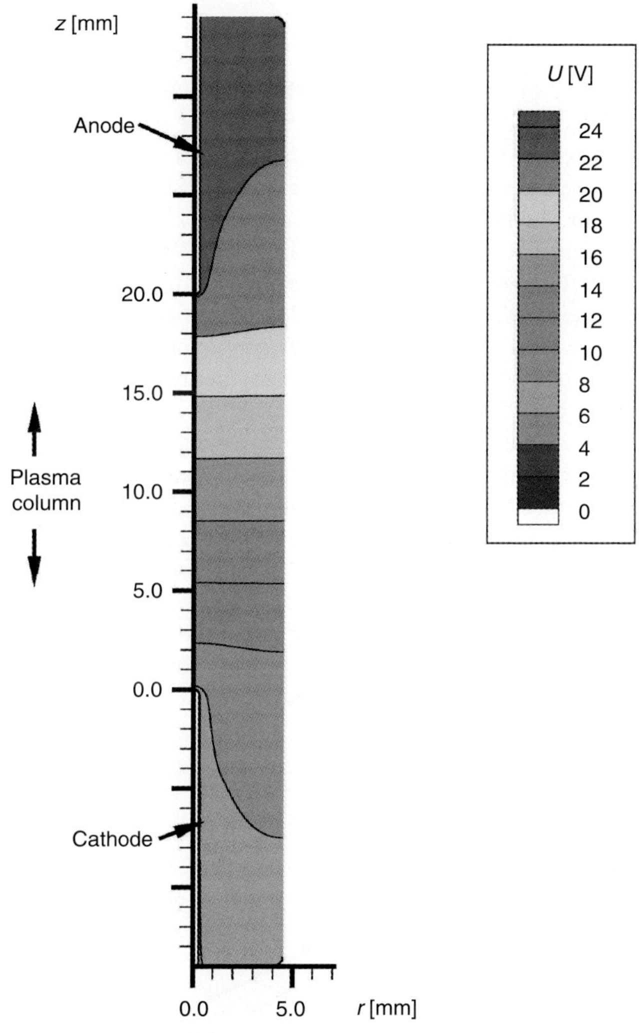

Fig. 4.28. Calculated electric potential distribution in the plasma (LS8-lamp, Xe, $p = 1$ MPa, $I = 4.5$ A (dc), $\varphi = 4.55$ eV, [56])

Bötticher and Schmitz do not connect the cathode layers to the actual high-pressure plasma region, the attachment of the arc on the cathode is influenced only by properties of the cathode, but not by the constriction zone of the plasma. The model used by Wendelstorf [167, 168] consists of a space charge layer and an ionization layer (so-called presheath). These layers are linked consistently to the LTE plasma column, so that the simulation area (cathode, anode, plasma with near-electrode plasma, hot plasma spots, constriction zone, and plasma column) is similar to the simulation area of the self-consistent electrode–plasma model described in Sects. 4.4.1 and 4.4.2. Both models ("Wendelstorf" and "Flesch") treat the electrodes and the plasma in two spatial dimensions, using rotational symmetry.

Cathode Surface Temperature

Figures 4.29 ($\varphi = 4.55$ eV) and Fig. 4.30 ($\varphi = 3.5$ eV) show modeling results for the surface temperature of the cathode along the surface path s (cathode tip at $s = 0$). The most obvious difference between the models is that Bötticher, Wendelstorf, and the self-consistent electrode–plasma models (Flesch) calculate temperature distributions along the surface path s, whereas the result of Schmitz is an average temperature of an "active" area. The size of this "active" area multiplied by the average electric current density j (Figs. 4.33 and 4.34) equals the lamp current I. Outside the "active" area, the electric current density is zero.

The differences in the predicted cathode tip temperatures are up to 400 K for $\varphi = 3.5$ eV (Fig. 4.31). At the boundary between hemispherical tip and cylindrical cathode body (marked by a solid line labeled "semisphere" in Figs. 4.29–4.34), these differences are already reduced to less than 300 K. For $\varphi = 4.55$ eV the differences between the models are even smaller than for $\varphi = 3.5$ eV. Comparing calculated cathode temperatures (Fig. 4.29) with measured cathode temperatures from Nandelstädt ([122], Fig. 4.32[39]), a strong raise in the measured temperature at the hemispherical tip starting at $x = 0.3$ mm (corresponding to $s = 0.47$ mm) is noticeable. This rise is probably caused by reflection of radiation originating from the hot plasma spot close to the cathode tip and by an enlarged emissivity ε due to a high surface roughness at the cathode tip, which influences the pyrometer reading. This renders the pyrometric measurements close to the cathode tip less reliable. The "real" cathode tip temperature is thus estimated by extrapolating the temperature

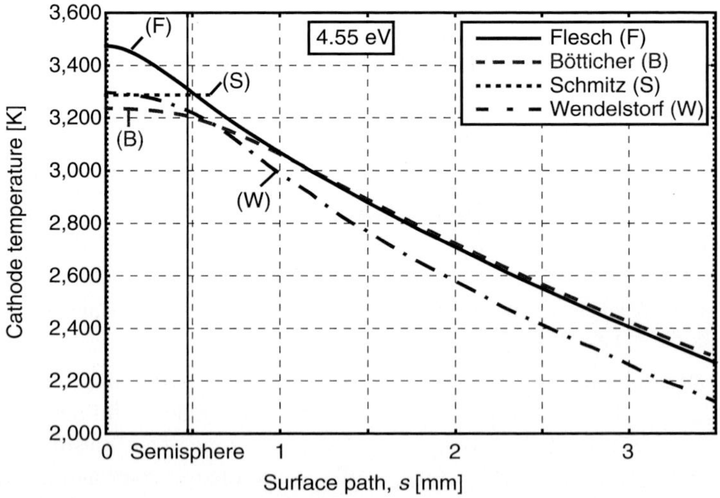

Fig. 4.29. Temperature on cathode tip surface calculated using four different approaches (LS8-lamp, Xe, $p = 1$ MPa, $I = 4.5$ A (dc), $\varphi = 4.55$ eV, [56])

[39] Cathode temperature depends on the distance x to the cathode tip instead of the surface path s, which is given on the upper space axis of Fig. 4.32.

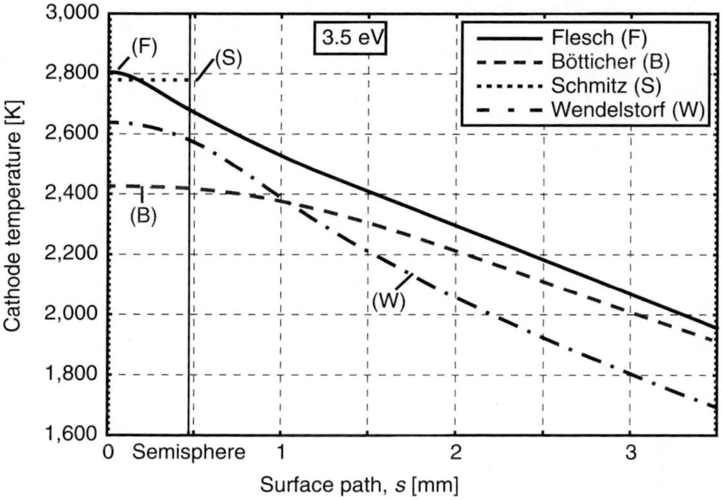

Fig. 4.30. Temperature on cathode tip surface calculated using four different approaches (LS8-lamp, Xe, $p = 1$ MPa, $I = 4.5$ A (dc), $\varphi = 3.5$ eV, [56])

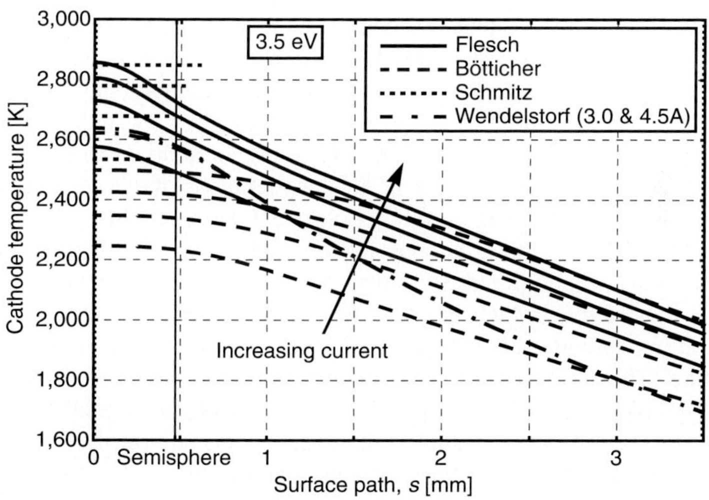

Fig. 4.31. Temperature on cathode tip surface calculated using four different approaches (LS8-lamp, Xe, p = 1 MPa, I = 1.5–6.0 A (dc), $\varphi = 3.5$ eV, [56])

readings on the last millimeters of the cathode[40] to $x = 0$, neglecting the temperature readings for $x < 1$ mm (cf. Sect. 3.1.4). The result of this extrapolation is 3,380 K,

---

[40] Comparing the data "Bötticher" and "Flesch" in Fig. 4.29 shows that even though the temperature distributions on the cylindrical part of the cathode (up to $s = 1$ mm) are very similar, they can differ by more than 200 K at the cathode tip. This shows that results of an extrapolation on the last millimeter of the cathode are questionable.

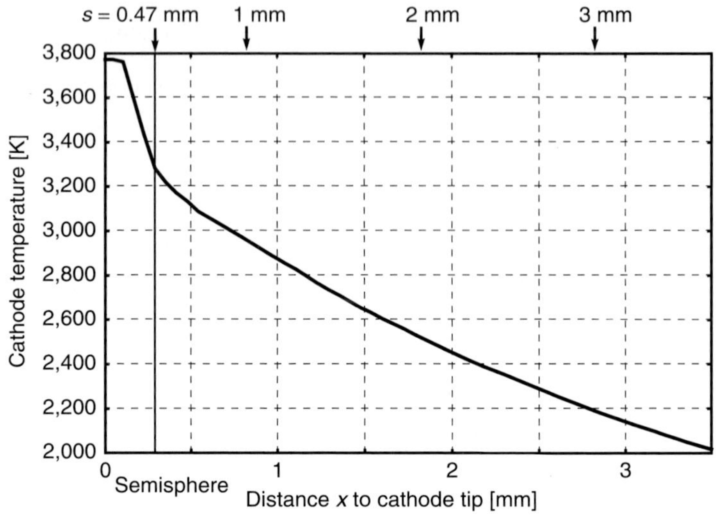

Fig. 4.32. Measured temperature on cathode surface depending on distance x to cathode tip from Nandelstädt ([122], LS8-lamp, Xe, $p = 1$ MPa, $I = 4.5$ A (dc), tungsten electrodes, [56])

lying midway between the calculated cathode tip temperatures ($T_{\mathrm{CathodeTip}} = 3{,}240$–$3{,}470$ K, Fig. 4.29) of the four models participating in the comparison. At the boundary between hemispherical tip and cylindrical cathode body ($s = 0.47$ mm), the measured temperature is 3,280 K, lying again between the calculated temperatures ($T_{\mathrm{tip\text{-}body}} = 3{,}200$–$3{,}300$ K, Fig. 4.29). In addition to the error at the cathode tip described earlier, there is an uncertainty in the measured values of cathode temperatures of about ±150 K (cf. Sect. 3.1.4). A conclusion as to which of the four models is preferable is not possible on the basis of available data. The work function presents an additional difficulty: All models need a definite value for φ (here 4.55 and 3.5 eV), which has a strong influence on the results (like cathode tip temperature, cathode fall voltage, etc.). However, work functions are difficult to determine. The only measurement of φ in an operating HID lamp has been published in [145, 146] (cf. Sect. 3.1.11). Normally, measurements are done using vacuum conditions. Moreover, the work function depends on the lifetime of the cathode, the crystal orientation, the preparation of the surface, the working gas, and so forth (cf. Sects. 3.1.11, 3.1.12, and 4.4.6). And experimental investigations reveal that macroscopically identical lamps might show a different behavior in spite of identical operation conditions (cf. e.g., Sects. 3.1.7 and 3.1.13). For this reason, the calculation of "exact" cathode temperatures or cathode fall voltages[41] is not as important and interesting as the prediction of the behavior of these quantities in the case of altering parameters like lamp current, working gas pressure, or work

[41] With an uncertain work function φ, there is no possibility to calculate exact cathode temperatures or cathode fall voltages.

function. Figures 4.29–4.31 show that all models predict, unsurprisingly, increasing cathode temperatures with increasing lamp current and increasing work function.

Electric Current Density

The electric current densities for $\varphi = 4.55$ eV (Fig. 4.33) and $\varphi = 3.5$ eV (Fig. 4.34) demonstrate a principal difference between the models "Wendelstorf" and "Flesch" on the one hand and the models "Bötticher" and "Schmitz" on the other hand: Wendelstorf and the self-consistent electrode–plasma model ("Flesch") include the plasma column, the hot plasma spots immediately in front of the electrodes, and the constriction zones, where the current density distribution changes from a broad cross section in the plasma column to a constricted cross section on the cathode or anode tip. This leads to a more constricted arc attachment on the cathode tip than the one predicted with the model used by Bötticher, which does not include the connection to the plasma column described earlier (good to see especially for $\varphi = 3.5$ eV in Fig. 4.34). The arc attachment on the cathode is not only more constricted with the models from Wendelstorf and Flesch, but also only weakly influenced by the work function of the cathode, different from the prediction using the models of Bötticher or Schmitz. The reason for this is again the constriction zone, which influences the size of the arc attachment more than the cathode parameters do. The model used by Schmitz does not include the constriction of the current profile from the plasma column to the arc attachment on the cathode tip, but here the size of the arc attachment is additionally influenced by the restricted treatment of the cathode (one-dimensional).

The maximum value of the electric current density (at $s = 0$ mm) is predicted by all models to rise with the lamp current I. The size of the arc attachment on the cathode tip increases in all models with the current, too. But a change in the cathode work function effects the predicted size of the arc attachment and the maximum value of the electric current density differently (compare Fig. 4.33 with 4.34): The results obtained by Schmitz show an increase in the size of the arc attachment and a corresponding decrease of the maximum value of the current density (lamp current fixed) for a larger work function, whereas all three other models predict the opposite effect. The reason for this is the calculation of the ion current density in the model used by Schmitz: The magnitude of the ion current density is not calculated in a self-consistent manner but results from a prescribed plasma temperature. As the ratio of ion to total lamp current increases with increasing work function (Figs. 4.35 and 4.36), the size of the arc attachment area must rise to assure the higher ion current (at a fixed ion current density calculated with the prescribed plasma temperature in the model by Schmitz). With a larger arc attachment area, the maximum value[42] of the electric current density at fixed lamp current must decrease. In the three other models (Bötticher, Wendelstorf, and Flesch) the ion current density is determined self-consistently, leading to a constriction of the arc

[42] In the model used by Schmitz, the maximum value equals the average value.

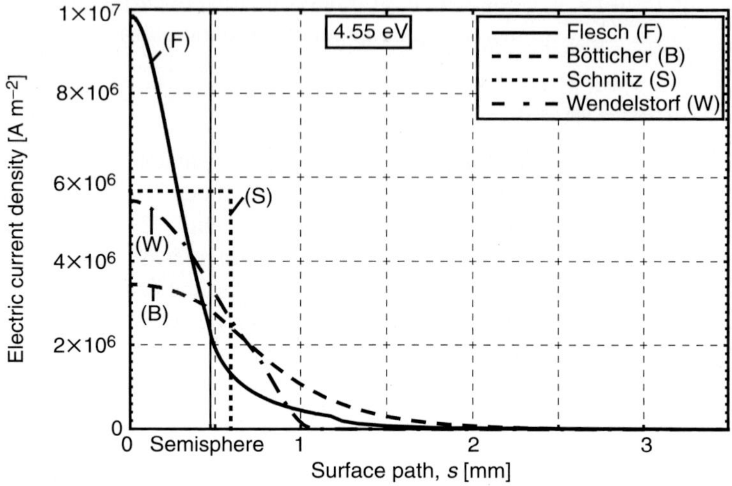

Fig. 4.33. Electric current density on cathode tip surface calculated using four different approaches (LS8-lamp, Xe, $p = 1$ MPa, $I = 4.5$ A (dc), $\varphi = 4.55$ eV, [56])

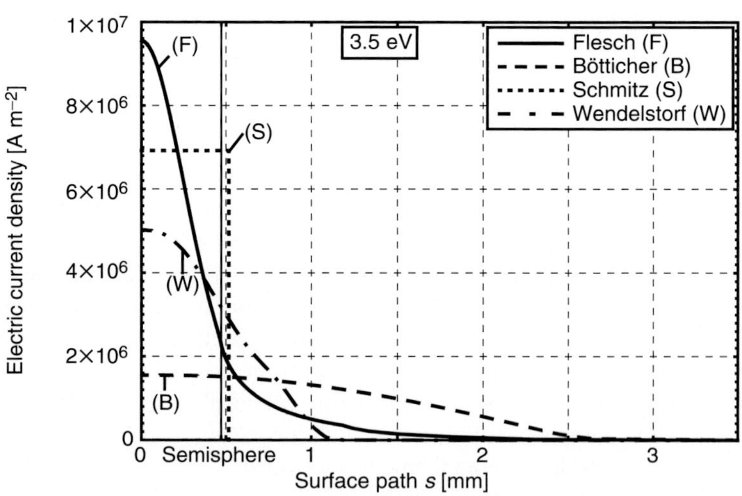

Fig. 4.34. Electric current density on cathode tip surface calculated using four different approaches (LS8-lamp, Xe, $p = 1$ MPa, $I = 4.5$ A (dc), $\varphi = 3.5$ eV, [56])

attachment and an increase of the magnitude of the electric current density at the cathode tip for an increasing work function φ.

Ion Current

The share of the ion current on the total lamp current at the cathode decreases with increasing lamp current and increases with increasing work function using the models Bötticher, Schmitz, and Flesch (Figs. 4.35 and 4.36). The ion current fraction predicted by the model used by Wendelstorf seems to be independent of lamp current but increases also with increasing work function. An experimental method to determine the ion current fraction has not yet been developed.

The models used by Bötticher and Schmitz have only one heating mechanism of the cathode, which is ion current. In the models of Wendelstorf and Flesch, an additional cathode heating by the thermal heat flux from the plasma exists. Thus, for a small work function (Fig. 4.36) the models used by Bötticher and Schmitz overestimate the ion current at the cathode compared to the predictions by the models used by Wendelstorf and Flesch, since they are constrained to the ion current to heat the cathode to sufficiently high temperatures. This consideration is confirmed by the following result: Simplified calculations with the model presented by Neumann (cf. [130] or Sect. 4.1.1) for a 4 MPa mercury HID lamp with a geometry comparable to a D2 automotive headlight lamp predict an ion current fraction of 43% ($\varphi = 2.6$ eV, $I = 0.75$ A), whereas the self-consistent electrode–plasma model described in Sects. 4.4.1 and 4.4.2 (Flesch) predicts an ion current fraction of 11% only, all other parameters being equal. This is due to the important heating of the electrodes by heat conduction from the hot plasma spots to the relatively cold electrodes (here cathode). This heating by the thermal flux from the plasma "replaces" the ion heating, so that realistic values for the ion current can be calculated.

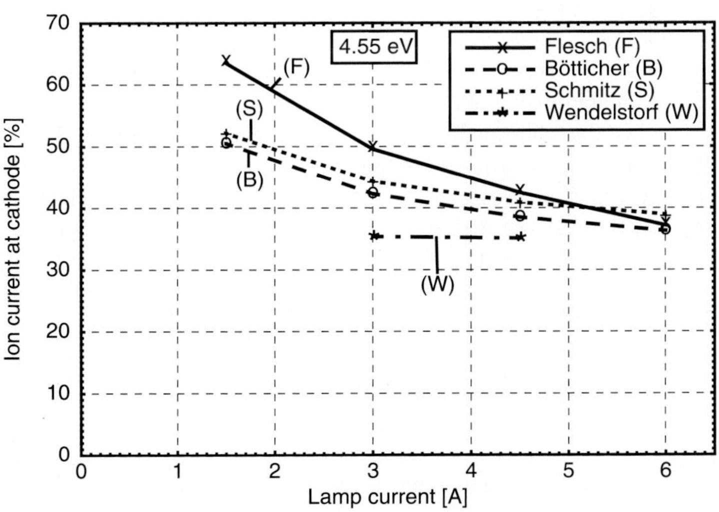

Fig. 4.35. Ratio of ion to total lamp current at the cathode vs. dc lamp current (LS8-lamp, Xe, $p = 1$ MPa, $\varphi = 4.55$ eV, [56])

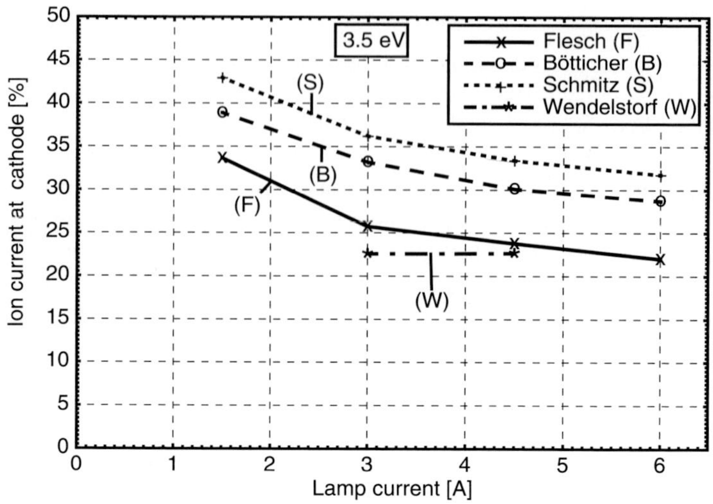

Fig. 4.36. Ratio of ion to total lamp current at the cathode vs. dc lamp current (LS8-lamp, Xe, $p = 1$ MPa, $\varphi = 3.5$ eV, [56])

Cathode Fall Voltage

The definition of the cathode fall voltage differs in the four models: The models of Bötticher and Schmitz, lacking an adjacent arc column, define the cathode fall voltage as the sum of the cathode sheath voltages. The models of Wendelstorf and the self-consistent electrode–plasma model ("Flesch") use a definition similar to the experimental determination of the sum of cathode and anode falls, i.e., the linear extrapolation of the electric potential of the plasma column to $z = 0$ (cathode) and $z = d_{elec}$ (anode), see Sects. 3.2.2 and 4.4.1 as well as Figs. 3.40, 4.44, 4.59, and 4.80. Using this procedure, the constriction zones in front of cathode and anode contribute to both anode and cathode fall voltage.

For all models, the cathode fall voltage decreases with increasing lamp current[43] (Figs. 4.37 and 4.38[44]). All models predict an increasing cathode fall voltage for an increasing work function. Despite the discussed differences in the definition of the cathode fall voltage and the differences in the current density distribution and ion current fraction, the calculated cathode fall voltages are quite similar for all models. The reason for this is the energy balance of the cathode, which mainly determines the cathode fall voltage. The work function φ and the cathode temperature T_c control the electron emission current,[45] i.e., the cathode needs a certain tip

[43] Nevertheless, the electrical power input into the near-cathode plasma due to the cathode fall voltage ($U_c \cdot I$) increases because of the increasing lamp current I.

[44] No data available for the cathode fall voltage for $\varphi = 3.5$ eV for the model of Wendelstorf.

[45] All models use comparable formulas for the emission current depending on work function, cathode temperature, and electric field at the cathode (cf. Sect. 2.4.4, especially Fig. 2.33).

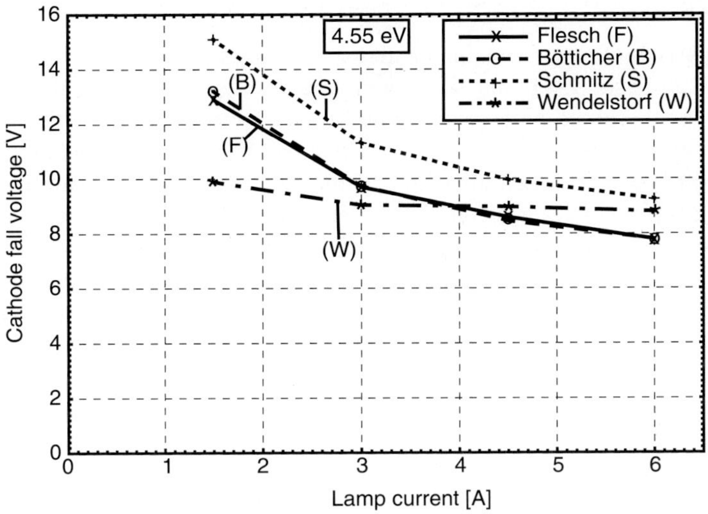

Fig. 4.37. Cathode fall voltage vs. dc lamp current (LS8-lamp, Xe, $p = 1$ MPa, $\varphi = 4.55$ eV, [56])

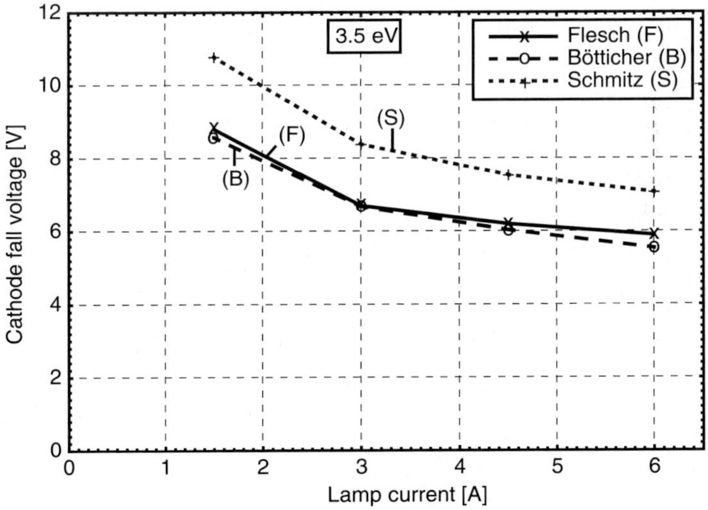

Fig. 4.38. Cathode fall voltage vs. dc lamp current (LS8-lamp, Xe, $p = 1$ MPa, $\varphi = 3.5$ eV, [56])

temperature to be able to emit sufficient electrons for the given lamp current I. This temperature is maintained by heating the cathode. The power necessary for upholding the cathode tip temperature is brought into the near-cathode plasma by the cathode fall voltage, independent of the different heating processes under consideration.

Summary

The comparison of results obtained with the self-consistent electrode–plasma model described in Sects. 4.4.1 and 4.4.2 with the results obtained using three other cathode models (based on layer or sheath models) shows differences and common characteristics of the models under consideration. The most important result of this comparison is that the energy balance of the cathode combined with the equation describing the electron emission current as a function of cathode temperature and local electric field is the most important part in modeling the cathode. But as soon as one is interested in details like, for example, the electric current density distribution on the cathode tip (arc attachment), it is important to consider the cathode at least in two dimensions and to include the plasma column and the constriction zone, where the current density distribution transforms from a rather broad profile in the plasma column to the small, constricted arc attachment at the cathode tip. Moreover, models neglecting the thermal heat flux from the hot plasma spots to the relatively cool cathode will overestimate the ion current, which is the only heating mechanism used in these models for the cathode.

4.4.4 Different Electrode Shapes

This section focuses on a comparison of different electrode shapes (cathode and anode) and the resulting changes in electrode and plasma hot-spot temperatures, electrode fall voltages, and electric current profiles at the electrode tips. This comparison is motivated, for example, by the investigation of Hartmann et al. (cf. [86] and Sect. 3.1.13), showing micrographs of new cathodes after 2 min, 30 min, 5 h and 10 h of operating time in a 0.26 MPa xenon lamp with a lamp current of 6.0 A. A substantial deformation of the cathode tip is already visible after 30 min and is increased after a couple of hours. This deformation changes electrode and plasma temperatures as well as the electrode fall voltages. A detailed analysis of the range of variation of computed results caused by different geometries of the electrode tip will be presented in this section (see especially Figs. 4.47–4.55).

There are two reasons for this investigation: First, comparing theoretical results with experimental data without knowing the exact shape of the electrode tip (which is often the case, e.g., due to erosion during operation [86]) may be difficult if one does not know the impact of different electrode tip geometries on the results of the simulations or the experimental data. This section will help to estimate the influence of the shape of the electrode tip on the theoretical results and will, therefore, improve the quality of comparisons between numerical results and experimental data. The second reason for this investigation is the better insight into the interaction between plasma and electrodes. This will improve the design of future electrodes. Knowing the influence of electrode tip geometry on plasma and on electrode temperature or on electrode fall voltages, one can optimize the shape of electrodes for better performance and/or longer lifetimes of lamps. Up to now, there is no other theoretical work known to the author which deals with the important impact of different geometries of the electrode tips taking the whole plasma and both electrodes into account.

The Lamp

The self-consistent electrode–plasma model described in Sects. 4.4.1 and 4.4.2 is applied within this section to a 0.6 MPa mercury discharge lamp considering six different electrode shapes (anode and cathode shape are always identical). This particular lamp has been chosen because of the many experimental data and theoretical experience available for this lamp, e.g., in [30, 84, 102]. Where experimental data are available, a comparison with the theoretical results is presented. The diameters of the different electrodes considered are 0.5 and 0.7 mm, with different shapes of the tip of anode and cathode (hemispherical, flat, or conical tip with 60° and 90° apex angle, see Fig. 4.39). Furthermore, an electrode with a larger diameter in the midsection of the rod, which approximates an idealized wire wound around the electrode, is considered (cf. Fig. 4.39). All electrodes are assumed to be pure tungsten electrodes, i.e., $\varphi = \varphi_{\text{emission}} = \varphi_{\text{cool}} = \varphi_{\text{heat}} = 4.55$ eV.

The electrode gap of the lamp is 10 mm, the electrodes have also a length of 10 mm, and the lamp fill is mercury with a working pressure of 0.6 MPa. In each numerical simulation, the geometry of anode and cathode is identical. Lamp currents of rectangular (400 Hz, $I = \pm 1.8$ A) and sinusoidal shape (400 Hz, $I_{\text{rms}} = 1.8$ A, $I_{\text{max}} = 2.55$ A) are considered as well as dc currents ($I = 0.4$–3.0 A). The inner diameter of the quartz bulb is 12.5 mm, the shape of the lamp is sketched[46] in Fig. 4.40, a picture of the lamp is shown in Fig. 4.56. According to experimental findings, the inner quartz wall temperature and the temperature of the electrode roots[47] are set to 1,000 K. Details of this lamp are the result of a discussion between experimentally and theoretically working research groups in the

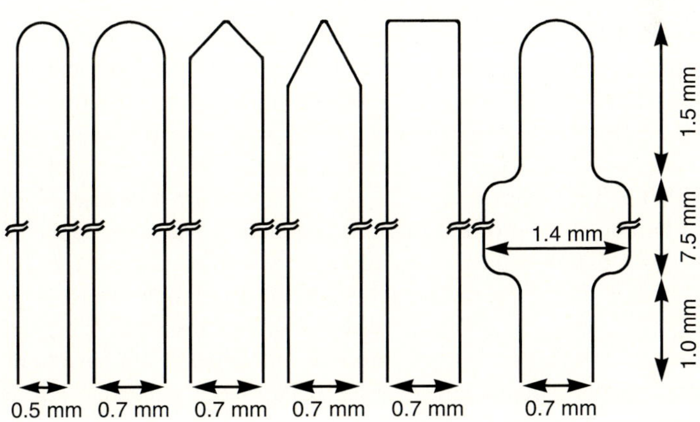

Fig. 4.39. Different electrode shapes (all electrodes have a length of 10 mm, [57])

[46] For a better understanding, the electrodes are labeled "cathode" and "anode", but this assignment is (of course) time dependent in the case of ac lamp currents.

[47] Bötticher uses the same value [30].

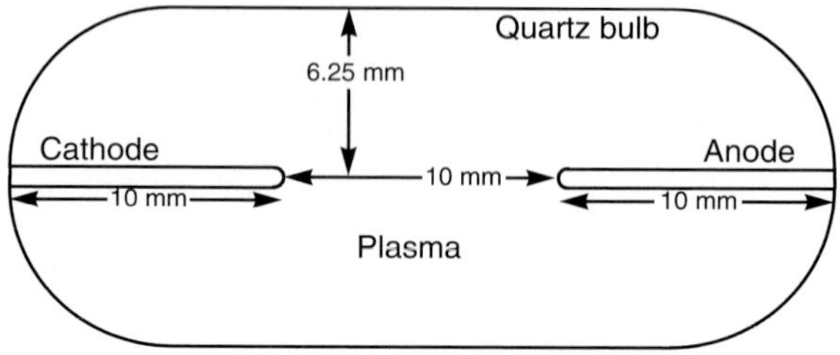

Fig. 4.40. Geometry of LS9hg-lamp [57]

context of the LS9-conference[48] (hence the lamp is called *LS9hg-lamp*) focusing on the comparison of modeling results and experimental data. Some results of this comparison can be found in this section, more comparisons with experimental results are shown in Sect. 4.4.5.

Plasma Properties

The plasma properties and the transport coefficients of the plasma are calculated according to the description given in Chap. 2. The radiation transport calculation for the thermal conductivity of the plasma is based on the temperature profile of the arc column for temperatures below the maximum temperature in the arc column (5,400–6,000 K depending on lamp current I). For higher temperatures, the radiation transport calculation is based on the temperature profile of the hot plasma spots (cf. Sects. 2.3.3 and 2.3.4, especially Fig. 2.25).

A paper of Kloss et al. [102] has been used to double check the plasma properties and transport coefficients of the plasma: The electrical conductivity σ_{LTE}, the thermal conductivity λ_{P} (in [102]: κ), the mass density ρ, and the specific heat at constant pressure c_p (all depending on plasma temperature T_{P}) are compared in Figs. 4.41 and 4.42 for a 0.6 MPa mercury plasma. Within the temperature range of 1,000–8,000 K, the differences for σ_{LTE}, λ_{P}, and ρ are less than 5%, for c_p less than 10%. The comparison of the thermal conductivity has been done without the contribution of radiation transport calculation, which is not used in [102] in the described manner. This is a rather good agreement. Furthermore, a paper from Hartel et al. [84], using a different method for radiation transport calculations, confirms that only mercury resonance lines at 185 and 254 nm have to be included for radiation transport calculation for a 0.6 MPa mercury plasma. For all material functions used in the present, the previous, and the following sections, constant pressure within the discharge is assumed.

48 The LS9-conference (9th International Symposium on the Science and Technology of Light Sources) took place at Cornell University from 12 to 16 August 2001.

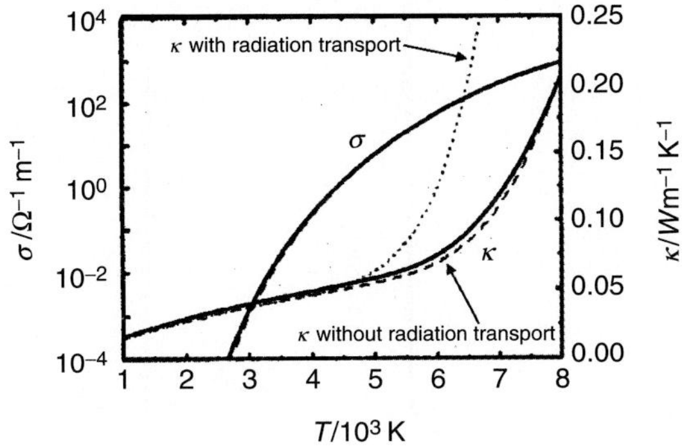

Fig. 4.41. Comparison of plasma properties: Electrical conductivity σ_{LTE} and thermal conductivity κ vs. plasma temperature for mercury, 0.6 MPa taken from [102] (*solid lines*) and calculated according to Sect. 2.3 and [171] (*dashed* and *dotted lines*). The thermal conductivity λ used for the numerical simulations is shown in Fig. 2.25

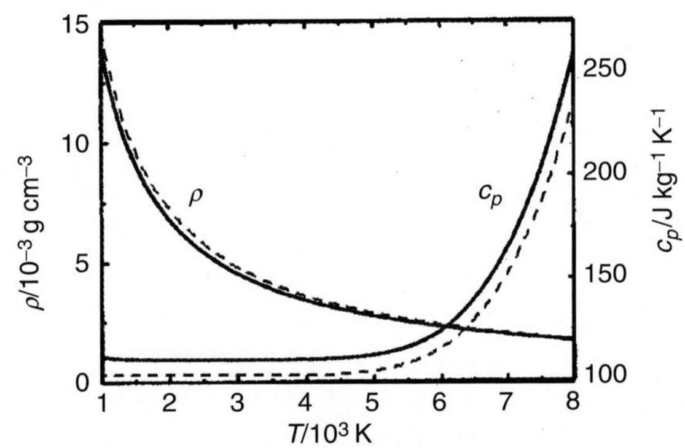

Fig. 4.42. Comparison of plasma properties: Mass density ρ and specific heat at constant pressure c_p vs. plasma temperature for mercury, 0.6 MPa taken from [102] (*solid lines*) and calculated according to Sect. 2.3 and [171] (*dashed lines*)

First Impression

As an example of the conditions in this high-pressure discharge lamp, Fig. 4.43 shows the temperature distribution on the lamp axis ($r = 0$ mm), starting from the outer cathode end ($z = -10$ mm) via the cathode tip ($z = 0$ mm) through the discharge plasma to the anode tip ($z = 10$ mm) all the way to the outer anode end ($z = 20$ mm). Clearly visible are the extremely hot plasma spots very close to the electrode tips (80–100 μm distance from the electrode tips). These hot plasma spots have temperature maxima between 7,000 and 8,000 K. The electrodes have tip temperatures of about 3,070 K (cathode) and 2,500 K (anode). This leads to extremely large temperature gradients between the hot plasma spots and the

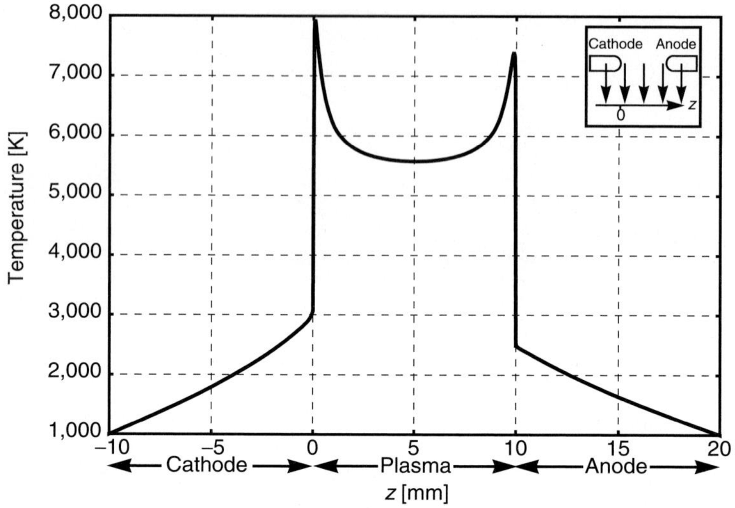

Fig. 4.43. Calculated electrode and plasma temperature on lamp axis (LS9hg-lamp, Hg, $p = 0.6$ MPa, $I = 1.0$ A (dc), $\varphi = 4.55$ eV, hemispherical tip, electrode diameter/length 0.7 mm/10 mm, [57])

electrode tips, which drive an intense thermal heat flux from the plasma to the electrodes. This power loss of the plasma together with the power loss due to the enhanced radiation by the hot plasma spots has to be balanced by an enlarged electric field near the electrodes. This enlarged electrical power density leads to the electrode fall voltages (cf. Fig. 4.44). Figure 4.44 shows the resulting electric potential distribution on the lamp axis between cathode and anode tip. The electrode fall voltages can be extracted easily by extrapolation of the linear part of the electric potential distribution toward anode and cathode as shown in Fig. 4.44 (dashed line). The dc lamp current for this temperature and electric potential distribution is 1.0 A, the electrodes have a diameter of 0.7 mm with a hemispherical tip.

Comparison to Experimental Results

Numerical results and experimental data are compared in Figs. 4.45 and 4.46. The experimental data were kindly supplied by the INP, Greifswald [154]. The lamp configuration (dimensions and geometry, electrode material, lamp current, etc.) is the same for measured and calculated data. The electrodes of the experimental lamps are cylindrical with a flat tip at the beginning of the experiments, changing to a hemispherical tip after some hours of burning time. The lamps are mounted vertically. Simulations of lamps with different shapes of the electrode tips showed that the principal agreement between measured and calculated data does not change with variation of the geometry of the electrode tip. The results of the simulations shown in Figs. 4.45 and 4.46 were obtained with hemispherical electrode tips. Later,

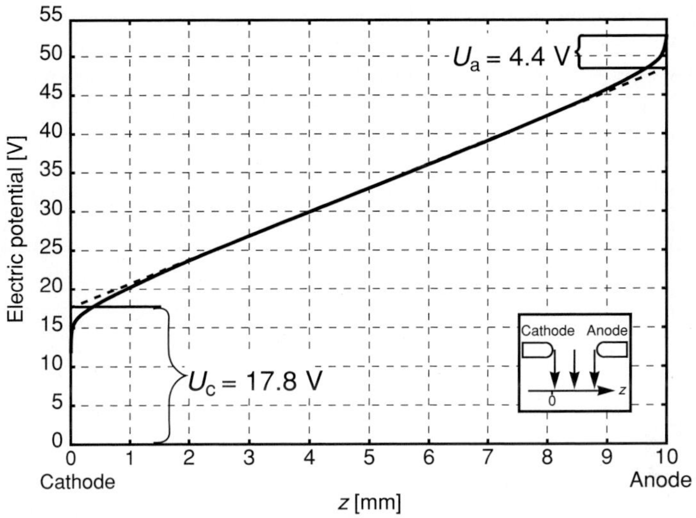

Fig. 4.44. Calculated electric potential on lamp axis and electrode fall voltages U_c and U_a (LS9hg-lamp, Hg, $p = 0.6$ MPa, $I = 1.0$ A (dc), $\varphi = 4.55$ eV, hemispherical tip, electrode diameter/length 0.7 mm/10 mm, [57])

it appeared that the electrodes used during the experimental investigations had electrode lengths of 13 mm instead of 10 mm. This difference can be seen comparing the calculated and the measured electrode temperature (see later).

Figure 4.45 shows the total lamp voltage for a 400 Hz sinusoidal lamp current of $I_{rms} = 1.8$ A ($I_{max} = 2.55$ A). There is a good agreement between calculated (dashed line) and experimental (solid line) results.[49] The point in time of the maximum of the total lamp voltage can be shifted a little bit by varying the shape of the electrode tips in the simulations, increasing the agreement of experimental and theoretical results. The computed values of the total lamp voltage are somewhat lower than the measured values in the decreasing part of the lamp voltage curve. A possible explanation is that the computed electric field in the cylindrical part of the discharge is too low. In [84] a model describing only this cylindrical part of a 0.6 MPa mercury plasma without effects caused by the electrodes (like constriction zones and electrode fall voltages) was developed, showing similar deviations of measured and computed values. As the electric field in the plasma column is mainly influenced by the temperature dependent values of σ_{LTE} (which are nearly the same in [84] and for the numerical simulations presented here), this might be the explanation for the small differences between experimental and theoretical data in Fig. 4.45. Bearing in mind that the simulations cover the whole discharge and both electrodes (the total lamp voltage consists of the voltage drop within the near-electrode plasma, the constriction zones, and the hot plasma spots in front of the

[49] In the Section “Experimental Data” in Sect. 4.4.5, a possible variation of the total lamp voltage from lamp to lamp will be described.

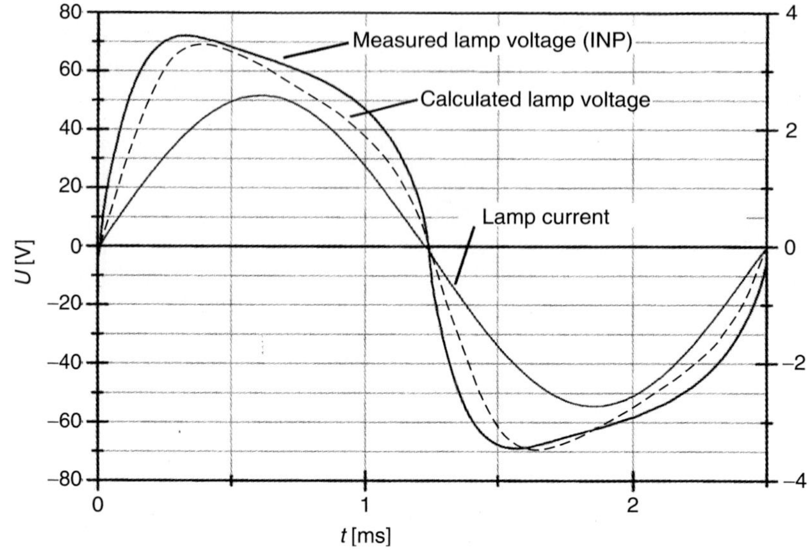

Fig. 4.45. Comparison of measured [154] and computed total lamp voltage (LS9hg-lamp, Hg, $p = 0.6$ MPa, $I_{\mathrm{rms}} = 1.8$ A, 400 Hz sinusoidal lamp current, $\varphi = 4.55$ eV, hemispherical tip, electrode diameter 0.7 mm, electrode length: 13/10 mm)

electrodes as well as the voltage drop across the cylindrical part of the plasma (cf. Fig. 4.44)), the agreement between simulations and experimental data can be regarded as quite satisfactory. This confirms that the self-consistent electrode–plasma model described in Sects. 4.4.1 and 4.4.2 is suitable for the simulation of high-pressure discharge lamps and delivers realistic results.

In Fig. 4.46, a comparison between measured and calculated electrode temperature depending on the distance z to the electrode tip ($z = 0$ mm) for a 400 Hz rectangular current of $I = \pm 1.8$ A is given. The measured electrode temperature is time averaged and has been measured using two different pyrometers (Impac IS10 and IGA100, cf. Sect. 3.1.4). In contrast, the computed electrode temperature belongs to a dc anode, a dc cathode, the end of the anode phase and the end of the cathode phase (400 Hz rectangular current of $I = \pm 1.8$ A). Taking uncertainties in temperature measurements into account (± 100 K would be a typical error) the results are very good. The slightly different slopes of the measured and calculated temperature profiles are due to the different electrode lengths: The numerical simulations were carried out assuming 10 mm long electrodes, the experimental values were obtained with 13 mm instead of 10 mm long electrodes, as it appeared later.[50]

[50] A comparison of measured and calculated electrode temperatures for an electrode with 0.5 mm diameter is shown in Fig. 4.69 (electrode length for calculated and measured temperatures: 13 mm).

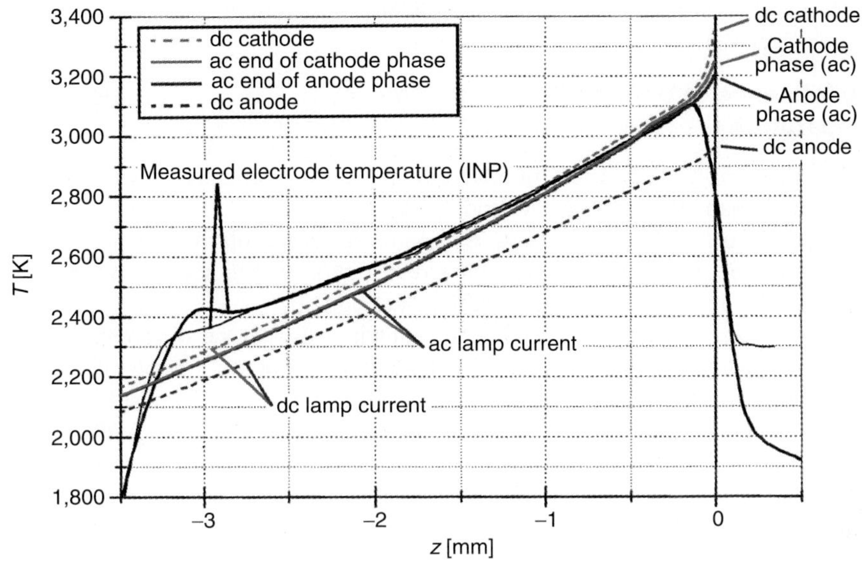

Fig. 4.46. Comparison of measured ([154], time averaged) and computed electrode temperature (LS9hg-lamp, Hg, $p = 0.6$ MPa, $I = \pm 1.8$ A, 400 Hz rectangular-wave lamp current, $\varphi = 4.55$ eV, hemispherical tip, electrode diameter 0.7 mm, electrode length: 13/ 10 mm)

The sudden drop of the measured temperature near the electrode tip ($z = -0.15$ to 0.15 mm) is caused by the size of the target spot (0.3 mm) so that only the measured temperatures below $z = -0.15$ mm are appropriate.[51] The measured temperatures below $z = -2.5$ mm are as well out of the range of confidence due to optical distortions. In addition, the electrode tip temperature is also influenced by the geometry of the electrode tip (see later), which was a flat tip at the beginning of the experiments, changing to a hemispherical tip after some hours of burning time. The results of the simulations were obtained with hemispherical electrode tips. Again, the agreement between experimental data and the simulations confirms that the self-consistent electrode–plasma model is suitable for the simulation of high-pressure discharge lamps.

Comparing the calculated dc and ac results (ac: 400 Hz rectangular current of $I = \pm 1.8$ A) in Fig. 4.46 shows that the electrode temperature does not settle just between anode and cathode dc temperatures. Instead, the ac temperature tends very clearly toward the dc cathode temperature. This is true for the experimental values

[51] This demonstrates one of the advantages of numerical simulations: The electrode tip temperature can be computed with much higher accuracy than its experimental determination can provide.

[153] as well as for the results from numerical simulations.[52] A discussion will be given in Sect. 4.4.5.

The Influence of Different Electrode Shapes on Plasma and Electrode Temperature

An overview of the differences in electrode and plasma temperatures for different shapes and diameters of the electrodes can be found in Fig. 4.47. The wide-stretching influence of the electrode tip geometry on the plasma temperature can clearly be seen. The plasma temperature distribution within a 0.4 mm "zone" is strongly influenced by the geometry of the electrode tip.

The dc lamp current for all cathodes shown is 1.0 A, the different electrode geometries correspond to Fig. 4.39. The three cathodes at the top row of Fig. 4.47 have a hemispherical tip but different diameters: 0.5 mm (left-hand side), 0.7 mm (middle), and 0.7/1.4 mm (right-hand side). The three cathodes at the bottom line of Fig. 4.47 have a diameter of 0.7 mm (i.e., the same diameter as the cathode in the middle of the top row) but different shapes of the tip: conical tip with 60° apex angle (left-hand side), conical tip with 90° apex angle (middle), and flat tip (right-hand side).

The cathode tip temperature depends on both the electrode diameter and the geometry of the cathode tip. Increasing electrode diameter decreases the cathode tip temperature. This is caused by an enlarged heat flux within the electrode body. Small, point-like electrode tips have increased cathode tip temperatures. This effect is due to a constriction of the arc attachment (and the related constriction of power transfer to the electrode tip) at the electrode tip for small, point-like electrode heads. The arc attachment at the flat cathode is much broader compared to that at a hemispherical or conical cathode tip. The most constricted arc attachment and the highest cathode tip temperature arises for the conical tip with 60° apex angle (Fig. 4.47, bottom row, left-hand side). A more detailed analysis of cathode tip temperatures depending on lamp current is shown in Fig. 4.48, and of anode tip temperatures in Fig. 4.49. The arc attachment is discussed in Fig. 4.55.

The temperature of the cathode body[53] is influenced by the electrode diameter in a similar way as the cathode tip temperature. An increasing electrode diameter decreases the cathode body temperature due to the enlarged heat flux within the electrode body (Fig. 4.47, top row). In contrast, the shape of the electrode tip has a different effect on the temperature of the cathode body compared to the effect on the cathode tip temperature. Small, point-like electrode tips have increased electrode tip temperatures but decreased electrode body temperatures (Figs. 4.47, 4.48 (cathode),

---

[52] As a consequence thereof, one cannot just take anode and cathode temperature from dc operation (calculated or measured) and calculate ac electrode temperatures by averaging the dc values. The (time dependent) ac electrode temperature depends nontrivial on lamp current, current frequency, current waveform, electrode diameter and length, work function, mode of arc attachment, etc.

[53] The electrode body temperature is defined within this section as the temperature 0.35 mm (half of the diameter) below the electrode tip.

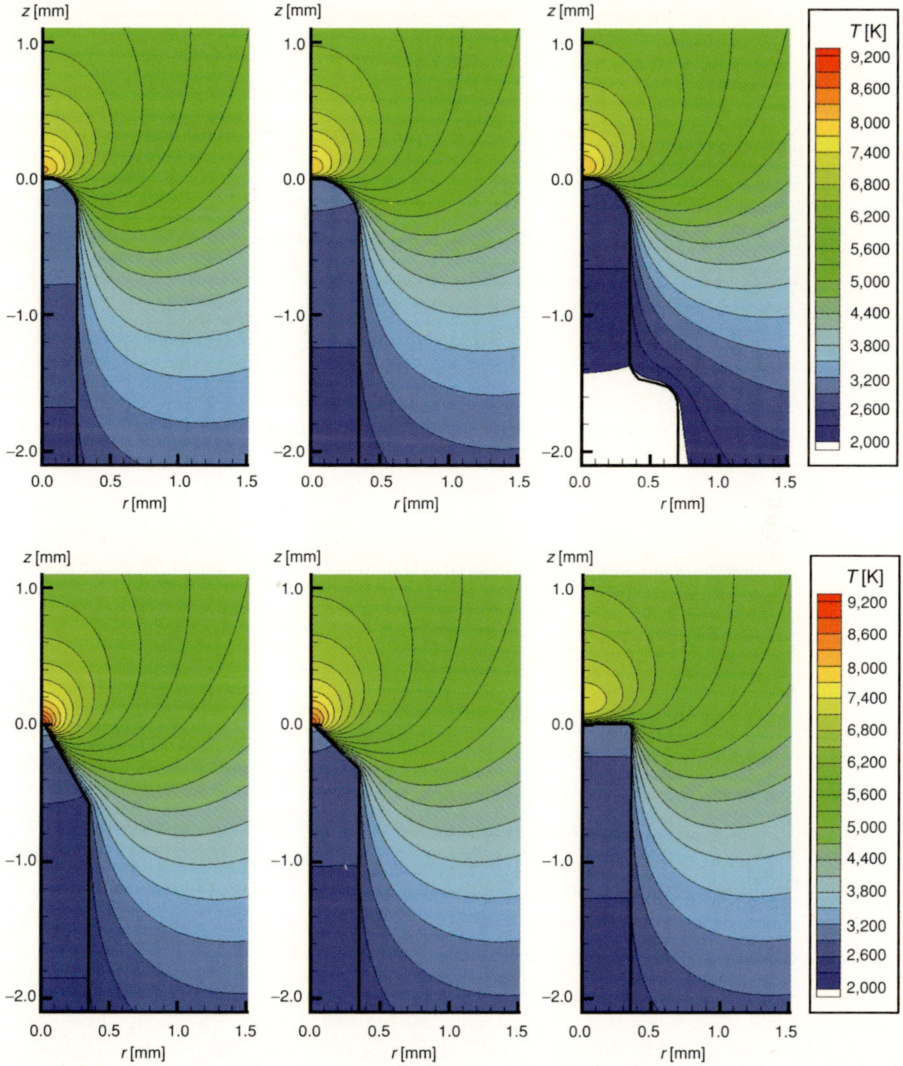

Fig. 4.47. Calculated plasma and cathode temperature for different electrode geometries (LS9hg-lamp, Hg, $p = 0.6$ MPa, $I = 1.0$ A (dc), $\varphi = 4.55$ eV, electrode length: 10 mm, [57])

and 4.49 (anode)). This is again caused by the constriction of the arc attachment (and the corresponding constriction of power transfer to the electrode) at the electrode tip: Small, point-like electrode heads have more confined arc attachments. This leads to a higher power density transfer within a smaller area (increasing the local electrode tip temperature). Nevertheless, the *total* power transfer (power density transfer times area) decreases for small, point-like electrode tips, leading to lower electrode body temperatures (exception: anode temperature for lamp currents above 1.6 A, see following text and Fig. 4.49).

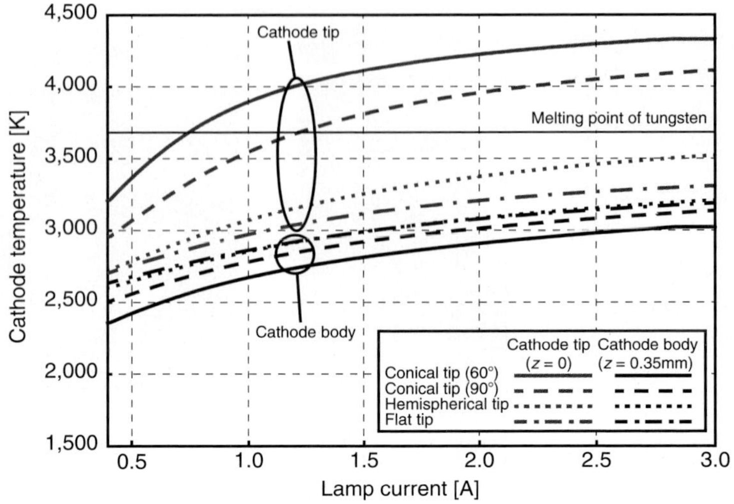

Fig. 4.48. Calculated temperature of cathode tip and cathode body vs. dc lamp current I for different electrode geometries (LS9hg-lamp, Hg, $p = 0.6$ MPa, $\varphi = 4.55$ eV, electrode diameter/length 0.7 mm/10 mm, [57])

The temperature of the hot plasma spots does not change considerably with the electrode diameter but significantly with the geometry of the electrode tip (cf. Figs. 4.47, 4.50 (cathode), and 4.51 (anode)). Small, point-like electrode tips have high plasma hot-spot temperatures and the hot plasma spots are closer to the electrode tip. This is again caused by the constriction of the arc attachment near the electrode tip for small, point-like electrode tips. This leads to a more constricted current density profile near the electrode tip and enlarges the local power input into the hot plasma spot. As a result, the plasma hot-spot temperature gets higher and the hot plasma spot profile gets smaller.

The electrode with the larger diameter in the midsection of the rod (0.7/1.4 mm diameter, see Fig. 4.47, top row, right-hand side) has a lower tip and body temperature compared to the "standard" electrode (0.7 mm diameter, hemispherical tip) due to an increased heat conduction within the electrode and an increased black body radiation from the surface. Depending on the lamp current, the tip temperature is reduced by 100–500 K, leading to higher cathode and anode fall voltages compared to the "standard" electrode. However, both electrodes have similar arc attachments, i.e., the arc attachment is influenced largely by the geometry of the electrode tip and the resulting arc attachment of the plasma, and not the heat conduction and radiation losses within the lower parts of the electrode.

A more detailed analysis of the electrode tip temperature and the electrode body temperature depending on electrode tip geometry and lamp current I is given in Figs. 4.48 (cathode) and 4.49 (anode). The effect of the shape of the electrode tip is much larger for the electrode tip temperature than for the electrode body

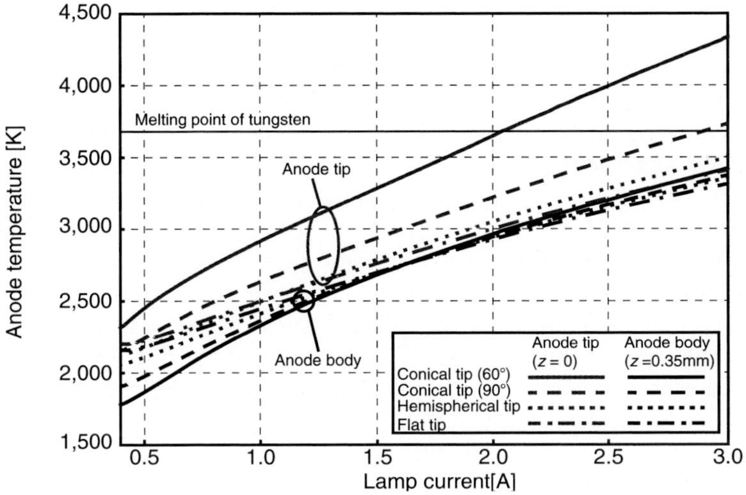

Fig. 4.49. Calculated temperature of anode tip and anode body vs. dc lamp current I for different electrode geometries (LS9hg-lamp, Hg, $p = 0.6$ MPa, $\varphi = 4.55$ eV, electrode diameter/length 0.7 mm/10 mm, [57])

temperature. The differences in the electrode tip temperature can be up to 1,000 K, the differences in the electrode body temperature are up to 300 K. The latter might be difficult to measure due to the uncertainty in the experimental data (±100 K would be a very small error, cf. Sects. 3.1.1–3.1.7). The differences in the electrode tip temperature are difficult to detect, because of the small geometrical dimensions of the electrodes compared to the size of the target spot of a commercial pyrometer (e.g., 0.3 mm for the temperature measurement in Fig. 4.46). Furthermore, the reflection of plasma radiation (especially from the hot plasma spots immediately in front of the electrodes) at the electrode tip increases the error.[54] An important advantage of the numerical simulations carried out with the self-consistent electrode–plasma model is the possibility to predict the electrode temperatures without any constraints concerning the resolution in time and space. This makes the theoretical results indispensable for analyzing and improving electrode characteristics.

The top of a cathode with a conical tip would melt[55] at a lamp current of about 0.8 A (60° apex angle) or 1.2 A (90° apex angle). The top of an anode with a conical tip would melt at about 2.0 A (60° apex angle). The melting of the top would change

[54] This error due to reflection of plasma radiation can be avoided if there is a "plasma correction" with measurements using a second or third wavelength (cf. Sects. 3.1.6 and 3.1.7).

[55] Melting point of tungsten: 3,680 K.

the geometry of the electrode tip. Differences between anode and cathode can clearly be seen comparing Figs. 4.48 (cathode) and 4.49 (anode): The anode temperature rises linearly with the lamp current *I*, the cathode temperature starts at currents below 1.0 A with a strong slope which decreases with increasing lamp current. This different behavior of cathode and anode tip temperature depending on lamp current *I* has indeed been observed in experimental setups (e.g., [43]).

A very interesting result can be observed concerning the temperature of the anode body around $I = 1.6$ A (cf. Fig. 4.49): For currents below 1.6 A, the body temperature of the anode with a flat or hemispherical tip is higher than the body temperature of the anode with a conical tip. For currents above 1.6 A, the opposite is true. This is related to the anode fall voltage, which shows the same characteristics as the anode body temperature (cf. Fig. 4.53). Above 1.6 A, the higher power density transfer to conical anode tips dominates the smaller arc attachment area (compared to hemispherical or flat anode tips), causing a higher total power transfer to the anode with a conical tip (power density transfer times arc attachment area). This leads to higher anode fall voltages and higher anode body temperatures.

Results for the plasma hot-spot temperature depending on electrode tip geometry and lamp current *I* are given in Figs. 4.50 (near cathode tip) and 4.51 (near anode tip). The differences in the plasma hot-spot temperature for different shapes of the electrode tip can be up to 2,500 K. Again, this might be difficult to measure due to the uncertainty in the experimental data (±500 K in plasma temperature measurements at the hot plasma spot would be a small error, cf.

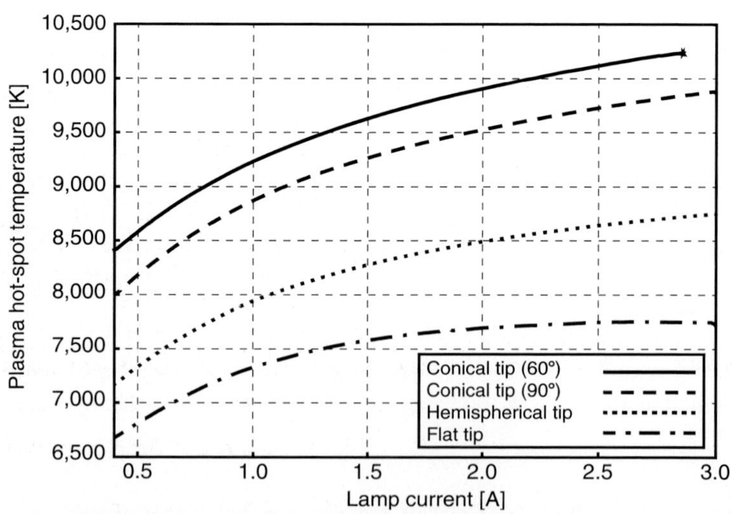

Fig. 4.50. Calculated plasma hot-spot temperature near cathode tip vs. dc lamp current *I* for different electrode geometries (LS9hg-lamp, Hg, $p = 0.6$ MPa, $\varphi = 4.55$ eV, electrode diameter/length 0.7 mm/10 mm, [57])

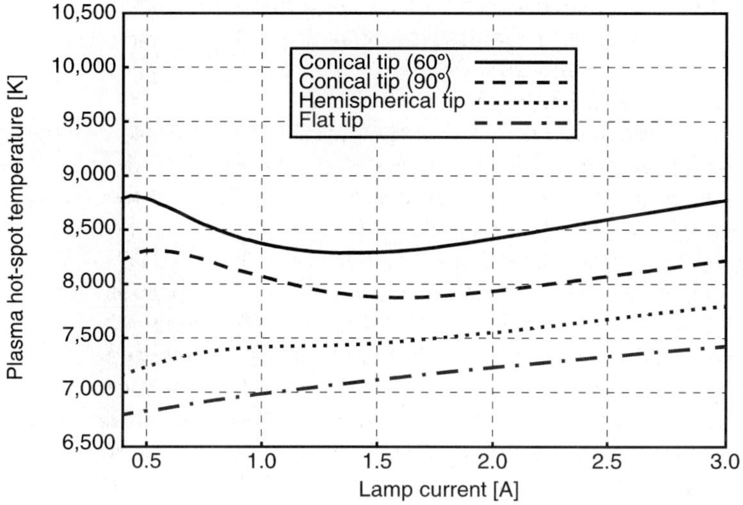

Fig. 4.51. Calculated plasma hot-spot temperature near anode tip vs. dc lamp current I for different electrode geometries (LS9hg-lamp, Hg, $p = 0.6$ MPa, $\varphi = 4.55$ eV, electrode diameter/length 0.7 mm/10 mm, [57])

Figs. 4.3–4.5 and error bars in Fig. 4.5) and the small distances of the hot plasma spot to the electrode tips (at $I = 1$ A: 160 μm for the flat cathode tip, 21 μm for the conical cathode tip (90° apex angle), 190 μm for the flat anode tip, 49 μm for the conical anode tip (90° apex angle)). Different plasma hot-spot temperatures in anode and cathode phase depending on the arc attachment and the tendency to lower plasma hot-spot temperatures during the anode phase are also presented in Sect. 4.1.3 (especially Figs. 4.3–4.5).

The dependence of the plasma hot-spot temperature in front of the anode tip on lamp current I (Fig. 4.51) showing a local maximum and minimum for conical anode tips might be explained by a different dynamic of the increase of the local power transfer to the anode tip and the increase in the arc attachment area on the anode tip with lamp current I. A more detailed analysis of local power density, arc attachment area, and distance of hot plasma spot to anode tip depending on lamp current I and anode tip geometry should be helpful to answer this question.

The Influence of Different Electrode Shapes on the Electric Potential

Different electrode shapes do not only alter electrode and plasma temperatures, but also influence the electric potential distribution in the near-electrode plasma. How wide-stretching this influence on the electric potential is can be seen in Fig. 4.52 for

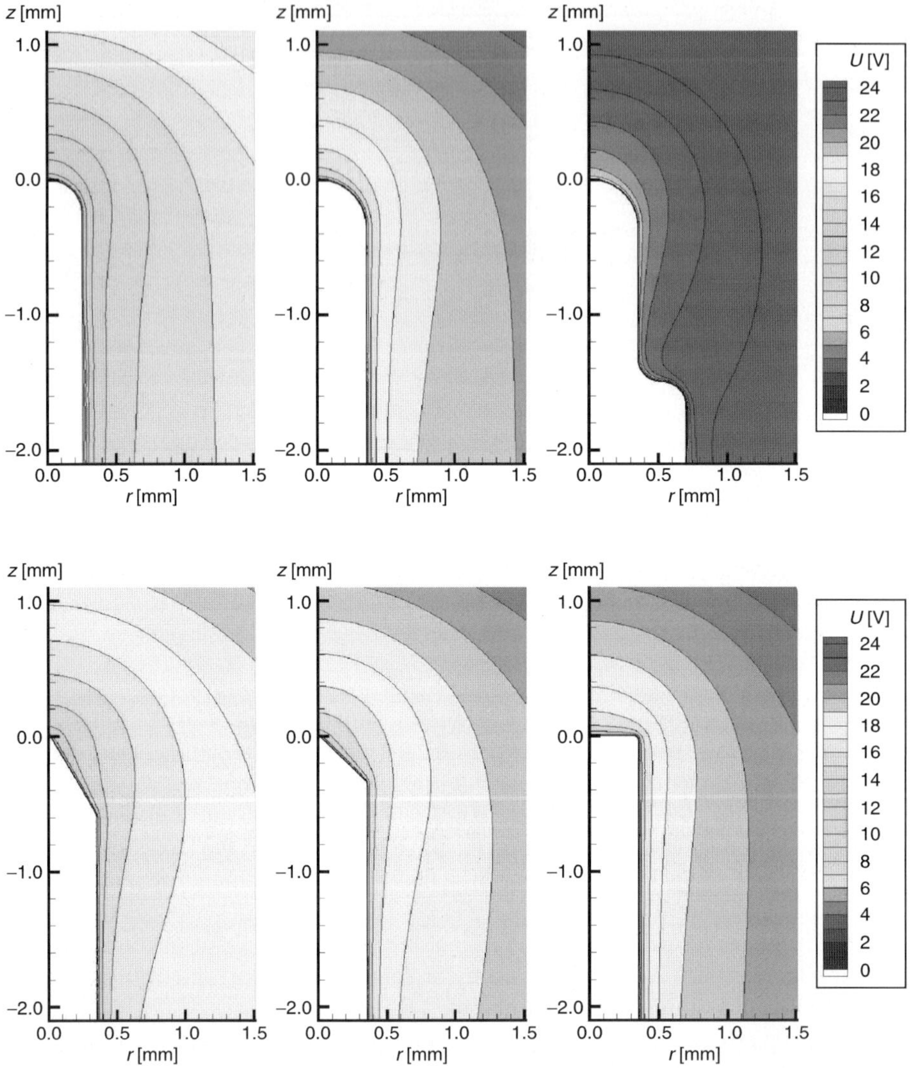

Fig. 4.52. Calculated electric potential during cathode phase for different electrode geometries (LS9hg-lamp, Hg, $p = 0.6$ MPa, $I = 1.0$ A (dc), $\varphi = 4.55$ eV, electrode length: 10 mm)

different shapes and diameters of the electrodes. At distances of about 0.2 mm (200 μm) from the cathode tip, the electric field is independent of the geometry (approximately 1 V/0.2 mm = 5 V mm^{-1}). The electric potential distribution within this 0.2 mm "zone" is strongly influenced by the geometry of the electrode tip.

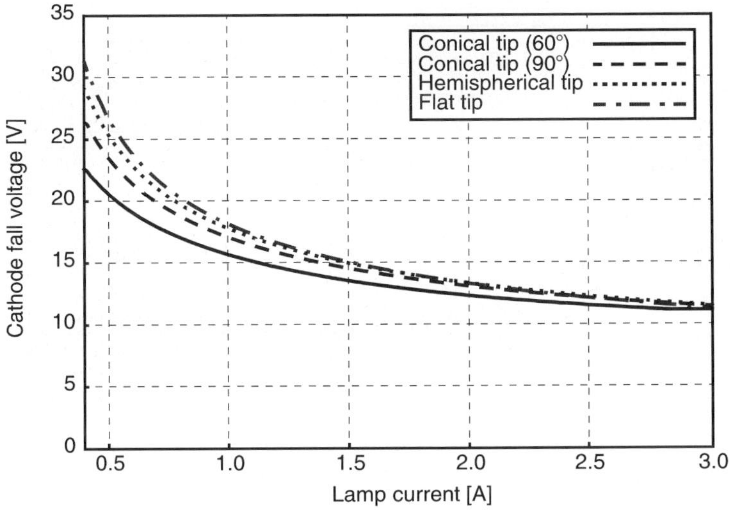

Fig. 4.53. Calculated cathode fall voltage vs. dc lamp current I for different electrode geometries (LS9hg-lamp, Hg, $p = 0.6$ MPa, $\varphi = 4.55$ eV, electrode diameter/length 0.7 mm/10 mm, [57])

The dc lamp current for all cathodes shown is 1.0 A, the different electrode geometries correspond to Fig. 4.39. The three cathodes at the top row of Fig. 4.52 have a hemispherical tip but different diameters: 0.5 mm (left-hand side), 0.7 mm (middle), and 0.7/1.4 mm (right-hand side). The three cathodes at the bottom line of Fig. 4.52 have a diameter of 0.7 mm (i.e., the same diameter as the cathode in the middle of the top row) but different shapes of the tip: conical tip with 60° apex angle (left-hand side), conical tip with 90° apex angle (middle), and flat tip (right-hand side).

The heating of the anode and cathode requires power which is supplied by anode and cathode fall voltages. These fall voltages decrease with increasing lamp current I (Figs. 4.53 (cathode fall voltage) and 4.54 (anode fall voltage)), but the product of electrode fall voltage times lamp current I (=electrical power input into the near-electrode plasma) increases with the lamp current I. The shape of the cathode tip has an important impact on the cathode fall voltage for currents less than 1.0 A, for higher currents the differences get smaller. Small, point-like cathode tips have lower cathode fall voltages. This corresponds with the lower cathode body temperatures for small, point-like electrode tips (cf. Fig. 4.48), leading to a lower thermal heat flux towards the cathode end. The same effect – reduction of cathode body temperature and cathode fall voltage for small, point-like cathode tips – has lately been observed for so-called "super spot modes" [86]. The electrode tip of these

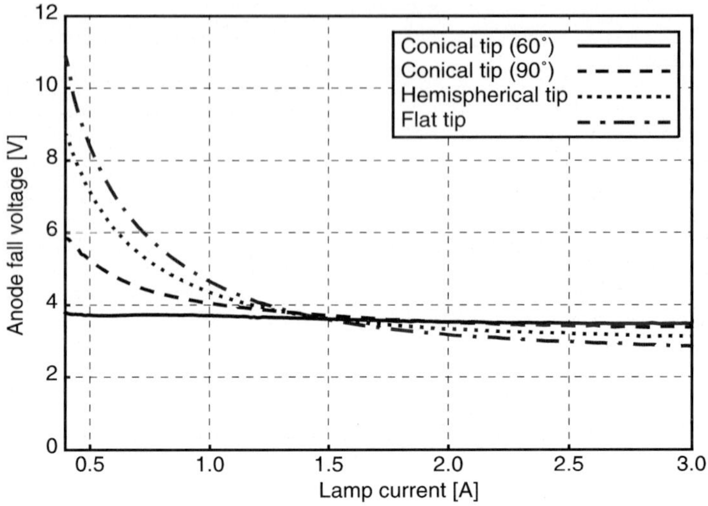

Fig. 4.54. Calculated anode fall voltage vs. dc lamp current I for different electrode geometries (LS9hg-lamp, Hg, $p = 0.6$ MPa, $\varphi = 4.55$ eV, electrode diameter/length 0.7 mm/10 mm, [57])

"super spot modes" is comparable to the conical electrode tips investigated in this section.

Similarly, the anode fall voltage corresponds with the body temperature of the anode. The observed effect of different dependences of anode body temperature on anode tip geometry for currents below or above 1.6 A, respectively (see text above and Fig. 4.49), is reflected in the behavior of anode fall voltage vs. lamp current I. As for the cathode, the anode tip geometry has an important impact on the anode fall voltage for currents below 1.0 A. For higher currents the differences get smaller.

The Influence of Different Electrode Shapes on the Arc Attachment

Finally, in Fig. 4.55 the electric current density on the cathode tip for different geometries is shown. This result may serve as a measure for the arc attachment area. The arc attachment at the conical cathode tip is much more confined compared to the arc attachment at the hemispherical or flat cathode tip. The very restricted arc attachment at a conical tip leads to high electrode tip temperatures but reduces the cathode fall voltage. Therefore, the choice of the best electrode tip geometry is a compromise between hot electrode tip temperature and low electrode fall voltage: If the electrode tip temperature gets near the melting point of tungsten, the melting of the electrode tip would contaminate the discharge, blacken the walls of the discharge vessel, and change the shape of the electrode tip. But the higher the electrode tip temperature, the lower the sum of anode and cathode fall voltage, leading to a better efficiency of the lamp.

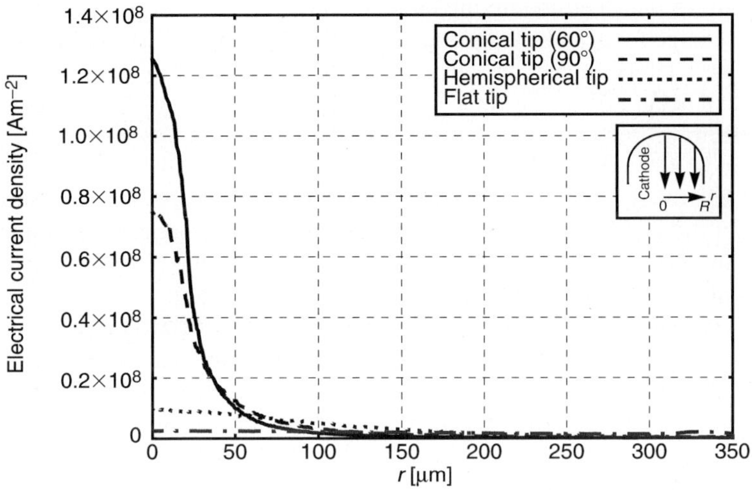

Fig. 4.55. Calculated electric current density on cathode tip for different electrode geometries (LS9hg-lamp, Hg, $p = 0.6$ MPa, $I = 1.0$ A (dc), $\varphi = 4.55$ eV, electrode diameter/length 0.7 mm/10 mm, [57])

Summary

The self-consistent electrode–plasma model is applied to a 0.6 MPa mercury discharge considering different (time dependent) lamp currents and different electrode shapes (electrode diameters 0.5, 0.7, and 0.7/1.4 mm with hemispherical, flat, and conical tip (60° and 90° apex angle)). The comparison of numerical results with experimental data for ac-operated lamps shows a good agreement of total lamp voltage and electrode temperature distribution.

Different results for different electrode shapes are compared and discussed (electrode and plasma hot-spot temperature, electrode fall voltage, arc attachment on electrode tip) with a special focus on the interaction of plasma and electrodes. An important result is that the shape of the electrode tip influences the lamp characteristics essentially. Electrode tip and electrode body temperature as well as plasma hot-spot temperature and electrode fall voltages are affected by the electrode tip geometry. Small, point-like electrode tips have high electrode tip and plasma hot-spot temperatures, but low electrode body temperatures and low electrode fall voltages (except for the anode for lamp currents above 1.6 A). This is important for the comparison of theoretical and experimental lamp data. Either one knows the "real" electrode tip geometry from the experimental setup (which is often not the case) or one has to know the impact of the shape of the electrode tip on the lamp characteristics. For the second alternative, the results presented in this section will be helpful. Furthermore, the analysis of different electrode tip geometries is important for the optimization of future electrodes. Choosing the best electrode tip geometry is a compromise between hot electrode tip temperatures and low electrode fall voltages: If the electrode tip temperature gets near the melting point of tungsten, the melting of the electrode tip might contaminate the discharge, blacken the walls of the discharge tube, and change the shape of the electrode tips. But the higher the electrode tip temperature, the lower the sum of anode and cathode fall voltage, leading to a better efficiency of the lamp. The electrode fall voltages are determined by the electrode body temperature, not by the electrode tip temperature. This result corresponds with the findings in [86] with respect to cathode temperature and cathode fall voltage after transitions between a diffuse arc attachment, a spot, and a super spot mode on the cathode tip. Furthermore, the numerical simulations show that the arc attachment is strongly influenced by the geometry of the electrode tip, which is also consistent with experimental observations [86].

An important point of these numerical simulations of HID lamps is that the interactions between electrodes and plasma are fundamental. The results on the basis of the self-consistent electrode–plasma model show that the use of one-dimensional electrode or plasma models or the use of models neglecting the hot plasma spots or the constriction zone might result in major errors concerning electrode and plasma hot-spot temperatures as well as electrode fall voltages. Further studies concerning the interaction among anode, cathode, and plasma and the dependence of this interaction on geometrical variations will be presented in Sect. 4.4.6 ("Different Electrode Geometries") and Sect. 4.4.7 ("Influence of Electrode Gap on Arc Attachment and Plasma Column")

4.4.5 External Laser Heating of Electrodes

HID lamps are usually operated on ac lamp current, so that the electrodes switch alternately from anode to cathode phase. Thus, time-dependent simulations including realistic and verified anode and cathode models allowing for the interaction between plasma and electrodes are essential. A recently published investigation of external laser heating of an electrode during anode and cathode phase in an operating HID lamp ([156] and Sect. 3.1.10) gives rise to the investigations presented in this section. These measurements reveal impressive influences of the external laser heating on electrode fall voltage and electrode temperature. Fortunately, the effects are very different during anode and cathode phase. Thus, by comparing the experimental findings with results obtained by using the self-consistent electrode–plasma model, we can learn much about the principles of electrode behavior and explain in detail the differences between anode and cathode phase. Furthermore, the self-consistent electrode–plasma model can be verified.

The Lamp and the Experimental Setup

The HID lamp under consideration contains a pure mercury plasma with a working pressure of about 0.6 MPa, the electrode gap is 10 mm. The electrodes (length 13 mm, diameter 0.5 mm) consist of pure tungsten, i.e., $\varphi = \varphi_{\text{emission}} = \varphi_{\text{cool}} = \varphi_{\text{heat}} = 4.55$ eV. The inner diameter of the quartz bulb is 12.5 mm, the shape of the lamp is sketched in Fig. 4.40,[56] a picture of the lamp is shown in Fig. 4.56. According to experimental findings, the inner quartz wall temperature and the temperature of the electrode roots[57] is set to 1,000 K [153, 156].

The lamp resembles the LS9hg-lamp under consideration in Sect. 4.4.4, where the effects of different electrode shapes have been analyzed and discussed. The only difference between the two lamps is the electrode length of 10 mm (Sect. 4.4.4) and 13 mm (this section). The plasma and electrode properties have been discussed in Sect. 4.4.4, double checking major plasma properties with [102]. The thermal conductivity λ_P includes radiative energy transport based on temperature profiles of the cylindrical plasma column as well as of the hot plasma spots as described in Sects. 2.3.3, 2.3.4, and 4.4.4. Considering the uncertainties in plasma and electrode properties discussed in Sect. 4.4.6 and the uncertainties in measured properties like working pressure [102, 156], we have a very good and reliable set of plasma and electrode properties for the numerical simulations.

The measurements were carried out on a vertically mounted lamp driven by a rectangular-wave current ($I = \pm 1.8$ A, 0.05 Hz). The lower electrode was heated using a diode laser ($\lambda = 808$ nm; maximum laser power P_L: 37.3 W), which was turned on for 3 s each half-cycle, starting 3 or 3.5 s after current reversal (half cycle length: 10 s). The laser was directed side-on at the upper part of the lower electrode just below the tip. Its beam diameter at the electrode surface was 0.5 mm [153, 156], and Fig. 4.57).

To determine the effective laser power $P_{L,eff}$ heating up the electrode (as opposed to the nominal laser power P_L, which is supposed to be provided by the laser), one electrode was heated by the laser using different laser powers without starting the discharge, i.e., the lamp was switched off. From the measured temperature profiles of the electrode, the effective laser power $P_{L,eff}$ could be determined. This has been done using a simple global power balance, namely $P_{L,eff} = P_{radiation} + P_{HeatConduction}$, where $P_{radiation}$ is the radiation cooling of the electrode and $P_{HeatConduction}$ the power transported by heat conduction within the electrode to the electrode root [156]. $P_{L,eff}$ is an important input parameter for the

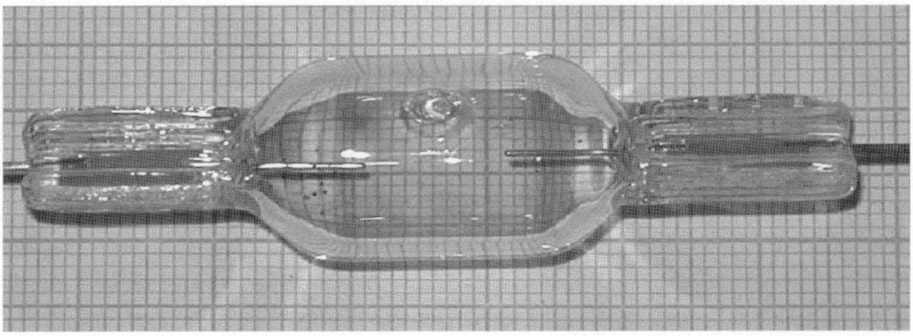

Fig. 4.56. Picture of LS9hg-lamp (courtesy of INP, Greifswald)

[56] Electrodes in Fig. 4.40 with a length of 10 mm, here the length of the electrodes is 13 mm.

[57] Bötticher uses the same value [30].

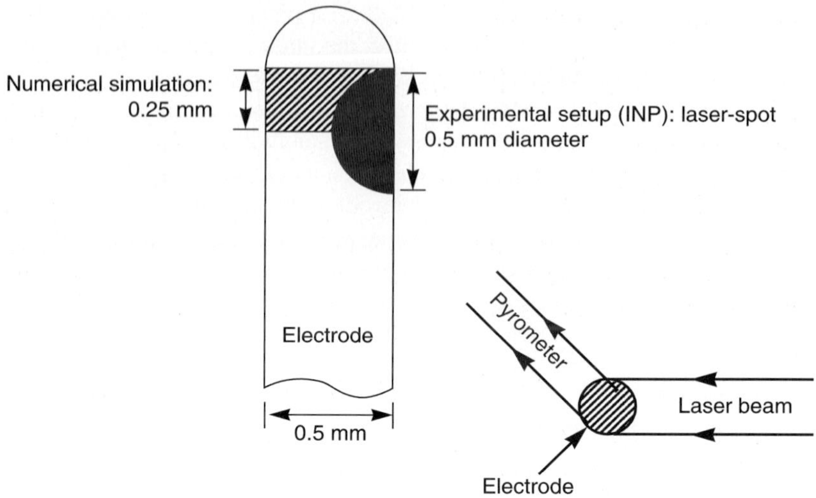

Fig. 4.57. Laser heating areas on the electrode surface in the experimental setup and for the numerical simulations with the self-consistent electrode-plasma model

numerical simulations. The problem with this procedure is that $P_{L,eff}$ might be different as soon as the discharge is turned on. Then, the electrode temperature will naturally be higher, resulting in a different emissivity $\varepsilon(\lambda,T)$ of tungsten at $\lambda = 808$ nm (cf. Fig. 2.30). If ε decreases for higher temperatures, the effective laser power should decrease as well, as ε is equal to the absorption coefficient. The result of this procedure is an efficiency η (effective laser power $P_{L,eff}$ to nominal laser power P_L) of 12–24% of the nominal laser power P_L [156]. The effective laser power $P_{L,eff}$ given for the experimental data in Figs. 4.70–4.73 corresponds to corrected[58] efficiencies (17% from 37.3 W = 6.3 W), the effective laser power $P_{L,eff}$ used for the numerical simulations corresponds to the efficiency of 12% as given in [156] (12% from 37.3 W = 4.5 W). The latter choice might account for the smaller ε for higher electrode temperatures as realized during the experiments in the operating lamp.

As described earlier and in the following, the laser power is coupled (asymmetrically) into the electrode across the spot area of the laser beam near the electrode tip, whereas the numerical simulations specify a rotational symmetric

[58] The efficiency η of 12% (= 4.5 W at maximum laser power of 37.7 W) in [156] was later corrected to values between 17% (high power = high electrode temperatures) and 24% (low power = lower electrode temperatures), confirming the above-proposed trend of ε as a function of temperature. Nevertheless, the higher electrode temperatures (smaller ε) in the operating lamp have not been taken into account for these corrected efficiencies.

input of the laser power into the electrode near the electrode tip (cf. Fig. 4.57). The electrode temperature near the electrode tip[59] and along the electrode was determined using a pyrometer (1–λ pyrometry, IMPAC IS10, cf. Sect. 3.1.4), the observation direction being at an angle of more than 90° compared to the laser beam. To avoid disturbance of the pyrometric measurements by the laser beam, a notch filter (10^{-4} at the laser wavelength) was used for the pyrometer [156]. The electrode temperature obtained from the numerical simulations can be given for every position and every point in time without any constraints in space and time. For good comparability, the electrode temperature from the numerical simulations is taken 0.5 mm (diameter of the electrode) below the electrode tip.

Adapted Boundary Conditions at the Electrode–Plasma Interface

At the boundary between cathode and plasma, the plasma temperature T_P and the electrode temperature T_E are coupled. Cooling by electron emission, heating by ion current, thermal heat flux from the plasma towards the cathode, and radiation cooling are considered. Additionally, the photons of the (external) laser beam are heating the cathode at certain times near the electrode tip. Thus, we cannot use (4.20) given in Sect. 4.4.1, but a different boundary condition has to be used, taking the additional laser heating into account:

$$\lambda_\mathrm{E}\frac{\partial T_\mathrm{E}}{\partial n} = \lambda_\mathrm{P}\frac{\partial T_\mathrm{P}}{\partial n} + L(z,t) + j_{\mathrm{t,i}}(\varphi_\mathrm{i} - \varphi_\mathrm{cool}) - j_{\mathrm{t,emission}}\varphi_\mathrm{cool} - \varepsilon\sigma_\mathrm{SB}T_\mathrm{E}^4. \qquad (4.61)$$

φ_i is the ionization energy of the gas (here 10.434 eV for mercury), φ_cool the work function for cathode cooling due to the release of an electron (cf. Sect. 4.4.6, here $\varphi = \varphi_\mathrm{emission} = \varphi_\mathrm{cool} = \varphi_\mathrm{heat} = 4.55$ eV), $j_{\mathrm{t,i}}$ the particle flux of ions, $j_{\mathrm{t,emission}}$ the particle flux of emitted electrons, ε is the emissivity of the electrode material (cf. Sect. 2.4.3), and σ_SB the Stefan–Boltzmann constant. The terms $\partial T_\mathrm{E}/\partial n$ and $\partial T_\mathrm{P}/\partial n$ denote the component of the temperature gradient normal to the electrode surface. The electron emission $j_{\mathrm{t,emission}}$ at the cathode is determined by equations given by Dyke and Dolan [47]. These equations include thermal electron emission, field-enhanced thermal electron emission, and field emission (cf. Sect. 2.4.4). The ion current density $j_\mathrm{i} = ej_{\mathrm{t,i}}$ at the cathode is set to $|j|$ - $|j_\mathrm{e}|$, if $|j| > |j_\mathrm{e}|$ and zero otherwise (cf. [144, 167, 168]).

The term $L(z,t)$ is used to couple the radiation power of the laser into the electrode (same function during anode and cathode phase). According to the experimental procedure (see earlier), $L(z,t)$ equals zero during the first three and the

[59] The measurement of the exact electrode tip temperature is not possible due to the finite spot size of the pyrometer of approximately 0.3 mm and interfering reflections of plasma radiation especially at the electrode tip (cf. Sect. 3.1.4).

last four seconds of each half period of ten seconds[60] (full period: 20 s, i.e., $f = 0.05$ Hz). During the three seconds of laser on, $L(z,t)$ is equal to zero everywhere outside $0.25\text{ mm} < z < 0.5\text{ mm}$, where $z = 0$ mm represents the electrode tip and $z = 13$ mm the electrode root. During the three seconds of laser on and $0.25\text{ mm} < z < 0.5\text{ mm}$, the function $L(z,t)$ is equal to a constant value representing the effective laser heating of the electrode. For example, for 1 W of effective laser heating this constant value would be 2.5465×10^6 W m^{-2}, i.e., 1 W of heating power spread over an area of 0.3926 mm^2 (=0.25 mm × 0.5 mm × π). This describes a rotational symmetric input of laser power into the electrode near the electrode tip. The experimental procedure is slightly different, as the laser power is coupled asymmetric into the electrode over the spot area of the laser beam near the electrode tip (cf. Fig. 4.57). The rotational symmetry of the numerical implementation of the self-consistent electrode–plasma model necessitates this difference. Considering the results, this approximation seems to be justified.

At the boundary between plasma and anode, thermal heat flux from the plasma towards the anode and radiation cooling are taken into account. The ion flux is zero at the anode and the electron flux $j_{\mathrm{t,e}}$ is heating the anode. Just as at the cathode tip, we have an additionally heating term caused by the photons of the (external) laser beam (see earlier). Thus, we cannot use (4.21) given in Sect. 4.4.1, but a different boundary condition has to be used, taking the additional laser heating into account:

$$\lambda_{\mathrm{E}} \frac{\partial T_{\mathrm{E}}}{\partial n} = \lambda_{\mathrm{P}} \frac{\partial T_{\mathrm{P}}}{\partial n} + L(z,t) + j_{\mathrm{t,e}} \varphi_{\mathrm{heat}} - \varepsilon \sigma_{\mathrm{SB}} T_{\mathrm{E}}^4 . \tag{4.62}$$

Recapitulating the above, the only difference between experimental setup and numerical simulations is the asymmetric/rotationally symmetric heating of the electrode and consequently a different "measurement" of the electrode temperature: In the experimental setup, the temperature is measured at an angle of more than 90° compared to the laser beam, the numerical simulations give a temperature at the lower edge of the rotational symmetric laser power input. Furthermore, there are experimental uncertainties concerning the effective laser power input into the electrode. Altogether, these differences or approximation are justified considering the results. The exact allocation of power input and temperature output and the exact value of the effective laser heating are apparently not crucial for the principal differences in anode and cathode behavior (change in electrode temperature and electrode fall voltage) as it will be described in the following.

[60] The exact experimental procedure is: 3.5 s laser off, 3 s laser on, 3.5 s laser off, i.e., the procedure for the numerical investigations differs by a shift of 0.5 s. Considering the numerical and experimental results, it is straightforward to see that this has no influence on the results and the comparability.

Experimental Data[61]

In the following, we will compare the experimental results with numerical simulations obtained from the self-consistent electrode–plasma model. The experimental data shown in the following are described in [156]. Nevertheless, the measured values shown here are not concordant with the values shown in [156], but originate from later, similar measurements with the same type of lamp [153]. One striking difference between the data in [156] and [153] is, for example, the absolute value of the electrode temperature in anode and cathode phase: In [156], the anode temperature (2,955 K without laser heating) is below the cathode temperature (2,995 K without laser heating), whereas for the later measurements ([153] and Fig. 4.70), the anode temperature (3,320 K without laser heating) is well above the cathode temperature (3,050 K without laser heating). The same is true for the total lamp voltage, which is above 58 V in [156] and below 53 V for the later measurements ([153] and Fig. 4.72). Consequently, without trying to discuss all possible reasons for these differences in the measured values using the same type of lamp, we should have at the back of our minds the fact that the absolute values of the measurements are not the crucial information, but the tendencies and the principal differences between anode and cathode behavior are what we are looking for in this section. The experimental uncertainties, the statistical spread of the experimental results (cf. Sect. 4.4.6 and especially Fig. 4.83), and the differences between numerical simulations and experimental realization do not influence or conceal these tendencies and the principal differences between anode and cathode behavior as can be seen in the following.

First Impression

Figure 4.58 shows the temperature distribution on the lamp axis ($r = 0$ mm), starting from the outer cathode end ($z = -13$ mm) via the cathode tip ($z = 0$ mm) through the plasma to the anode tip ($z = 10$ mm) all the way to the outer anode end ($z = 23$ mm). The extremely hot plasma spots very close to the electrode tips are clearly visible. These hot plasma spots have temperature maxima between 7,500 and 8,500 K, whereas the electrodes have tip temperatures of about 3,520 K (cathode) and 3,570 K (anode). This demonstrates again the extremely large temperature gradients between the hot plasma spots and the electrode tips, driving an intense thermal heat flux from the plasma to the electrodes. The resulting power loss can be seen in the additional voltage drop in the near electrode plasma (cf. Fig. 4.59).

The corresponding electric field is shown in Fig. 4.60 (logarithmic scale). The increase of the electric field in the near-electrode plasma can be explained using two different points of view: One possibility is to think of the power loss of the plasma due to the heat flux from the plasma to the electrodes and due to the enhanced

---

[61] Many thanks to M. Sieg, M. Kettlitz, and H. Hess from the INP Greifswald for providing the experimental data shown within this section.

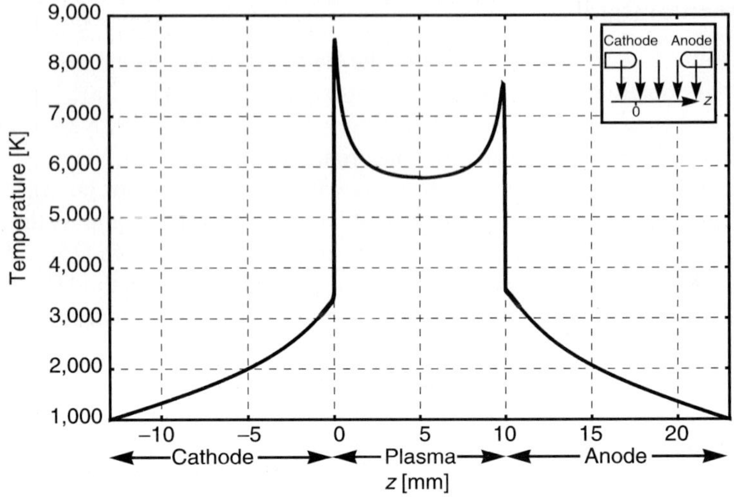

Fig. 4.58. Calculated electrode and plasma temperature on lamp axis (LS9hg-lamp, Hg, $p = 0.6$ MPa, $I = 1.8$ A (dc), $\varphi = 4.55$ eV, hemispherical tip, electrode diameter/length 0.5 mm/13 mm)

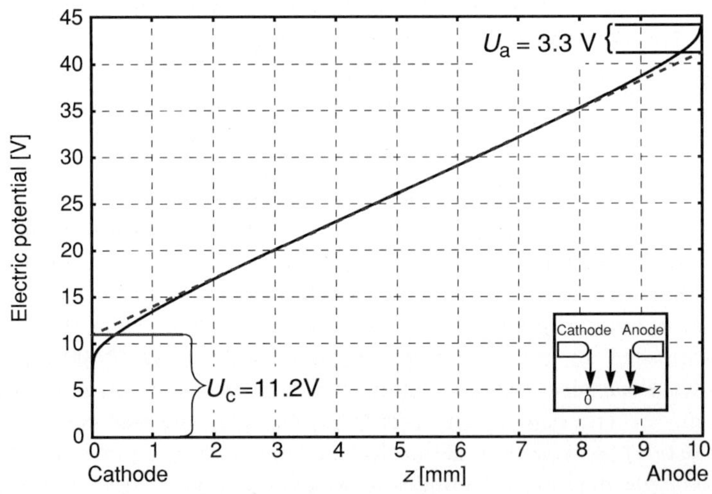

Fig. 4.59. Calculated electric potential on lamp axis and electrode fall voltages (LS9hg-lamp, Hg, $p = 0.6$ MPa, $I = 1.8$ A (dc), $\varphi = 4.55$ eV, hemispherical tip, electrode diameter/length 0.5 mm/13 mm)

radiation by the hot plasma spots (cf. Fig. 4.67). This power loss is balanced by an increased electrical power input (σE^2), thus an enlarged electric field can be observed near the electrodes.

A different point of view is given by the current continuity equation: The lamp current must be passed through the constriction zones and the near-electrode

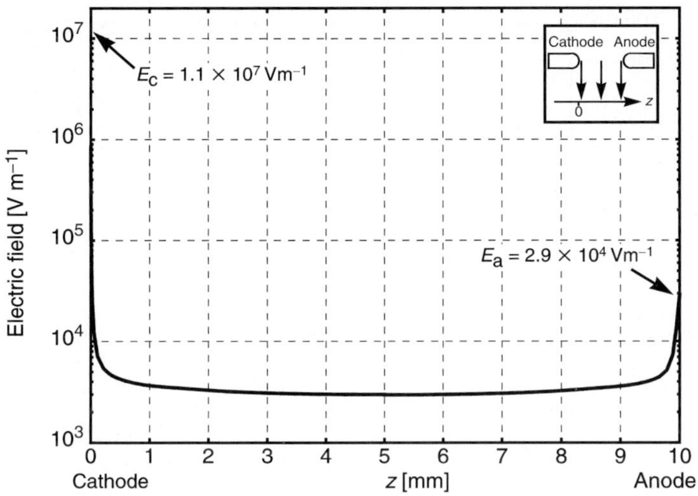

Fig. 4.60. Calculated electric field (logarithmic scale) on lamp axis (LS9hg-lamp, Hg, $p = 0.6$ MPa, $I = 1.8$ A (dc), $\varphi = 4.55$ eV, hemispherical tip, electrode diameter/length 0.5 mm/13 mm)

plasma. In the constriction zones, the relatively broad electric current profile (cf. Fig. 4.61, full width at half maximum: approximately 3.6 mm) must be "squeezed" to the size of the arc attachment on the electrode tips, which is smaller or comparable to the diameter of the electrodes (here 0.5 mm). The result is an increase of the electric current density on the lamp axis as shown in Figs. 4.62 (logarithmic scale), 4.63 (double logarithmic scale, near-cathode plasma), and 4.64 (double logarithmic scale, near-anode plasma). This increase of the electric current density $j = \sigma E$ causes the electric field to raise if the electrical conductivity would remain constant. So, we have to take a look at the electrical conductivity on the lamp axis: Figures 4.65 and 4.66 show the calculated electrical conductivity (solid line) on the lamp axis in a double logarithmic scale in the near-cathode and near-anode plasma, respectively. Comparing Figs. 4.63 and 4.65 and Figs. 4.64 and 4.66 respectively, it appears that the increase in the electric current density in the constriction zone is largely balanced by the increase of the electrical conductivity. This increase of the electrical conductivity is due to the increase of the plasma temperature caused by the hot plasma spots (cf. Fig. 4.58). Thus, the electric field increases only moderately due to the increase in the electric current density near the electrodes. But in the close vicinity of the electrode tips (a plasma layer with thickness of about 0.1 mm on the electrodes), the electrical conductivity decreases whereas the electric current density even increases getting closer to the electrode tips. This is responsible for the steep increase of the electric field in the near-electrode plasma.

Figures 4.65 and 4.66 also demonstrate the "working principle" of the self-consistent electrode–plasma model used within Sect. 4.4: The calculated non-LTE electrical conductivity is equal to the LTE electrical conductivity nearly in the whole

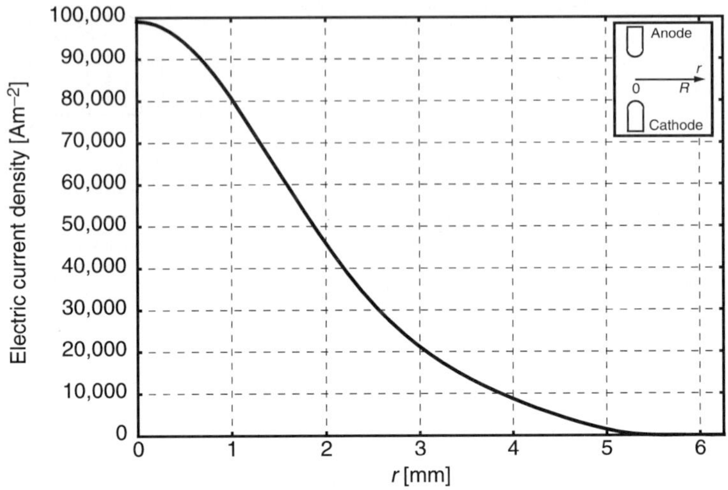

Fig. 4.61. Calculated electric current density midway between electrodes (LS9hg-lamp, Hg, $p = 0.6$ MPa, $I = 1.8$ A (dc), $\varphi = 4.55$ eV, hemispherical tip, electrode diameter/length 0.5 mm/13 mm)

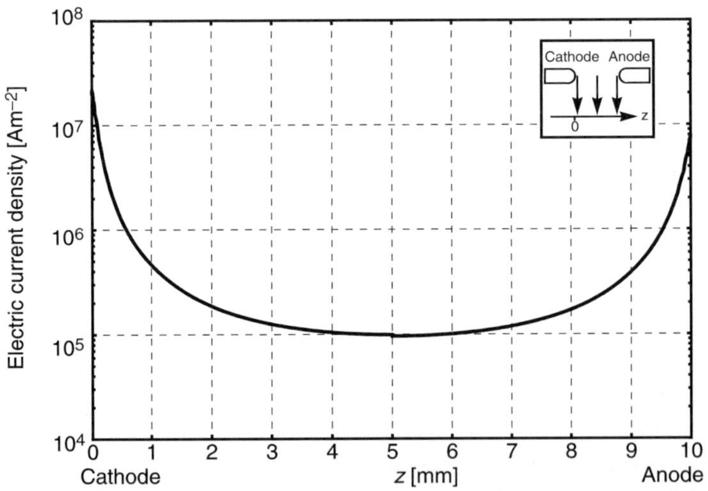

Fig. 4.62. Calculated electric current density (logarithmic scale) on lamp axis (LS9hg-lamp, Hg, $p = 0.6$ MPa, $I = 1.8$ A (dc), $\varphi = 4.55$ eV, hemispherical tip, electrode diameter/length 0.5 mm/13 mm)

plasma. Only in the close vicinity of the electrode tips, the non-LTE electrical conductivity is strongly enhanced compared to the LTE electrical conductivity due to the diffusion of electrons from the hot plasma spots to the relatively cold near-electrode plasma (cf. Sect. 4.4.1). In the near-cathode plasma, this diffusion is

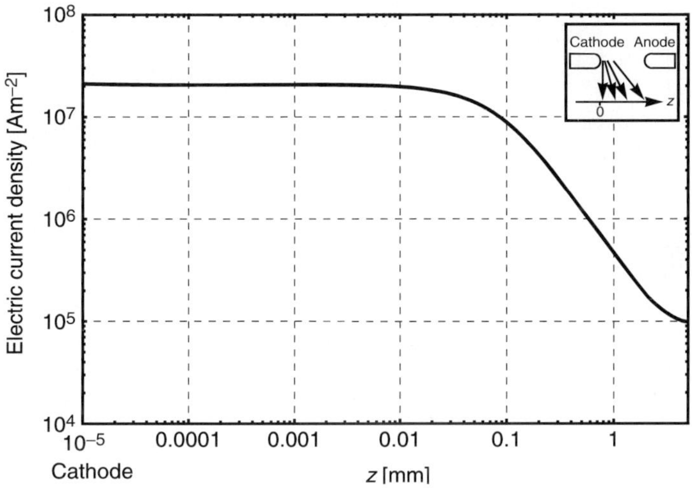

Fig. 4.63. Calculated electric current density (double logarithmic scale) on lamp axis near cathode (LS9hg-lamp, Hg, $p = 0.6$ MPa, $I = 1.8$ A (dc), $\varphi = 4.55$ eV, hemispherical tip, electrode diameter/length 0.5 mm/13 mm)

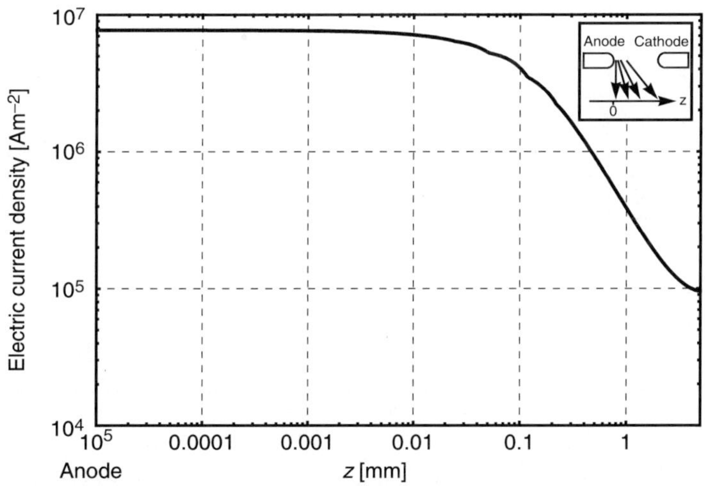

Fig. 4.64. Calculated electric current density (double logarithmic scale) on lamp axis near anode (LS9hg-lamp, Hg, $p = 0.6$ MPa, $I = 1.8$ A (dc), $\varphi = 4.55$ eV, hemispherical tip, electrode diameter/length 0.5 mm/13 mm)

against the electric field, whereas in the near-anode plasma, the diffusion is enhanced by the electric field.

Concerning the power balance, Figs. 4.67 and 4.68 are useful visualizations for the conditions in this high-pressure mercury discharge lamp. Figure 4.67 depicts the power densities on the lamp axis: The electrical power input is transformed into

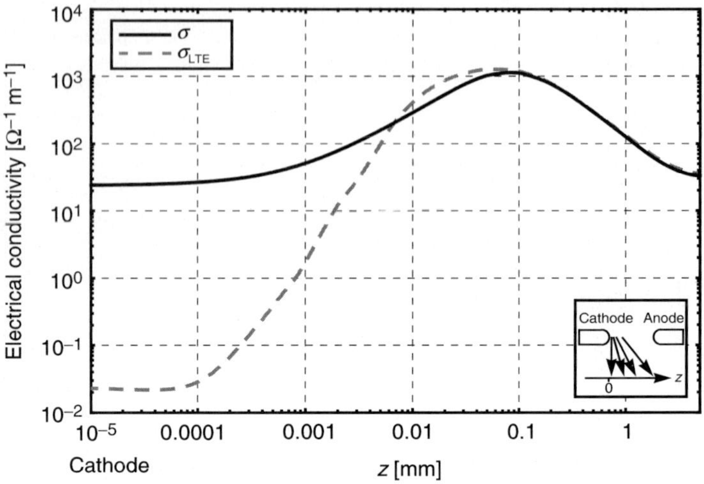

Fig. 4.65. Calculated electrical conductivity (LTE and non-LTE, double logarithmic scale) on lamp axis near cathode (LS9hg-lamp, Hg, $p = 0.6$ MPa, $I = 1.8$ A (dc), $\varphi = 4.55$ eV, hemispherical tip, electrode diameter/length 0.5 mm/13 mm)

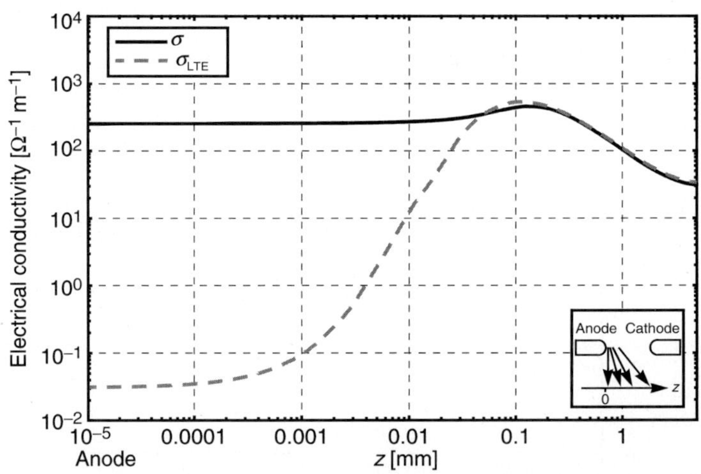

Fig. 4.66. Calculated electrical conductivity (LTE and non-LTE, double logarithmic scale) on lamp axis near anode (LS9hg-lamp, Hg, $p = 0.6$ MPa, $I = 1.8$ A (dc), $\varphi = 4.55$ eV, hemispherical tip, electrode diameter/length 0.5 mm/13 mm)

radiation, the rest is drawn off to the tube walls and the electrodes. In the plasma column, the electrical power input is largely transformed to radiation. This radiation gets even stronger at the hot plasma spots but decreases to very small values in the relatively cold near-electrode plasma. In the close vicinity to the electrode tips, nearly all of the high electrical power input is transported via heat conduction to the electrodes. A similar picture for the power balance in the plasma column vs. radius is given in Fig. 4.68. The electrical power input decreases from the plasma core to the tube walls. In the outer regions of the plasma, nearly all of the electrical power input is transformed to a heat flux to the tube walls.

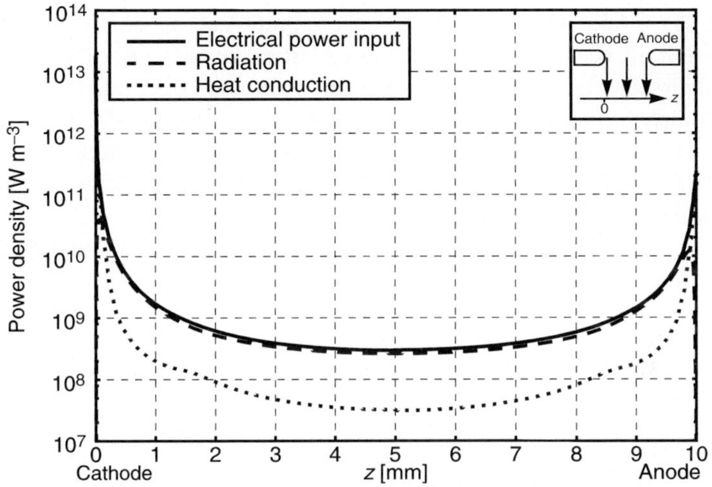

Fig. 4.67. Calculated power densities (logarithmic scale) on lamp axis (LS9hg-lamp, Hg, $p = 0.6$ MPa, $I = 1.8$ A (dc), $\varphi = 4.55$ eV, hemispherical tip, electrode diameter/length 0.5 mm/13 mm)

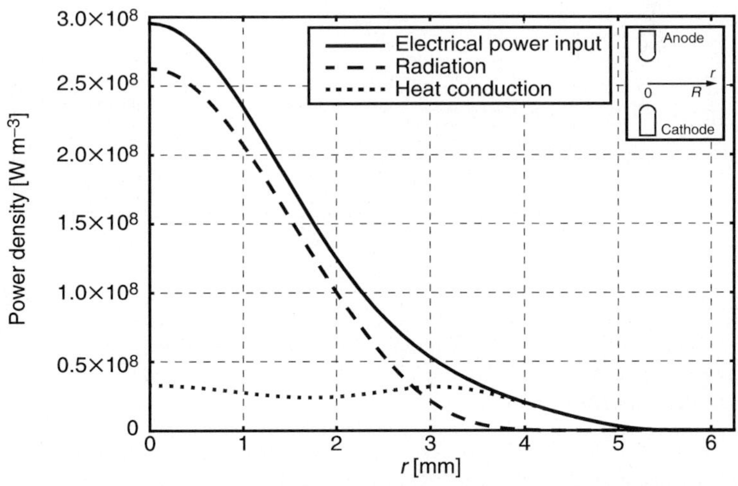

Fig. 4.68. Calculated power densities midway between electrodes (LS9hg-lamp, Hg, $p = 0.6$ MPa, $I = 1.8$ A (dc), $\varphi = 4.55$ eV, hemispherical tip, electrode diameter/length 0.5 mm/13 mm)

Electrode Temperature Distribution

A first impression of the electrode temperature in the lamp under consideration is shown in Fig. 4.69. The electrode temperature in anode and cathode phase without laser heating is given (quasistationary). The experimental values[62] (solid lines) agree very well with the results from the numerical simulations (dashed lines), especially if we take the experimental uncertainties into account (see earlier). The $z = 0$ position of the experimental data has been chosen according to the maximum value of the measured values. Due to the finite size ($d = 0.3$ mm) of the target spot of the pyrometer, this does not correspond to the very electrode tip.[63] The maximum value ought to be half the diameter of the target spot (=0.15 mm) away from the electrode tip (whole target spot on the electrode). This means that the experimental values can be shifted by 0.15 mm to the left.

Increasing the ac frequency to 50 or 400 Hz (rectangular-wave current), the electrode temperature does not settle just between anode and cathode dc temperatures, but tends very clearly toward the dc cathode temperature (cf. numerical results shown in Fig. 4.46). This is true for the experimental values [153]

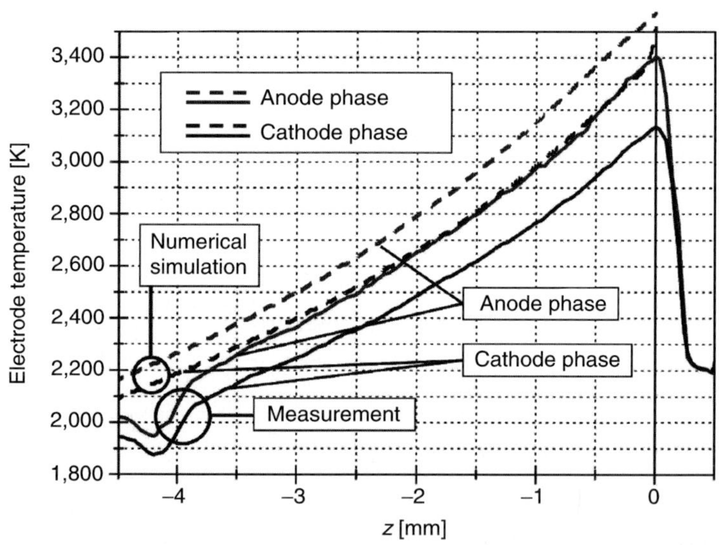

Fig. 4.69. Comparison of measured ([153, 156], *solid lines*) and calculated (*dashed lines*) electrode temperatures in anode and cathode phase without laser heating (LS9hg-lamp, Hg, $p = 0.6$ MPa, $I = \pm 1.8$ A (quasistationary), $\varphi = 4.55$ eV, hemispherical tip, electrode diameter 0.5 mm, electrode length:13 mm, [63])

[62] The deviations near $z = -4$ mm are due to optical distortions.

[63] See for example Fig. 4.83.

as well as for the results from the numerical simulations with the self-consistent electrode–plasma model.[64] Furthermore, the anode temperature is influenced much more by the lamp current I than the cathode temperature (cf. e.g., [43, 153], or Sect. 4.4.4). A discussion will be given later in this section. It should be noted that Fig. 4.46 shows numerical results for a 10 mm long electrode with a diameter of 0.7 mm. Compared to the 13 mm long and 0.5 mm thick electrode dealt within this section,[65] the dc cathode temperature for this electrode lies well above the dc anode temperature. This means that anode and cathode phase not only depend differently on lamp current I, but also the geometry of the electrode has a different effect during anode and cathode phase.

Electrode Tip Temperature during External Laser Heating

Figure 4.70 shows the comparison of measured and calculated electrode tip temperature during anode and cathode phase for different effective laser powers heating the electrode. Bearing all the differences and approximations described earlier in mind, the agreement is very good. The basic result is that the electrode temperature during the anode phase reacts much stronger to the external laser heating compared to the reaction during the cathode phase. Moreover, numerical simulations and experimental finding agree as well with respect to the response time: At current reversal ($t = 0$ and 10s), laser on ($t = 3/3.5$ and 13/13.5s), and laser off ($t = 6/6.5$ and 16/16.5s), the temperature adjustment is much faster during the cathode phase.

During the anode phase, the electrode was only heated with moderate laser powers up to $P_L = 10.5$ W ($P_{L,eff} = 2.4$ W) to prevent the melting of the anode tip [153]. The numerical simulation also predicts the melting[66] of the electrode tip during the anode phase for effective laser powers above 4 W ($T_{ElectrodeTip} > T_{MeltTungsten} = 3{,}680$ K). A more detailed presentation of the electrode temperature during the cathode phase is shown in Fig. 4.71.

Electrode Fall Voltages during External Laser Heating

The influence of the external laser heating on the lamp voltage is shown in Fig. 4.72. The measured value is the total lamp voltage (please note the negative voltage scale during the cathode phase). As the laser heating has only influence on the near-

[64] As a consequence thereof, one cannot just take anode and cathode temperature from dc operation (calculated or measured) and calculate ac electrode temperatures by averaging the dc values. The (time dependent) ac electrode temperature depends nontrivial on lamp current, current frequency, current waveform, electrode diameter and length, work function, mode of arc attachment, etc.

[65] All other parameters like lamp current, work function, etc. are identical.

[66] The latent heat of melting was not taken into account in this section.

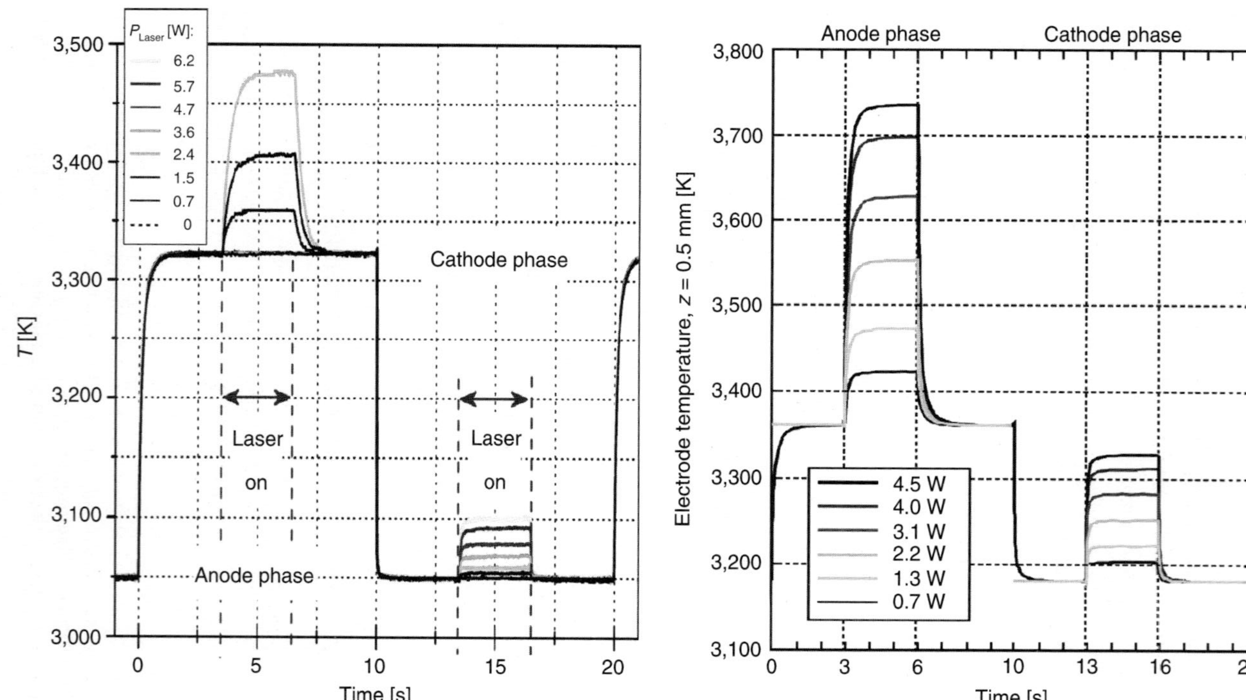

Fig. 4.70. Comparison of measured ([153, 156], left-hand side) and calculated (right-hand side) electrode temperatures in anode and cathode phase during laser heating (LS9hg-lamp, Hg, $p = 0.6$ MPa, $I = \pm 1.8$ A, 0.05 Hz rectangular-wave lamp current, $\varphi = 4.55$ eV, hemispherical tip, electrode diameter/length: 0.5 mm/13 mm, [63])

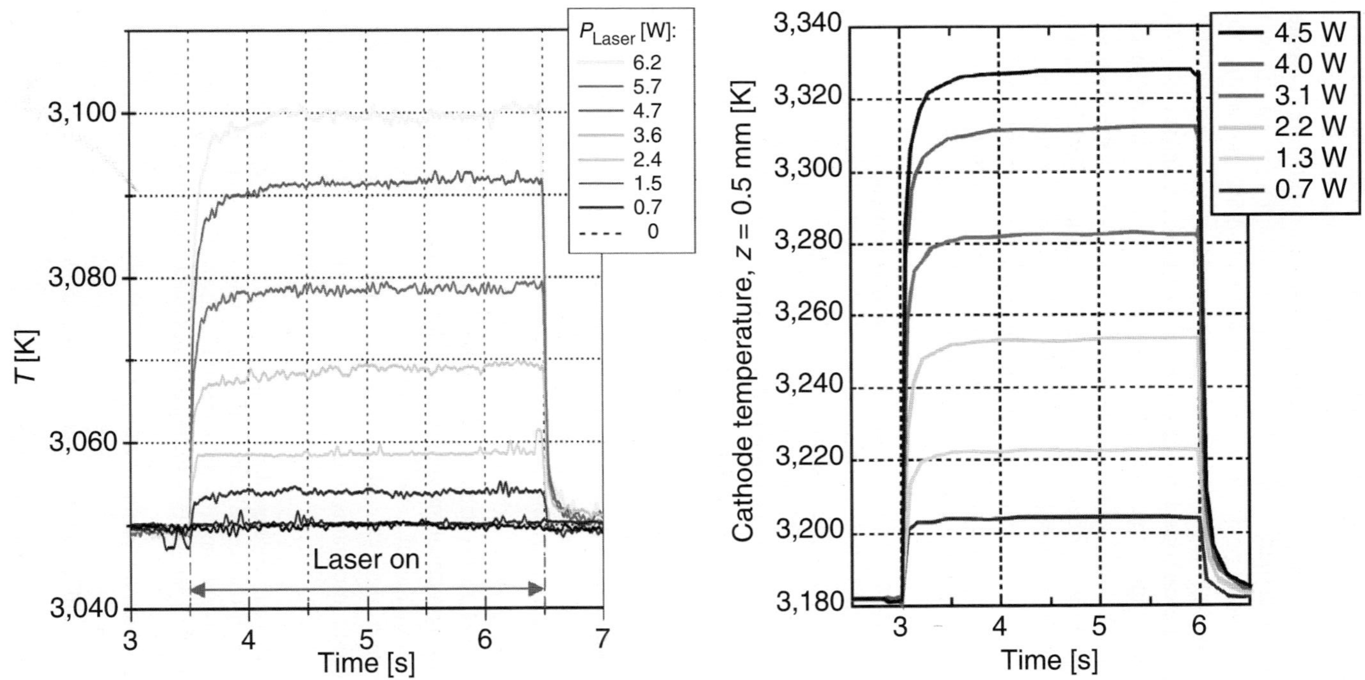

Fig. 4.71. Comparison of measured ([153, 156], left-hand side) and calculated (right-hand side) cathode temperatures during laser heating (LS9hg-lamp, Hg, $p = 0.6$ MPa, $I = \pm 1.8$ A, 0.05 Hz rectangular-wave lamp current, $\varphi = 4.55$ eV, hemispherical tip, electrode diameter/length: 0.5 mm/13 mm, [63])

electrode plasma,[67] the measured change in total lamp voltage is probably due to a change in anode or cathode fall voltage. This is confirmed by the numerical simulations using the self-consistent electrode–plasma model described in Sects. 4.4.1 and 4.4.2, where anode and cathode fall voltages U_a and U_c are computed by extrapolation of the linear part of the electric potential distribution on the lamp axis toward anode or cathode position (cf. Figs. 3.40, 4.44, 4.59, and 4.80). The difference between this extrapolation and the electric potential at the cathode surface (= 0 V by definition) is the cathode fall voltage (here between 9.5 and 11.2 V), the difference between the extrapolation and the electric potential at the anode surface (= total lamp voltage) is the anode fall voltage (here 3.2–3.3 V). This definition is similar to the typical experimental determination of cathode and anode fall voltages. The electrode fall voltages correspond to the additional power consumption in the near-electrode plasma,[68] i.e., the extra demand of electrical power input due to the losses to anode and cathode and due to the increased radiation output by the bright hot plasma spots (cf. (4.18)). The experimental measurement of electrode fall voltages is difficult (cf. Sects. 3.2.2 and 3.2.3), so only values for the total lamp voltage have been measured. Our numerical simulations deliver total lamp voltage as well as anode and cathode fall voltage, the latter even separated into the part of the cathode fall voltage necessary for the production of ions near the cathode surface (= V_0, cf. (4.23)) and the part of the cathode fall voltage necessary to balance losses due to heat conduction from plasma to cathode and due to the additional radiational "losses" by the hot plasma spot in front of the cathode (= $U_c - V_0$). The given electrode fall voltages are the anode fall voltage during the anode phase and the cathode fall voltage during the cathode phase, i.e., we do not give the so-called "electrode-sheath voltage," which is the sum of anode and cathode fall voltage (cf. [102]).

The striking feature comparing measured lamp voltage and calculated electrode fall voltages is the strong influence of the external laser heating on the cathode fall voltage during the cathode phase and the hardly observable influence on the anode fall voltage during the anode phase (Figs. 4.72 and 4.73). This is contrary to the observation of the electrode temperature, which reacts strongly during the anode phase but weakly during the cathode phase. Also, numerical simulations and experimental finding agree well with respect to the response time (cf. especially Fig. 4.73).

The strong influence of the external laser heating on the cathode fall voltage is due to the self-regulating mechanism of the cathode. The laser heating of the

---

[67] This is not the absolute truth, because the lamp pressure will rise a little bit due to the higher plasma temperature in the near-electrode plasma. Because the cathode temperature rises only slightly during the laser heating (thus the near-cathode plasma temperature rises only slightly) and because the change in the cathode fall voltage is quite high, the reason for this change in the measured total lamp voltage is almost exclusively due to the change in the cathode fall voltage. The situation is different during the anode phase, as it will be explained in the following.

[68] In contrast, the electrode fall voltages in sheath or layer models (normally cathode fall voltage, cf. Sect. 4.1) correspond to the total voltage drop over the sheaths or layers under consideration without taking the "normal" power consumption per unit length (in the arc) into account and in general without consideration of constriction zones or hot plasma spots.

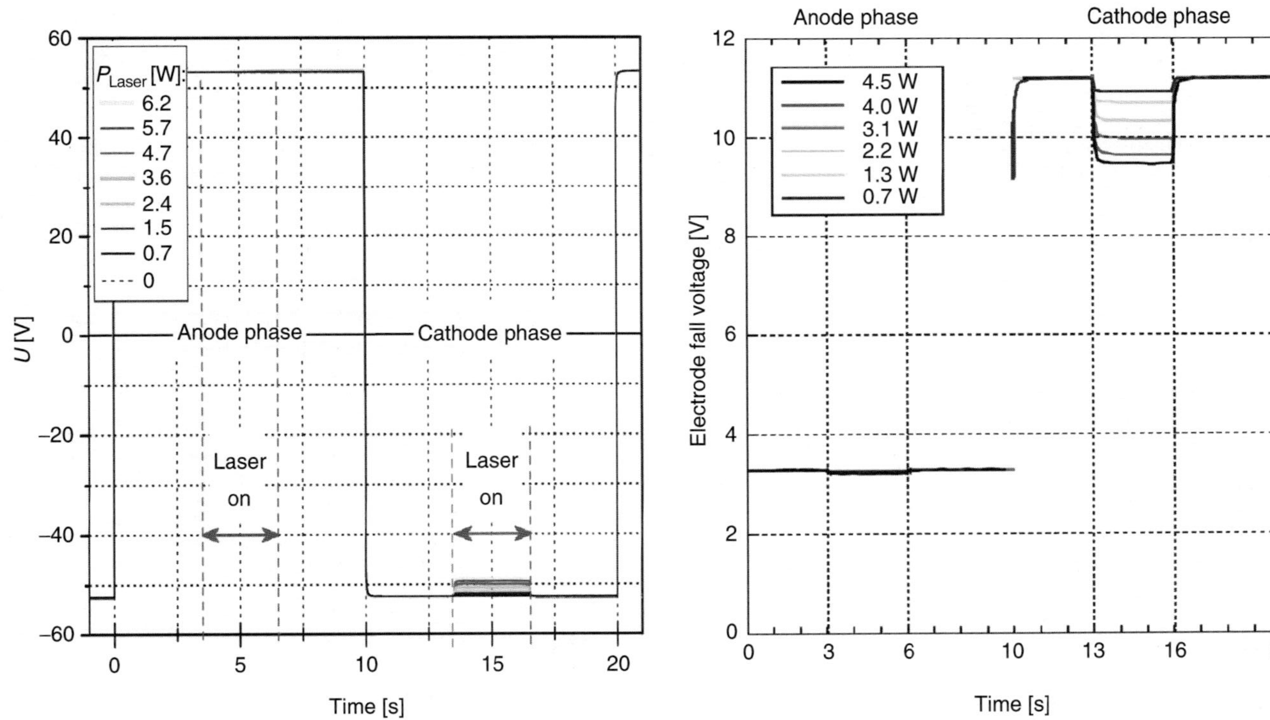

Fig. 4.72. Comparison of measured lamp voltage ([153, 156], left-hand side) and calculated electrode fall voltage (right-hand side) in anode and cathode phase during laser heating (LS9hg-lamp, Hg, $p = 0.6$ MPa, $I = \pm 1.8$ A, 0.05 Hz rectangular-wave lamp current, $\varphi = 4.55$ eV, hemispherical tip, electrode diameter/length 0.5 mm/13 mm, [63])

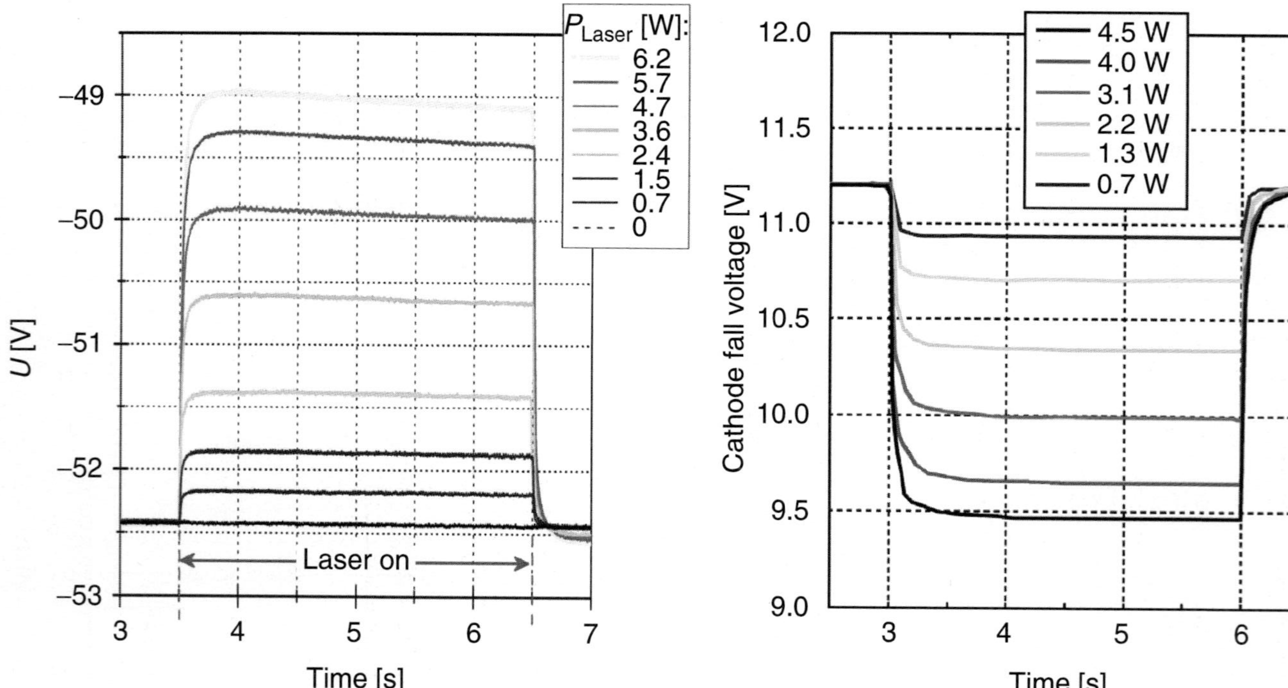

Fig. 4.73. Comparison of measured lamp voltage ([153, 156], left-hand side) and calculated cathode fall voltage (right-hand side) during laser heating (LS9hg-lamp, Hg, $p = 0.6$ MPa, $I = \pm 1.8$ A, 0.05 Hz rectangular-wave lamp current, $\varphi = 4.55$ eV, hemispherical tip, electrode diameter/length 0.5 mm/13 mm, [63])

electrode during the cathode phase increases the electron emission current. Thus, the ion current is reduced immediately by the same amount. Consequently, the part of the cathode fall voltage necessary for the production of ions near the cathode surface (= V_0) is reduced, which can be seen in Fig. 4.73 (please note the negative voltage scale during the cathode phase for the experimental values, i.e., the absolute value of the voltage decreases during the cathode phase with increasing laser power). The approximately 1.7 V reduction in the calculated cathode fall voltage for $P_{L,eff} = 4.5$ W is cause by a reduction of 1.5 V in V_0. This corresponds to an increase of the electron current at the cathode of approximately 0.26 A (without laser heating 1.04 A electron current and 0.76 A ion current) and the corresponding decrease of the ion current at the cathode. Only 0.2 V of the 1.7 V cathode fall voltage is caused by the part of the cathode fall voltage necessary to balance losses due to heat conduction from plasma to cathode and due to the additional radiational "losses" by the hot plasma spot in front of the cathode. Apparently, the interplay between ion current and electron emission is "adjusting" cathode fall voltage, cathode temperature, and the ratio of electron to ion current.

In contrast to the cathode fall voltage, which is reduced by laser heating, the anode fall voltage is reported to raise somewhat due to the laser heating, but the measurement of these small changes is very difficult ([153] and Sect. 3.2.4). We calculate a small reduction of anode fall voltage due to the external laser heating during the anode phase (approximately 0.04 V for $P_{L,eff} = 2.2$ W), but constant pressure is assumed for the numerical simulations. Due to the increasing temperatures in the near anode plasma, the pressure in the experimental setup will rise slightly. By keeping track of the total amount of mercury in the discharge (which is easily possible) during the numerical simulations, we can predict a (not measurable) raise in the operating pressure of about 0.4% for $P_{L,eff} = 2.2$ W. According to the results presented in Sect. 4.4.7, this slightly higher pressure would result in an increase of the total lamp voltage of approximately 0.12 V. Together with the calculated decreasing anode fall voltage for constant pressure, we would get a net change in anode fall voltage of +0.08 V for $P_{L,eff} = 2.2$ W, which is comparable to the experimental findings. One of the many advantages of the numerical simulations using the self-consistent electrode–plasma model is the ability to separate different effects, which are inherently superimposed in the experimental setup.

Differences Between Anode and Cathode Phase

The additional laser heating induces a new "adjustment" of the electrode temperature. This new adjustment is very fast during the cathode phase due to the fast assimilation of ion and electron current. Such a process does not exist at the anode or during the anode phase, where the total lamp current at the electrode surface is given by the electron current.

When the electrode is heated by, for example, $P_{L,eff} = 4.5$ W, the reaction in anode and cathode phase is in principle equal. Due to the strong additional heating the electrode gets hotter. This is a fast process, which gets slower near the "end"-

temperature (see anode phase in Fig. 4.70). During the anode phase, this “end”-temperature is finally reached, but during the cathode phase, the fast (immediate) decrease of the ion current (and the corresponding decrease of the heating of the cathode by this ion current) results in a lower final temperature compared to the “end”-temperature which would be reached without the reduction of the ion-current-heating. This process takes place at the beginning of the laser heating, where the cathode temperature rises rapidly. Thus, due to the decreasing ion current heating for increasing laser heating, the cathode temperature rise is smaller and faster compared to the temperature rise during the anode phase, i.e., the immediate adjustment of the ratio of electron to ion current is responsible for the small time constant during the cathode phase. This can be seen at laser on ($t = 3/3.5$ and $13/13.5$s) and laser off ($t = 6/6.6$ and $16/16.5$s), but also at current reversal ($t = 0$ and $t = 10$s). Furthermore, in the case of ac lamp currents as shown in Fig. 4.46, this faster response time during the cathode phase is responsible for the tendency of the ac temperature to the dc cathode temperature.

The electron emission current, which is essential at the cathode or during the cathode phase of the electrode, seems to have no impact during the anode phase. Despite the strong increase in the anode tip temperature (and the correlated increase in electron emission current) due to the external laser heating, the anode fall voltage is hardly altered (compare Figs. 4.70 and 4.72). This is true for the measured values as well as for the numerical results. Also, the response time of the electrode during the anode phase (compared to the response time during the cathode phase) indicates a thermal process without the influence of the electron emission current. In the self-consistent electrode–plasma model described in Sects. 4.4.1 and 4.4.2, the electron emission current is only included in the boundary conditions at the cathode (4.20) or (4.61) but not at the anode (4.21) or (4.62), which appears to be justified by the results.

As a result of the different characteristics of anode and cathode behavior described earlier, it is not surprising that lamp current or electrode length and diameter have different impacts during anode and cathode phase. The lamp current, for example, affects the anode much more than the cathode (see also [43, 153], and Sect. 4.4.4). The cathode with its self-regulating mechanism must have a sufficiently high temperature to emit enough electrons. The difference between electron emission current and total lamp current is provided by ion current which heats the cathode. Thus, the cathode temperature is adjusted to a value where the ratio of ion to electron emission current corresponds to a stable power balance of plasma and cathode. The anode is a more “passive” component, which is heated by the incoming electron current. The higher the lamp current, the larger this heating and, therefore, the higher the anode temperature. The different influence of electrode diameter and length during anode and cathode phase can be studied comparing Fig. 4.46 ($d_{\mathrm{electrode}} = 0.7$ mm, $l_{\mathrm{electrode}} = 10$ mm) and Fig. 4.69 ($d_{\mathrm{electrode}} = 0.5$ mm, $l_{\mathrm{electrode}} = 13$ mm), where only electrode diameter and length are varied (same lamp current, work function, etc.). For the 13 mm long and 0.5 mm thick electrode (Fig. 4.69), the electrode temperature during the anode phase is higher than the electrode temperature during the cathode phase, but for the 10 mm long and 0.7 mm thick electrode (Fig. 4.46), the electrode temperature during the cathode phase is higher than the electrode temperature during the anode phase.[69]

The explanation for this is the following: The heat transfer by the incoming electron current is independent of the electrode geometry, i.e., a narrow anode will be hotter than a thick anode. The change in the heat transfer from the plasma to the anode tip as a function of the anode diameter does not compensate for this effect. In contrast, the cathode tip has to have a sufficiently high temperature to emit enough electrons. The increase in the electron emitting area and the increase of the heat conduction to the cathode root due to the increase of the diameter has only a minor influence on the cathode tip temperature. Instead, the ratio of ion to electron emission current is adjusting the cathode temperature to ensure a stable power balance of plasma and cathode. The cathode tip temperature is a bit lower for a thick electrode, but the difference is much smaller compared to the temperature difference observed during the anode phase.

Summary

The comparison of the numerical results obtained with the self-consistent electrode–plasma model with different experimental data shows a very good agreement. The calculated temperature distribution along the electrodes (Fig. 4.69) and the tendency of the ac electrode temperature to the dc cathode temperature (Fig. 4.46) agree very well with the experimental findings. The investigations including the external laser heating of the electrodes revealed even more interesting results: The electrode temperature reacts strongly on the external laser heating during the anode phase but weakly during the cathode phase (Figs. 4.70 and 4.71). In contrast, the striking feature comparing measured lamp voltage and calculated electrode fall voltages is a strong influence of the external laser heating on the cathode fall voltage during the cathode phase and a barely observable influence on the anode fall voltage during the anode phase (Figs. 4.72 and 4.73). Also, numerical simulations and experimental findings agree with respect to the response time (Figs. 4.70–4.73).

The numerical simulations are able to explain all these effects in great detail. The strong influence of the external laser heating on the cathode fall voltage is due to the self-regulating mechanism of the cathode: The power brought into the electrode by the external laser heating partly "replaces" the heating by ion current. This leads to a reduced cathode fall voltage (less power needed for the generation of ions), but results also in an only moderately increased electrode temperature (more power input due to the external laser heating but less power input due to reduced ion current). The anode reacts to the external laser heating with an increased electrode temperature, but the anode fall voltage is hardly influenced. The observed small

[69] Figure 4.46 ($d_{\text{electrode}} = 0.7$ mm, $l_{\text{electrode}} = 10$ mm): $T_{\text{a,tip}} = 2{,}950$ K, $T_{\text{c,tip}} = 3{,}350$ K, and Fig. 4.69 ($d_{\text{electrode}} = 0.5$ mm, $l_{\text{electrode}} = 13$ mm): $T_{\text{a,tip}} = 3{,}550$ K, $T_{\text{c,tip}} = 3{,}450$ K (all values taken from the numerical simulations).

increase of the anode fall voltage can be explained with the help of the numerical simulations and is due to a small increase of the operating pressure (<1%).

Furthermore, the numerical simulations explain the different response times during anode and cathode phase at current reversal or laser on/off: The cathode reacts faster because its self-regulating mechanism is determined by the fast adjustment of the ratio of electron to ion current, whereas the longer response time during the anode phase is determined by the thermal inertia of the anode. Moreover, from the comparison of experimental results and numerical simulations we can deduce that the electron emission current, which is essential at the cathode or during the cathode phase of the electrode, has no impact during the anode phase. A further result of the numerical simulations is the explanation of the different impact that lamp current or electrode length and diameter have during anode and cathode phase: The anode, being a more "passive" component, reacts more strongly to variations in lamp current or electrode length and diameter. The cathode with its self-regulating mechanism is able to "compensate" such variation to a large extend by adjusting the ratio of electron to ion current. An interesting open question for further investigations is the usage of external laser heating on cathodes with no or very small ion current (cathode heated by heat conduction from plasma to temperatures high enough for sufficient electron emission to supply the lamp current I, cf. e.g., Sect. 4.4.6). In this case, the laser heating cannot "replace" the ion current and the cathode might react differently. Alternatively, the laser heating during the cathode phase might be further enhanced, overcompensating the ion current heating.

Together with the analysis of the impact of different electrode tip geometries on electrode and plasma temperature and electrode fall voltages (Sect. 4.4.4), the differences between anode and cathode phase described within this section can be used for the optimization of electrode design. In the case of a dc lamp current, anode and cathode can be optimized independently, whereas the electrodes have to be optimized for anode and cathode phase simultaneously in the case of an ac lamp current. In the latter case, the different response times during anode and cathode phase might be of particular importance.

4.4.6 D2 Automotive Headlight Lamp

The Lamp

The first generation of Xenon automotive headlight lamps (so called D1 and D2 lamps) was introduced in 1991/1992 (cf. Sect. 1.3.8). The D2 lamp is a prominent example of modern short arc high-intensity discharge lamps (metal halide lamp). It brought a quantum leap with respect to light intensity and lumen output (about 3,200 lm) compared to halogen lights (about 1,500 lm) with power consumption of only 35 W instead of about 50 W in the case of halogen lights (H7). This increase in light intensity can be used, for example, to offer different light distributions depending on the traffic situation like town light, crosscountry light, and motorway light (adaptive headlamp technologies [67]). The geometry of the D2 automotive headlight lamp is given in Fig. 4.74. The electrodes have a length of only 1.2 mm in

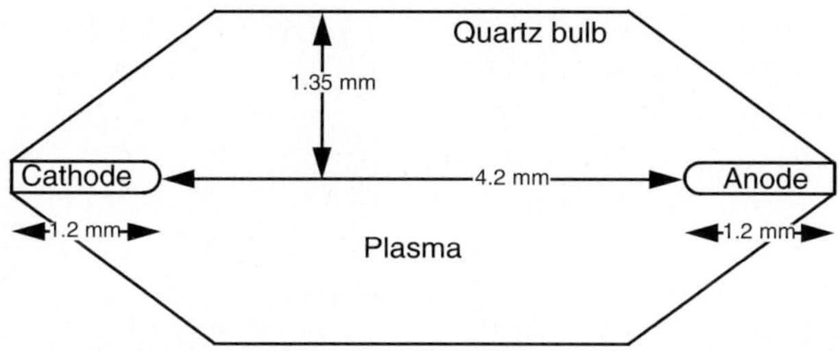

Fig. 4.74. Geometry of D2 automotive headlight lamp (electrode gap 4.2 mm, quartz diameter 2.7 mm, electrode length 1.2 mm, electrode diameter 0.25 mm). Electrodes change their "function" (anode/cathode) with time [59]

the discharge space, their diameter is 0.25 mm. The electrode gap is 4.2 mm and the inner diameter of the discharge tube 2.7 mm. For a better understanding, the electrodes are labeled "cathode" and "anode", but this assignment is of course time dependent due to the ac current. A polished cut image of the D2 lamp is shown in Fig. 4.75 (left-hand side). According to experimental findings in [85, 86], the electrode temperature T_{E} at the outside connections of the electrodes (electrode roots) is set to 1,900 K for thoriated tungsten electrodes and to 2,050 K for pure tungsten electrodes. The inner quartz wall temperature is set to 1,200 K. The default choice for the work function for electron emission $\varphi_{\mathrm{emission}}$ is 2.6 eV, corresponding to thoriated tungsten electrodes. Additionally, pure tungsten electrodes with $\varphi_{\mathrm{emission}} = 4.5$ eV are considered.

The filling of a D2 lamp consists of xenon (several bar in the cold lamp, about 5 MPa = 50 bar operating pressure), mercury (approximately 2 MPa = 20 bar operating pressure), NaI (operating pressure about 0.003 MPa = 0.03 bar), and ScI_3 (about 0.005 MPa = 0.05 bar operating pressure). Xenon is responsible for the so-called "Xenon-Instant-Light" at the cold start of the lamp to provide the required minimum output of 800 lumen one second after switch on. After a couple of seconds the mercury and the metal halides are vaporized. Mercury increases the resistance of the discharge, making it possible to operate the lamp at low currents (0.4 A, 400 Hz rectangular-wave current) and high voltages (85 V), which is desirable with respect to low strain and long lifetime of the electrodes. The excitation energies of sodium and scandium lie mainly in the visible part of the electromagnetic spectrum, rendering light output largely in the visible. Metal halides are used because their vapor pressure is higher than the vapor pressure of the metals themselves, i.e., the metal halides are more volatile. Moreover, metal iodides are in general less aggressive than their metals, preventing the quartz wall being attacked (cf. Sect. 1.3.8).

Considering the small dimensions of the D2 lamp (cf. right-hand side of Fig. 4.75), a detailed knowledge of the (time-dependent) interaction of arc plasma and electrodes is very important for lamp efficacy, lamp stability, and lifetime of electrodes. Unfortunately, experimental investigations of such lamps are rather difficult. Their smallness and optical distortions caused by the quartz bulb are not the only challenges experimental setups have to face. Thus, numerical simulations are again an important tool for the understanding, improvement, and development

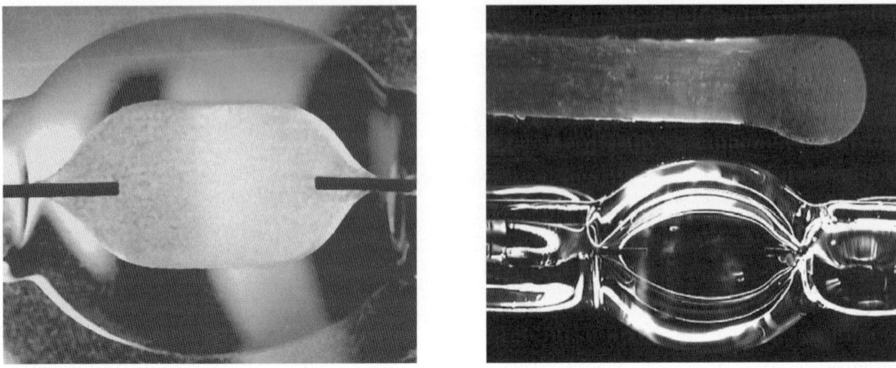

Fig. 4.75. D2 automotive headlight lamp: polished cut image (left-hand side) and comparison to standard match (courtesy of Osram, Berlin, [85])

of these HID lamps. To adequately describe these lamps, a detailed knowledge of the time-dependent properties of plasma and electrodes and their interactions is necessary. Important features to be determined are, for example, temperature distributions within plasma and electrodes, heat fluxes to electrodes and to quartz wall, electrode losses, and type of arc attachment to the electrodes.

All these desired data can be delivered by the self-consistent electrode–plasma model described in Sects. 4.4.1 and 4.4.2. Time-dependence, anode and cathode phase, interaction between plasma and electrodes, complex plasma compositions, and different electrode geometries can be addressed by the model. The results of the numerical simulations with the self-consistent electrode–plasma model and comparisons with experimental data concerning electrode and plasma temperature will be presented in the following. Because electrode and plasma properties play a decisive role for the numerical simulations, they will be discussed in detail as well in the following.

Electrode Properties

As in all other numerical simulations shown in Sect. 4.4, the thermal conductivity $\lambda_E(T_E)$ of tungsten is taken from [94] (cf. Fig. 2.27 and Sect. 2.4.1), the emissivity $\varepsilon(T_E)$ corresponds to that given in [174] (cf. Fig. 2.29 and Sect. 2.4.3). The mass density ρ of tungsten is 19,300 kg m^{-3}, the specific heat c is described by $129 + 0.0175T$ according to [31] (T in K, c in J kg^{-1}K^{-1}).

The emissivity ε might be a problem, because it depends on the surface conditions of the electrodes ([43] and Sect. 2.4.3). A rough surface has a considerably higher emissivity ε than a polished one (black body effect). This problem arises especially at the electrode tips, where the surface conditions might change with time. For the numerical simulations, this effect is not important if we assume polished surfaces. But comparing to experimental data, we have to keep in mind that the real surface might have a higher emissivity ε. If the front surface of an electrode gets deeply structured with large bumps as shown in [86], not only the emissivity ε might change, but also the emitting surface itself is increased.

The main challenge of electrode properties, however, is the work function φ (cf. Sect. 2.4.4). Within Sect. 4.4 concerned with the self-consistent electrode–plasma model, we distinguish between two work functions [144]: $\varphi_{\text{emission}}$ determines the thermionic electron emission current given for example by Richardson–Schottky. $\varphi_{\text{emission}}$ can be lowered by thoria (thorium dioxide ThO_2) at the surface of tungsten electrodes, resulting in $\varphi_{\text{emission}} = 2.6$ eV instead of $\varphi_{\text{emission}} = 4.5$ eV in the case of pure tungsten. The "second" work function is $\varphi_{\text{heat/cool}}$ and determines the heating of the anode[70] by electron absorption (φ_{heat}) and the cooling of the cathode by electron emission (φ_{cool}). Mostly, the energy gained/lost by the anode/cathode per absorbed/emitted electron is assumed to be equal to the work function for thermionic emission. But Sansonnens et al. [144] propose that the total energy gain/loss of the electrode will equal the sum of components from electrons passing through the thoria–plasma interface and the tungsten–thoria interface (and vice versa). For small percentages of thoria in tungsten, the net energy loss should be given by the work function of tungsten. So the work functions proposed in [144] are $\varphi_{\text{emission}} = \varphi_{\text{heat/cool}} = 4.5$ eV for pure tungsten electrodes and $\varphi_{\text{emission}} = 2.6$ eV and $\varphi_{\text{heat/cool}} = 4.5$ eV for thoriated tungsten electrodes (2.6 eV corresponds to the work function of thoria, since this represents the energy needed for extraction of one electron from the activated surface of the electrode). In addition to the arguments given in [144], energy conservation suggest this choice of work functions: If we have one thoriated tungsten electrode (lets say the cathode) and one pure tungsten electrode (the anode), the emitted electron would cool the cathode by 2.6 eV (if we take $\varphi_{\text{emission}} = \varphi_{\text{heat/cool}} = 2.6$ eV for thoriated tungsten at the cathode) and heat the anode by 4.5 eV, which would violate energy conservation. The energy we "pay" at the cathode for electron emission is "paid back" at the anode.[71] With the choice $\varphi_{\text{emission}} = 2.6$ eV and $\varphi_{\text{heat/cool}} = 4.5$ eV we solve this problem. For the numerical simulations of the D2 automotive headlight lamp, the default choice of work functions is $\varphi_{\text{emission}} = 2.6$ eV and $\varphi_{\text{heat/cool}} = 4.5$ eV, but we also try other combinations for comparison.

Measured values of the work function for electron emission $\varphi_{\text{emission}}$ at the cathode in an operating (!) lamp can be found in [145, 146] as described in Sect. 3.1.11. A very interesting result concerning the D2 lamp (metal halide lamp) is the lowering of the work function by the adsorption of, for example, sodium [145, 146]. If the metal halide lamp (not a D2 lamp) is operated at half the nominal power, nearly no salts are evaporated and the ("undisturbed") work function is measured to be 4.45 eV (±0.25 eV). Under nominal operating conditions, the partial pressures of the metal halides are significant, leading to adsorption of (for example) sodium at the cathode tip. The work function measured under nominal operating conditions is thus reduced to 2.41 eV (±0.2 eV). A low work function for electron emission $\varphi_{\text{emission}}$ at the cathode can thus arise from the thoria of the electrodes or (for example) the sodium of the discharge. Possibly, in D2 automotive headlight lamps, the thoria is

---

[70] In the following "anode" means "anode" for dc lamp currents and "electrode during anode phase" for ac lamp currents (likewise for the cathode).

[71] This is also reflected in anode and cathode fall voltages.

responsible for the low work function $\varphi_{\mathrm{emission}}$ at low power load (e.g., starting) and under nominal operating conditions the sodium is responsible for the low work function. This is also suggested by measurements and calculations described in [1] (cf. Sect. 3.1.12), showing a lowering of the work function for electron emission for increasing partial pressure of sodium in a sodium high-pressure discharge[72] from 4.3 to 2.7 eV (pure tungsten cathodes). In [145, 146] the work function $\varphi_{\mathrm{emission}}$ in an operating mercury HID lamp was measured to be 4.52 eV (pure tungsten electrode) and 3.7 eV (thoriated tungsten electrode), respectively.

A proposal in [112] (based on [173]) that the heating of the cathode by ion current is given by $j_{\mathrm{t,i}}\varphi_{\mathrm{i}}$ instead of $j_{\mathrm{t,i}}(\varphi_{\mathrm{i}} - \varphi_{\mathrm{cool}})$ is not employed here (cf. (4.20) or (4.61), φ_{i}: ionization energy of the gas). The same effect would have an increase in φ_{i} by φ_{cool}. This approach is questionable because the electrons "extracted" from the cathode by ions heat the anode.[73] The energy balance of the lamp would be out of equilibrium if we could heat the anode by an electron, which has not cooled the cathode beforehand. Nevertheless, for the numerical simulation of the D2 automotive headlight lamp these considerations are superfluous because there is no ion current at the cathode (see later).

In [85] (in parts also in [86]) electrode temperature measurements at different electrodes in D2 lamps are presented (see also Fig. 4.83). Three different types of electrodes are investigated: thoriated tungsten electrodes (type A), pure tungsten electrodes glowed at 1,200°C for cleaning (type B), and pure tungsten electrodes glowed at 2,400°C for cleaning (type C). For each of the three types of electrodes at least five lamps have been investigated to reduce statistical fluctuations. Thoriated tungsten electrodes should have lower electrode temperatures compared to pure tungsten electrodes due to the enhanced electron emission by a lower work function $\varphi_{\mathrm{emission}}$ for thoriated tungsten electrodes. This is true if one compares the electrode temperatures of type A with type C electrodes. But surprisingly, the pure tungsten electrodes of type B have practically the same temperature as type A electrodes (thoriated tungsten). This might have many reasons including work function $\varphi_{\mathrm{emission}}$, type of arc attachment (spot or diffuse), adsorption of, for example, sodium at electrode tip, increased front surface of deeply structured electrode tip by large bumps as shown in [86], etc. One of many possible and imaginable scenarios responsible for the large difference between type B and C pure tungsten electrodes could be the following: Due to the different "history" of the electrodes in the production process, the arc attachment on type B electrodes might be more confined than the arc attachment on type C electrodes (cf. left-hand side of Fig. 3.35 (type C electrode) and right-hand side of Fig. 3.35 (type B electrode) after 1,000 h of operating time). The sodium available in the discharge might be enough to produce a monolayer of sodium over the (small) arc attachment area of the type B electrodes with a coverage of 30–50% for instance. According to [1], where work functions

[72] Sodium pressure between 11 and 84 kPa, i.e., between 0.11 and 0.84 bar.

[73] Of course, it is not the same electron which neutralizes the ion at the cathode and heats the anode. The electron reaching the anode originates from the ionization of an atom and has been replaced many times by other electrons during the "journey" through the discharge.

depending on sodium coverage are calculated, this would lead to a work function for electron emission of around 2.5 eV for the type B electrodes. For the type C electrodes (with larger arc attachment area), the sodium available in the discharge might not be enough to create a sufficiently high sodium coverage to lower the work function for electron emission considerably, i.e., for a coverage of, for example, 10% the work function would be about 4.0 eV [1]. These two different work functions would lead to different electrode temperatures as measured and presented in [85]. Another possible scenario would be that the ability to adsorb sodium on the electrode surface might be influenced due to the different "history" of the electrodes in the production process, leading to the same result as described above.

What we have to keep in mind is that knowing material and (initial) geometry of the electrode is not always enough to characterize the electrode and its behavior. Even small differences in the production process of the electrodes [85] or different plasma compositions or electrical power inputs [146] might change the electrode properties considerably (see also Sects. 3.1.4 and 3.1.7). Especially the work function is very sensitive to plasma composition, electrode temperature, surface structure, etc. This is important to know because the work function is essential for numerical simulations of electrodes.

In [130, 168] (quoting [2]) concrete values for $\varphi_{\mathrm{emission}}$ and Richardson constant A_{R} can be found. For pure tungsten we find values for $\varphi_{\mathrm{emission}}$ between 4.2 and 6.0 eV with A_{R} equal to 0.75×10^6 AK^{-2}m^{-2} [130] and 600×10^6 AK^{-2}m^{-2} [168]. For thoriated tungsten $\varphi_{\mathrm{emission}}$ is 2.6 eV [130, 168] with A_{R} equal to 0.5×10^6 AK^{-2} m^{-2} [130] and 0.03×10^6 AK^{-2}m^{-2} [168]. In [146] the work function $\varphi_{\mathrm{emission}}$ in an operating mercury HID lamp was measured to be 4.52 eV (pure tungsten electrode) and 3.7 eV (thoriated tungsten electrode), respectively. So, although the electron emission is very important for the numerical simulation of HID lamps, we have to live with poor data concerning $\varphi_{\mathrm{emission}}$ and A_{R}. Besides the effects altering $\varphi_{\mathrm{emission}}$ stated earlier, there are many more factors which might be important for the ability to emit electrons. For pure tungsten $\varphi_{\mathrm{emission}}$ might depend, for example, on the surface state (crystal orientation, polycrystalline tungsten), for doped tungsten the local surface concentration of the dopant is important, the ambiance influences $\varphi_{\mathrm{emission}}$ (in general measurements are done in vacuum, but the real electrodes are operated in different plasmas), etc. For the numerical simulations of the D2 automotive headlight lamp, $\varphi_{\mathrm{emission}} = 4.5$ eV for pure tungsten electrodes and 2.6 eV for thoriated tungsten is chosen. A_{R} is set to the theoretical value of approximately 1.2×10^6 A K^{-2}m^{-2} (as, e.g., in [30 and 168]).

As soon as the work function $\varphi_{\mathrm{emission}}$ is known, the calculation of the electron emission current is straightforward. Thermal electron emission and field-enhanced thermal electron emission are important in HID lamps. Only for electric fields above 10^9 Vm^{-1}, pure field emission gets important, but these high electric fields are not reached in D2 lamps. A comparison of three different approaches for the calculation of the electron emission current is shown in Fig. 2.33. The Richardson–Schottky approach includes thermal electron emission and field-enhanced thermal electron emission. The approaches of Dyke and Dolan ([47], used in Sect. 4.4) and Paulini et al. [135] include additionally pure field emission. All three approaches agree on the electron emission current for a given temperature (in Fig. 2.33: 3,000 K) and an electric field up to 10^8 Vm^{-1}. Only for electric fields above 10^9 Vm^{-1} the effect of pure field emission comes into play. The work function $\varphi_{\text{emission}}$ is (besides the electrode temperature) the most important factor determining the electron emission current.

Plasma Properties

The plasma composition, the thermodynamic properties, and the transport coefficients of the plasma strongly vary with plasma temperature. The composition of the plasma and the thermodynamic properties are obtained using chemical equilibrium ([171], Sect. 2.2.2). The calculation of transport coefficients is based on the work of Hirschfelder, Curtiss, and Bird [90] using rigorous kinetic theory according to Chapman–Enskog theory ([171], Sect. 2.3). Constant pressure in the discharge can be assumed in a D2 lamp driven by a rectangular-wave current of 400 Hz.

To get a general idea of the plasma, the particle densities of the different plasma constituents are shown in Fig. 2.8 (cf. Sect. 2.2.2). According to the high partial pressures of 50 bar xenon and 20 bar mercury, the xenon and mercury atoms dominate the plasma. The overall ionization rate (electrons or ions compared to neutral particles) is well below 1% even at the highest temperatures in the hot plasma spots immediately in front of the electrodes (<8,000 K). ScI_3, Sc_2I_6, NaI, and Na_2I_2 are present at temperatures up to 3,000 or 4,000 K, then they dissociate and the Sc, Na, and I atoms outstrip the molecules (3,000–6,000 K). For even higher temperatures (6,000 K), the Sc^+ (ionization energy: 6.54 eV) and Na^+ ions (ionization energy: 5.1 eV) get ahead of the neutrals Sc and Na.

The radiation of the plasma is presented in Fig. 2.16 (cf. Sect. 2.2.3). The line radiation of sodium and scandium dominates all other radiation mechanisms. Only for very high temperatures of 6,000 K and more (as in the hot plasma spots), the line radiation of mercury matters. The net radiation emission coefficient u is calculated according to [160], taking the radiation due to recombination and bremsstrahlung additionally to line radiation into account. The mercury resonance lines at 185 and 254 nm are not included in u. These lines are treated separately for radiative energy transport, using the approximation of radiation diffusion (cf. Sect. 2.3.4 and below). For spectral lines which are partially absorbed within the plasma, a radiation transport calculation yields spectral line weakening factors χ [160]. According to [160], χ is set to 0.4 for the mercury atomic lines at 365.02, 365.48, and 546.08 nm, and $\chi = 0.5$ for the line at 435.83 nm. For the xenon infrared lines at 820.86, 828.24, 823.39, and 834.91 nm the spectral line weakening factor χ is set to 0.7.

The thermal conductivity λ_P, which is shown in Fig. 2.26, is composed of transport of kinetic energy, transport of reaction energy (here: ionization and dissociation), and radiative energy transport. The latter is important at high temperatures (above 4,000 K in the D2-discharge). It is estimated by radiation transport calculation for the resonance lines at 185 and 254 nm (cf. Sect. 2.3.4). The computation of the radiation transport is based on a given temperature profile that can be extracted from the numerical simulations to improve λ_P iteratively. It turns out that depending on the chosen temperature profile (of the cylindrical plasma column or of the hot plasma spots in front of the electrodes), contributions of radiative energy transport to λ_P can be quite different (compare $\lambda_\mathrm{rad,col}$ and $\lambda_\mathrm{rad,hs}$ in Fig. 2.26). To solve this problem, a new and successful method is to use $\lambda_\mathrm{rad,col}$ and $\lambda_\mathrm{rad,hs}$ in different temperature regions (cf. Sects. 2.3.4 and 4.4.4). For temperatures below the maximum temperature of the arc column (here 5,100 K), the radiation transport calculation is based on the temperature profile of the arc column. For higher temperatures the temperature profile of the hot plasma spots is taken into account to calculate radiation transport. This method delivers good results for the temperature profile of the plasma column as well as for the temperature distribution in the hot plasma spots (cf. Fig. 4.78). A radiation transport calculation based only on the plasma column temperature profile would result in too low plasma hot-spot temperatures, a radiation transport calculation based only on the temperature distribution within the hot plasma spots would result in too high plasma column temperatures.

Mass density ρ and specific heat at constant pressure c_p are given in Fig. 4.76. The peak in c_p at around 2,700 K is due to the dissociation of NaI and ScI_3 molecules, the increase of c_p at temperatures above 6,000 K is caused by the ionization of xenon and mercury (see also particle densities in Fig. 2.8).

An important point of the self-consistent electrode–plasma model is the calculation of a non-LTE electrical conductivity σ, based on the LTE electrical conductivity σ_LTE (cf. (4.42a)). One major assumption in the derivation of the non-LTE electrical conductivity equation is that σ is proportional to n_e/n (cf. Sect. 4.4.1). Figure 4.20 shows σ_LTE and the ratio $n_\mathrm{e,LTE}/n$. Both curves are very close to each other, which complies with the model assumption. Furthermore, it is obvious that σ_LTE varies strongly with temperature (logarithmic scale in Fig. 4.20) and is very low for temperatures below 3,000–4,000 K (typical plasma temperatures immediately in front of the electrodes). This makes the concept of a non-LTE electrical conductivity important (as for the other HID lamps discussed in Sect. 4.4).

The diffusion coefficient between ions and atoms ($D_\mathrm{i,n}$, see Fig. 4.77) has been defined as the mean value of diffusion coefficients of most frequently occurring positive ions Xe^+, Hg^+, Na^+, and Sc^+ and the mixture of neutrals of the plasma [129, 158]. The rate coefficient for ionization (q^I, see Fig. 4.23) is the weighted average value[74] of partial ionization rate coefficients q^I_Xe, q^I_Hg, q^I_Na, and q^I_Sc of most frequently occurring neutral atoms. These coefficients have been obtained beforehand from cross section data for ionization of atoms in collision with electrons [129, 158]. The ionization energy of the gas, φ_i, necessary for the boundary condition at the cathode (4.20) or (4.61), can also be calculated as the weighted average value of most frequently occurring positive ions. However, the question comes up if the corresponding ions can be supplied in a sufficient quantity in the near-cathode plasma. Thus, the ionization energy used for the numerical simulations described in this section has been set to 10.434 eV corresponding to mercury with the operating pressure of 2 MPa (xenon with the operating pressure of

[74] A similar approach as described for the calculation of q^I and $D_\mathrm{i,n}$ is used by Benilov et al. [22].

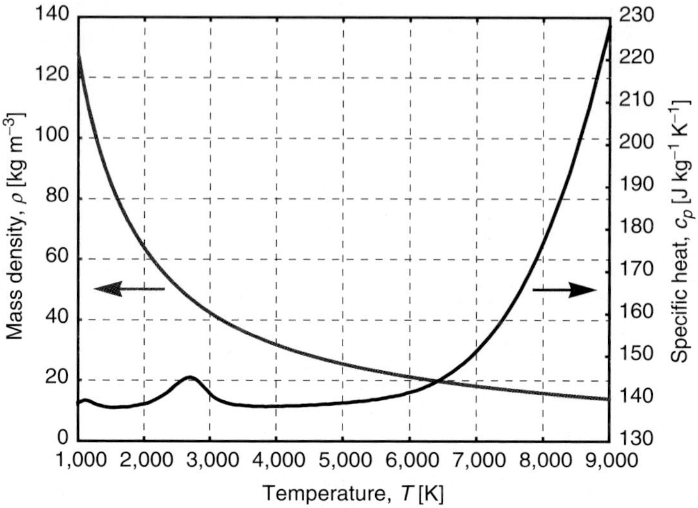

Fig. 4.76. Mass density ρ (left axis) and specific heat at constant pressure c_p (right axis) for a D2 lamp plasma (Xe, Hg, NaI, ScI_3) vs. plasma temperature [59]

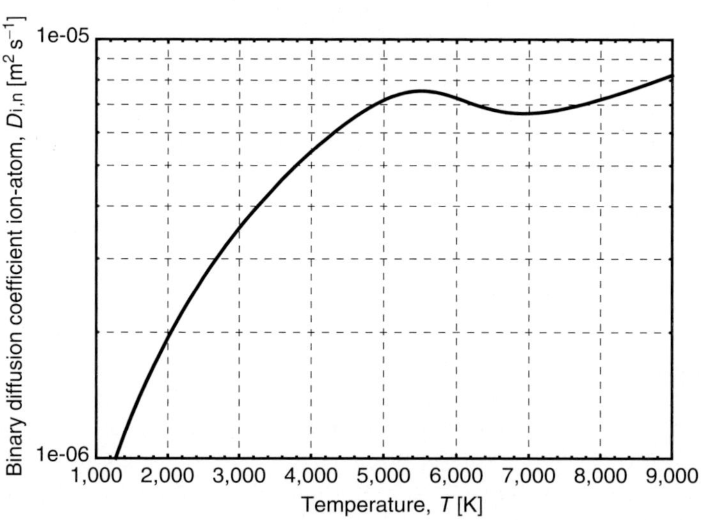

Fig. 4.77. Binary diffusion coefficient between ions and atoms $D_{i,n}$ for a D2 lamp plasma (Xe, Hg, NaI, ScI_3) vs. plasma temperature [59]

5 MPa has a much higher ionization energy of 12.13 eV and is thus ionized with lower probability). Nevertheless, for the numerical simulation of the D2 automotive headlight lamp these considerations are superfluous, because there is no ion current at the cathode (see later).

All plasma properties have been double checked with plasma properties calculated at Philips (where available). In the case of deviations (net radiation

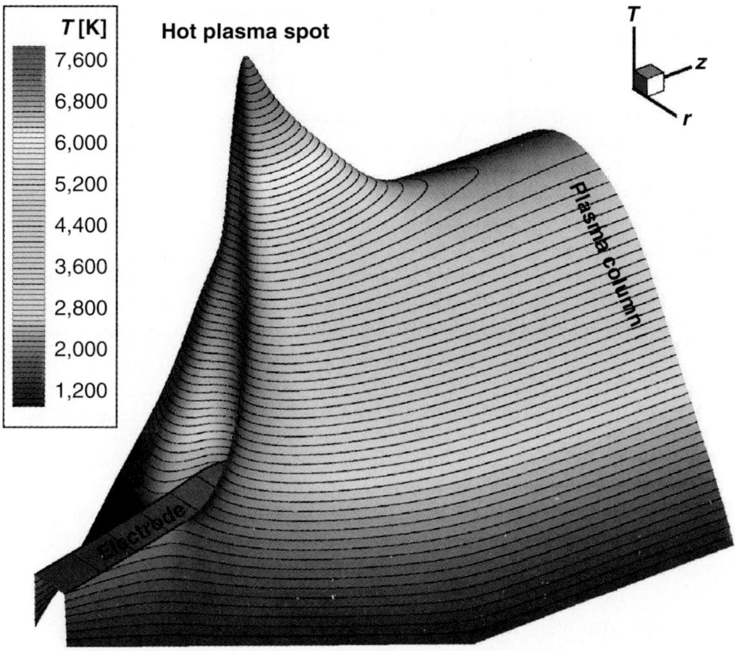

Fig. 4.78. Calculated temperature distribution (plasma and electrodes) with hot plasma spot and plasma column during anode phase (D2 automotive headlight lamp, Xe/Hg/NaI/ScI_3, $p = 7$ MPa, $I = 0.4$ A, 400 Hz rectangular-wave lamp current, $\varphi_{\text{emission}} = 2.6$ eV, $\varphi_{\text{cool/heat}} = 4.5$ eV, [59])

emission coefficient and thermal conductivity) the plasma properties have been averaged between the two values at the disposal. In general, there is a very good agreement on the plasma properties.

First Impression

A first impression of the conditions within plasma and electrodes is given by the temperature distribution shown in Fig. 4.78 (see also Fig. 4.24 for a closer look on the near-electrode temperature distribution). The temperature (depending on r and z) is used as the third coordinate to illustrate the tremendous temperature drop from the hot plasma spots to the electrodes. It is obvious that the electrodes are enormous heat (and power) sinks for the plasma, so that an additional power input into the plasma must compensate for these losses (electrode fall voltages). A different way to look at the temperature distribution is to focus on the lamp axis ($r = 0$), as done in Fig. 4.79. The large temperature gradient of around 5,000 K within 30–60 μm is responsible for the corresponding large electron density gradients near the electrode tips. Driven by these gradients, diffusion of electrons increases the electron density in the close proximity of the electrode tips, leading to the increased non-LTE electrical conductivity discussed in Sect. 4.4.1 and used for the self-consistent

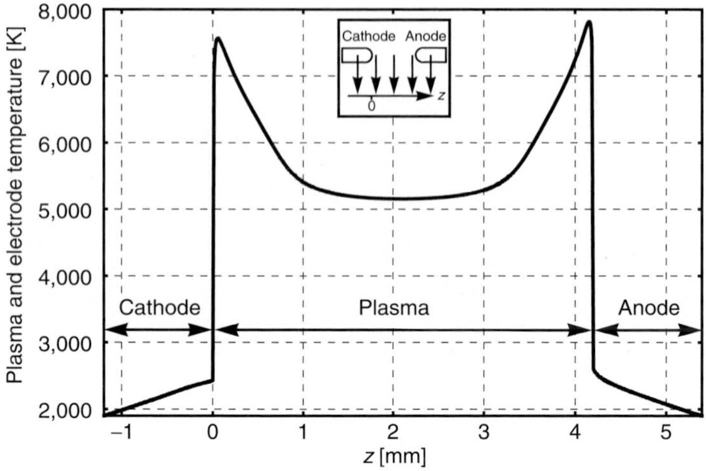

Fig. 4.79. Calculated electrode and plasma temperature on lamp axis before current reversal (D2 automotive headlight lamp, Xe/Hg/NaI/ScI_3, $p = 7$ MPa, $I = 0.4$ A, 400 Hz rectangular-wave lamp current, $\varphi_{\text{emission}} = 2.6$ eV, $\varphi_{\text{cool/heat}} = 4.5$ eV, [59])

electrode–plasma model. The electric potential distribution on the lamp axis (Fig. 4.80) is determined by the non-LTE electrical conductivity (cf. (4.22)). The electric field, which is proportional to the gradient of the electric potential, is constant nearly everywhere (constant power input per unit length). Only close to the electrodes the electric field increases, leading to the so-called electrode fall voltages. The anode and cathode fall voltages are defined as the extrapolation of the linear part of the electric potential distribution on the lamp axis to the electrode tip positions (cf. Fig. 4.80). The difference between this extrapolation and the electric potential at the cathode surface (=0 V by definition) is the cathode fall voltage (in Fig. 4.80: 6.5 V), the difference between the extrapolation and the electric potential at the anode surface (=total lamp voltage) is the anode fall voltage (in Fig. 4.80: 3.6 V). This definition is similar to the experimental determination of cathode and anode fall voltages. The electrode fall voltages correspond to the additional power consumption in the near-electrode plasma,[75] i.e., the extra demand of electrical power input due to the losses to anode and cathode and due to the increased radiation output by the bright hot plasma spots (cf. (4.18), Sects. 3.2.2 and 4.1).

We have different possibilities to interpret the increased electric field in the near-electrode plasma: One way to look at the electric field is to have Poisson's equation at the back of our mind (cf. (4.10)). Then, the increased electric field is due to sources or drains. These sources and drains are ions and electrons, i.e., the

[75] In contrast, the electrode fall voltages in sheath or layer models correspond to the total voltage drop over the sheaths or layers without consideration of the "normal" power consumption per unit length in the arc and in general without consideration of constriction zones and hot plasma spots.

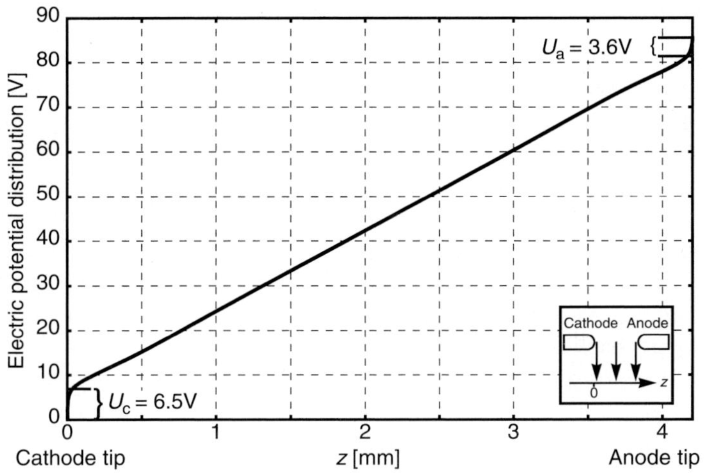

Fig. 4.80. Calculated electric potential on lamp axis and electrode fall voltages U_c and U_a before current reversal (D2 automotive headlight lamp, Xe/Hg/NaI/ScI_3, $p = 7$ MPa, $I = 0.4$ A, 400 Hz rectangular-wave lamp current, $\varphi_{emission} = 2.6$ eV, $\varphi_{cool/heat} = 4.5$ eV, [59])

space charge regions in front of cathode and anode. This space charge shields the electrodes. This interpretation is preferable, if particle densities are calculated in the first place, resulting in an electric field with the help of Poisson's equation. If the electrical conductivity is known/calculated first, then the current continuity equation (4.22) is used to calculate the electric field. A high electric field corresponds then to low electrical conductivities,[76] i.e., one gets high electric fields in plasma areas where it is "tough" to get the given current through. The force (= electric field strength) must consequently be enlarged to ensure current continuity. This second point of view can easily be combined with the power balance of the near-electrode plasma to include the important self-regulating mechanisms. If the electric field is very high (low electrical conductivity), the corresponding electrical power input (σE^2) elevates the temperature of electrode and plasma. An increased plasma temperature results in an increased electrical conductivity, a higher cathode temperature implies stronger electron emission, resulting as well in an enlarged electrical conductivity in the near-cathode plasma. So, the electric field can decrease to a value where power balance and current continuity are in balance. The constriction of the arc near the electrode tips is also responsible for an enlargement of the electric field due to the increase in the electric current density j (same current I through smaller area A). This effect cannot be accounted for in a one-dimensional model.

[76] Low electrical conductivities caused, for example, by low temperatures in the near electrode plasma.

Power Balance

The electrode tip temperature in the anode phase is higher (about 2,600 K) than the electrode tip temperature in the cathode phase (ca. 2,450 K), which does not correspond to the higher cathode fall voltage compared to the anode fall voltage ($U_c = 6.5$ V, $U_a = 3.6$ V before current reversal). At first sight, a high cathode fall voltage should generate high cathode tip temperatures, but we have to take into account that anode and cathode phase are different. The electrons released from the cathode (as electron emission current and to neutralize incoming ions, cf. (4.20)) have a cooling effect on the cathode, whereas the electrons reaching the anode have a heating effect. Accordingly, the energy we put in the cathode to release an electron, we get back at the anode. Therefore, equal anode and cathode fall voltages do not have to correspond to equal anode and cathode tip temperatures. In the case of pure tungsten electrodes ($\varphi_{\text{emission}} = 4.5$ eV), the cathode tip temperature (about 3,300 K) is higher than the anode tip temperature (approx. 3,150 K) with a cathode fall voltage of 17 V and an anode fall voltage of 3 V.

The total lamp voltage calculated for the D2 lamp is 85 V (corresponding to 34 W electrical power consumption) as imaged in Fig. 4.80. The specification of the D2 lamp (voltage 85 ± 17 V, light source power 35 ± 3 W) matches perfectly with these results. The power input of 34 W into the plasma is converted into radiation and heat. About 6.3 W is calculated to be heat losses to the quartz walls, about 4.0 W is heat losses through the electrode roots, and 23.7 W is radiated (IR + visible + UV). The radiation originates mainly from the plasma, only 0.7 W of the 23.7 W is caused by the radiation of the hot electrodes.

The power fluxes to/from the electrodes (mainly to the tips) are very time dependent. Looking at the electrodes just before current reversal, we have the following situation: electron cooling at cathode 1.80 W, thermal heat flux from plasma to cathode 2.76 W, radiation from cathode 0.33 W. The corresponding values at the anode are: electron heating 1.80 W, thermal heat flux from plasma 1.87 W, radiation 0.34 W. Adding/subtracting all these heating/cooling terms we end up with the 4.0 W heat losses through the electrode roots (see earlier), i.e., the heat flux through the outside connections of one electrode (electrode root) corresponds to 2.0 W on average.

For the thoriated tungsten simulations ($\varphi_{\text{emission}} = 2.6$ eV), the heating of the cathode by thermal heat flux from the plasma is sufficient to obtain high enough cathode temperatures to emit the total lamp current of 0.4 A by electron emission. Therefore, there is no additional heating by an ion current ($I_e = 0.4$ A, $I_i = 0$ A). Moreover, for $\varphi_{\text{emission}} = 2.6$ eV the anode tip temperature is higher than the cathode tip temperature, i.e., the electrode is "preheated" in the anode phase, increasing the electron emission current in the cathode phase. For pure tungsten electrodes ($\varphi_{\text{emission}} = 4.5$ eV), the situation is different: The cathode tip temperature (about 3,300 K) is higher than the anode tip temperature (about 3,150 K), but even this high cathode tip temperature is not sufficient to emit the total lamp current of 0.4 A by electron emission. This results in an ion current of

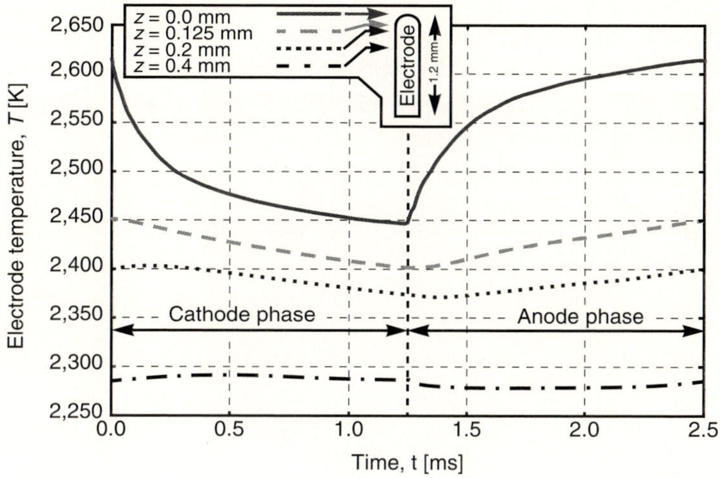

Fig. 4.81. Calculated time-dependent electrode temperature at certain distances to the electrode tip (D2 automotive headlight lamp, Xe/Hg/NaI/ScI_3, $p = 7$ MPa, $I = 0.4$ A, 400 Hz rectangular-wave lamp current, $\varphi_{emission} = 2.6$ eV, $\varphi_{cool/heat} = 4.5$ eV, [59])

0.28 A and an electron emission current of 0.12 A (averaged over period) for pure tungsten electrodes.

Electrode Temperature

The electrode tip temperature changes over a range of 2,450–2,600 K within one period of the rectangular-wave current of 400 Hz (cf. Figs. 4.81 and 4.82 with $\varphi_{emission} = 2.6$ eV). But already 0.125 mm below the tip (at the changeover from hemispherical electrode tip to the rod), the temperature varies only between 2,400 and 2,450 K, which is hard to determine experimentally. Going further down the electrode, the maximum and minimum temperatures move to later times within the period and further approach each other. A time-dependent measurement of the electrode temperature is very difficult because (apart from other problems) reflections of the plasma especially at the hemispherical electrode tip disturb the measurements there (cf. Sects. 3.1.1–3.1.7). Moreover, below the electrode tip the amplitude of the temperature variation is very small. Therefore, numerical simulations are an important tool to learn more about and to further improve HID lamps.

Comparison to Experimental Results

Due to their smallness, HID lamps in general and especially the D2 automotive headlight lamp are difficult to conduct experimental measurements with. Fortunately, there are two publications presenting measurements concerning plasma temperature [100] and electrode temperature [85, 86]. The results of the numerical

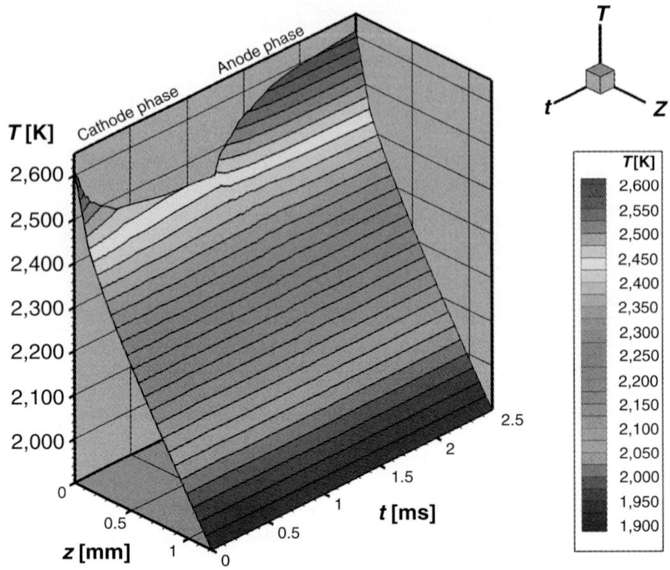

Fig. 4.82. Calculated variation of axial electrode temperature within one period (D2 automotive headlight lamp, Xe/Hg/NaI/ScI_3, $p = 7$ MPa, $I = 0.4$ A, 400 Hz rectangular-wave lamp current, $\varphi_{emission} = 2.6$ eV, $\varphi_{cool/heat} = 4.5$ eV, [64], © 2004 IEEE)

simulations obtained with the self-consistent electrode–plasma model cover all plasma and electrode properties, thus we can compare the outcome of the calculations with the experimental findings. All numerical results originate from one numerical simulation, i.e., we do not have one calculation optimized for the comparison with plasma temperature measurements and another numerical simulation optimized for the comparison with electrode temperature measurements. A first agreement between numerical simulation and real lamp concerning lamp voltage has already been mentioned.

Figure 4.83 shows the measured[77] ([85], time averaged) electrode temperature in comparison with the (time-dependent) electrode surface temperature resulting from the numerical simulation.[78] The agreement is very good, especially if the variation of the measurements at different (but identical) lamps is taken into account (Fig. 4.83 shows measurements on the electrodes of six different lamps). The measurements shown here were carried out at thoriated tungsten electrodes, accordingly the work function $\varphi_{emission}$ for the numerical simulations was set to 2.6 eV.

Numerical simulations of a D2 lamp with pure tungsten electrodes ($\varphi_{emission} = 4.5$ eV) result in electrode temperatures about 300 K higher than the measured temperatures for pure tungsten electrodes (about 3,200 K compared to

---

[77] The method used is described at the end of Sect. 3.1.4.

[78] At the end of anode phase and cathode phase, respectively.

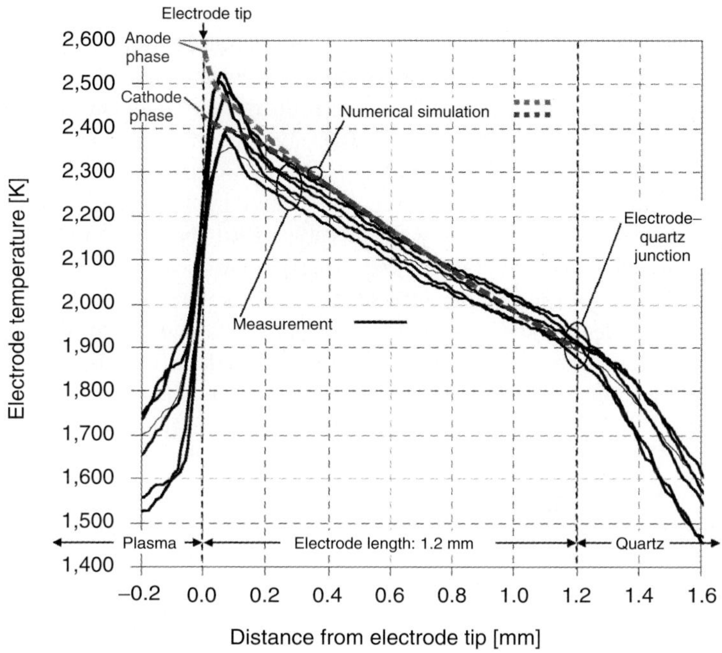

Fig. 4.83. Comparison of measured ([85], time averaged, six different lamps) and calculated (time-dependent) electrode temperature (D2 automotive headlight lamp, Xe/Hg/NaI/ScI_3, $p = 7$ MPa, $I = 0.4$ A, 400 Hz rectangular-wave lamp current, $\varphi_{emission} = 2.6$ eV, $\varphi_{cool/heat} = 4.5$ eV, [59])

circa 2,900 K measured in [85]). Nevertheless, in [85] there are also temperature measurements on pure tungsten electrodes resulting in electrode temperatures comparable to those of thoriated tungsten electrodes. These two different series of pure tungsten electrodes (with at least five lamps in each series) are different with respect to the cleaning (glowing) temperature in the production process (1200 and 2400°C, respectively) and with respect to the arc attachment. Together with the in situ measurements of the work function in metal halide lamps presented in [145, 146], this might be a hint that not the thoria of the thoriated tungsten electrodes lowers the work function, but the salts of the discharge adsorbed on the cathode tip[79] (supported maybe by the type of arc attachment).

[79] One type of pure tungsten electrodes (type B) might have a work function for electron emission of around 2.5 eV (due to a high coverage of sodium on the arc attachment area of 40% for instance), the other pure tungsten electrode (type C) might have a work function of 4.0 eV (i.e., smaller than the pure tungsten value of 4.5 eV due to a low coverage of sodium on the arc attachment area of 10% for instance).

Furthermore, Fig. 4.83 reveals the difficulties in measuring the temperature at the very tip of the electrode (top 0.1 mm). The temperature decreases already 0.05–0.1 mm below the electrode tip due to the finite diameter of the target spot. Reflections of plasma radiation at the electrode tip might be the reason for the increased temperature gradient between 0.05 mm and 0.2 mm as mentioned, for example, for similar measurements in [43]. At 1.2 mm distance from the electrode tip, the electrode enters the quartz vessel. The change in slope near 1.2 mm is due to distortions arising from the glass wall curvature [86]. Needless to say that these effects (finite diameter of target spot, plasma reflections on electrode tip, distortions) do not appear in the results of the numerical simulations using the self-consistent electrode–plasma model described in Sects. 4.4.1 and 4.4.2, so that a detailed analysis of the most interesting part of the electrode, the very tip of the electrode, is possible without any constrains. This makes the numerical simulations very important for the investigation and development of HID lamps.

A comparison of measured [100] and calculated plasma temperature on the lamp axis (Fig. 4.84 with $\varphi_{\text{emission}} = 2.6$ eV) shows a good agreement of the arc column temperature as well as the hot plasma spot temperature. According to [100], there is no significant difference in the measured plasma temperature between cathode and anode phase, which is confirmed by the numerical simulation. The error in the temperature measurement can be estimated to be between ± 300 and ± 500 K (cf.

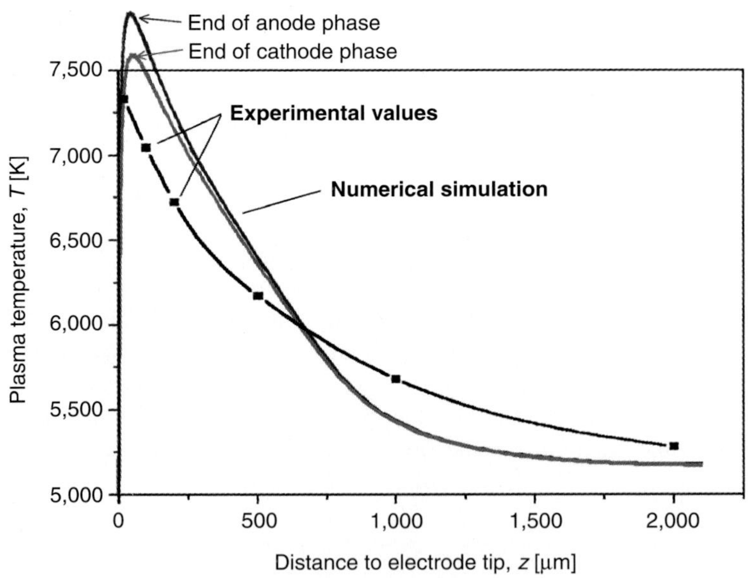

Fig. 4.84. Comparison of measured [100] and calculated plasma temperature on the lamp axis (D2 automotive headlight lamp, Xe/Hg/NaI/ScI_3, $p = 7$ MPa, $I = 0.4$ A, 400 Hz rectangular-wave lamp current, $\varphi_{\text{emission}} = 2.6$ eV, $\varphi_{\text{cool/heat}} = 4.5$ eV, [59])

[99] or Fig. 4.5), furthermore the determination of the distance to the electrode tip might be difficult for distances of only 20 or 100 μm.

The calculated arc temperature of the hot plasma spot near the electrode tips is different in the anode and cathode phases. But the difference is not very large compared with the temperature difference between hot plasma spot and electrode tip (about 7,500 K to about 2,500 K). For the heat flux from plasma to electrode, the temperature difference between hot plasma spot and electrode tip is not the best measure because the power transported to the electrode tip by heat conduction (heat flux from plasma to electrode) is compensated/balanced by the increased electrical power input between hot plasma spot and electrode tip.

A different presentation of the same measurements compared with the results of the numerical simulations is given in Fig. 4.85. Here, the measured radial temperature profiles for different distances to the electrode tip from [100] are compared to the corresponding calculated temperature profiles. Again, the agreement is very good. The different profiles of the arc column and the hot plasma spot can be clearly distinguished.

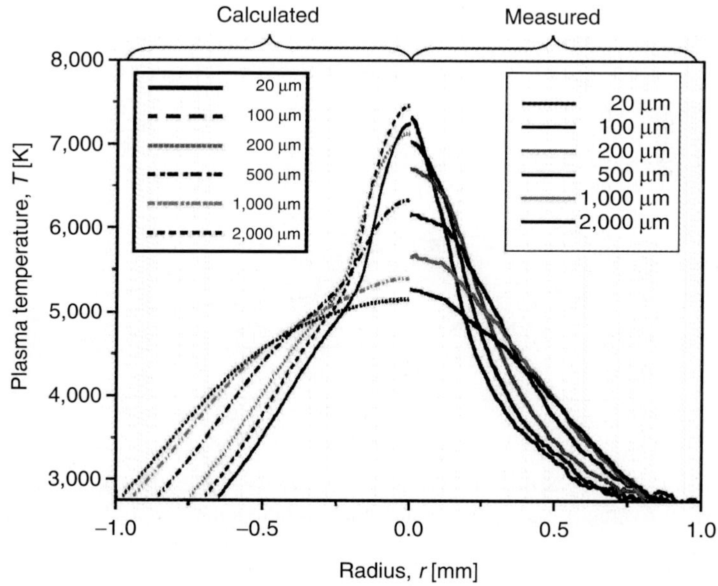

Fig. 4.85. Comparison of measured [100] and calculated radial plasma temperature for different distances to the electrode tip (D2 automotive headlight lamp, Xe/Hg/NaI/ScI$_3$, $p = 7$ MPa, $I = 0.4$ A, 400 Hz rectangular-wave lamp current, $\varphi_{\text{emission}} = 2.6$ eV, $\varphi_{\text{cool/heat}} = 4.5$ eV, [59])

Different Electrode Geometries

The influence of different electrode geometries on the electrode temperature is studied in Fig. 4.86. The motivation for this investigation lies in the published pictures of used electrodes ([86] and Sect. 3.1.13) and in the interesting results concerning different electrode shapes in mercury HID lamps described in Sect. 4.4.4. Figure 4.86 shows the temperature distribution resulting from two different electrode tips: A (standard) hemispherical tip (dashed lines) as used for all other numerical results shown within this section, and a structured electrode tip (solid lines) with some kind of bump at the very top of the electrode (motivated by pictures in [86] and Sect. 3.1.13, especially Figs. 3.35 and 3.36). The temperatures are given at the end of anode (higher temperatures) and cathode phase, respectively. The arc attachment is much more confined in the case of the structured electrode tip, leading to a higher temperature at the very tip of the electrode and lower temperatures outside the bump compared to the hemispherical electrode tip. The effect is more pronounced during the anode phase.[80] The effect of the temperature drop at the cylindrical part of the electrode is not as strong as measurements show for two different tungsten electrodes with different arc attachments (cf. [85], discussion of Fig. 4.83). This means that the cultivation of an extra bump as studied in Fig. 4.86 cannot be the only cause for the effects observed in real lamps, but we

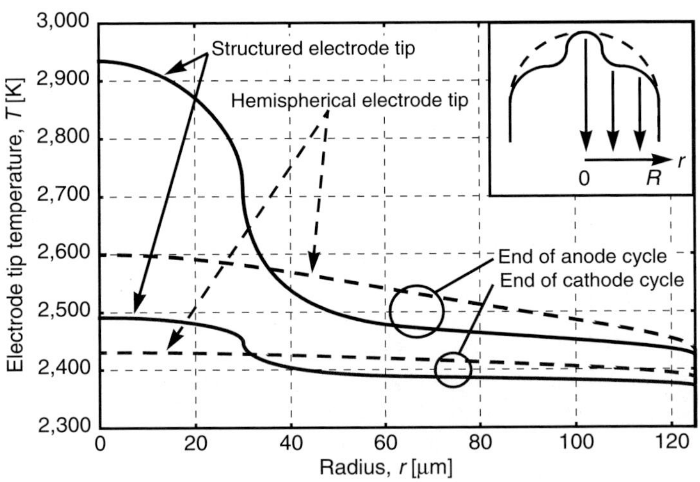

Fig. 4.86. Influence of electrode tip geometry on calculated electrode tip temperature at the end of anode and cathode phase (D2 automotive headlight lamp, Xe/Hg/NaI/ScI_3, $p = 7$ MPa, $I = 0.4$ A, 400 Hz rectangular-wave lamp current, $\varphi_{emission} = 2.6$ eV, $\varphi_{cool/heat} = 4.5$ eV, [59])

[80] This cannot be regarded as a rule of thumb: in Sect. 4.4.4, the temperature change during the cathode phase is more pronounced than during the anode phase.

can propose some explanations for these observed effects: We have already mentioned that plasma components are able to change the work function for electron emission at the electrode tip (cf. [1, 146], Sects. 3.1.11, 3.1.12). This influence might depend on the arc attachment, i.e., the smaller the arc attachment area, the better the adsorption of, for example, sodium might work. Furthermore, for a smaller arc attachment fewer atoms are needed for the same coverage than for larger arc attachments. This would mean that a structured electrode tip leads to a smaller arc attachment (as shown in Fig. 4.86), increasing the adsorption of, for example, sodium (coming from the discharge plasma), resulting in a lowering of the work function for electron emission (as reported in [1] or [146]), leading to notedly smaller electrode temperatures (as observed in [85]). Further important issues concerning the different electrode temperatures of pure tungsten electrodes as observed in [85] are the extreme increase in the surface area of the electrode tip (cf. Fig. 3.36) and the possibility of (to a large extent) thermally isolated structures on the electrode tip. The extreme increase in the surface area of the electrode tip by many extra bumps and some kind of swelling of the electrode tip as shown in Fig. 3.36 increases the electron emitting surface, which is not encountered in the geometries studied in Fig. 4.86. This increase of electron emission by the larger surface would result in lower electrode temperatures (as a lower work function would do). Furthermore, spherical structures with little contact surface to the electrode body might be formed that are (to a large extend) thermally isolated but with a good electrical contact to the electrode body. These structures might have a much larger temperature than the electrode body without transferring a large amount of heat to the electrode body (due to the thermal isolation). These very localized high temperatures would be very difficult to detect by the means used in [85, 86] and would result in much lower electrode body temperatures compared to "normal" electrodes without these structures. Figures 3.35 and 3.36 themselves suggest that the melting temperature of tungsten (about 3,680 K) is reached at least temporarily and localized, although the temperature measurements indicate smaller electrode tip temperatures of around 2,500 K. One of the many advantages of numerical simulations with the self-consistent electrode–plasma model is the possibility to separate different effects (here caused by the geometry of the electrode tip, the variation of the work function for electron emission, etc.), which are superimposed (and not separable) in real lamps.

Summary

The self-consistent electrode–plasma model described in Sects. 4.4.1 and 4.4.2 yields many interesting results concerning overall lamp performance, temperature profiles of plasma and electrodes, electrode fall voltages, etc. Only one single numerical simulation is necessary for all these results, i.e., the lamp is treated as a whole, as a system with many interactions leading to the final (time dependent) state of the lamp. It is easily possible to investigate the impact of the variation of different parameters like work function for electron emission, geometry of the electrode, plasma composition, etc. The results can be analyzed straightforward without any constraints in space and time, i.e., the temperature of the electrodes, for example, can be studied at every position and at every point in time requested.

The only input parameters required for the numerical simulations are electrode and plasma properties, like thermal conductivity or heat capacity, which are thoroughly discussed within this section. Especially the D2 automotive headlight lamp under consideration within this section has a very complex plasma composition resulting in complex plasma properties. But the electrodes have challenging features as well. Especially the work function for electron emission needs the greatest care and attention. Specifying only the temperature at the inside walls and the outside connections to the electrodes, everything else can be calculated for any given time dependent lamp current. The fundamental principles determining the outcome of the time-dependent simulations are the power balance and interactions between electrodes and plasma, electrical conductivity and temperature, etc. These basic principles are discussed within this section.

Beyond the presentation of the numerical results, a comparison with different experimental data is presented. The time-dependent numerical results for the D2 automotive headlight lamp agree very well with the measured total lamp voltage, the measured plasma temperature profiles, and the measured electrode temperature profile. Numerical simulations carried out for mercury-free lamps for automotive lighting applications show very good agreement with experimental results as well [129, 158].

4.4.7 Mass, Pressure, and Electrode Gap

Within this and the following section, a specially built high-pressure mercury lamp with different electrode gaps, operating pressures, and lamp currents will be discussed. The numerical results will be compared to experimental findings.[81] Furthermore, comparing the numerical results obtained for lamps with different electrode gaps, operating pressures, and lamp currents, the influence of these parameters can be studied. Thus, we can learn a lot about the behavior of HID lamps by using numerical simulations.

The Lamp

The lamp under consideration within this and the following section was specially built by Philips to investigate the influence of different parameters like operating pressure, electrode gap, and lamp current on lamp operation. Pictures of the lamp are shown in Sect. 3.1: Figure 3.3 shows a picture of a precursor lamp with a coil at the cathode, Fig. 3.21 depicts one of the lamps used for the experimental investigations, and Figs. 3.23 and 3.24 show measured electrode temperature

[81] Many thanks to C. Bauer for providing the experimental data shown within this and the following Sect. [8, 9, 11, 128].

distributions (two-dimensional and one-dimensional, respectively). A characteristic feature of the lamp is the Apollo-shaped anode,[82] thus we will call this lamp in the following *ApolloHg-lamp* to simplify matters.

Altogether, 25 different high-pressure mercury discharge lamps similar to this shown in Fig. 3.21 were built. They differ in electrode gap (2, 3, and 4 mm), operating pressure[83] (2, 4, and 6 MPa, i.e., 20, 40, and 60 bar), and shape of the anode (Apollo-shaped or rod-shaped as shown in Fig. 3.25). Besides the investigation of the influence of different parameters like lamp pressure, electrode gap, and lamp current on lamp operation, the lamp was used for the comparison of different methods of electrode temperature measurements (cf. [9, 128]).

The Apollo-shaped anode has a length of 8 mm and a maximum diameter of 3.5 mm. The rod-shaped cathode has a length of 4, 5, or 6 mm depending on the electrode gap, and a diameter of 0.7 mm. The electrodes are made of pure tungsten, thus the work function was set to $\varphi = \varphi_{\mathrm{emission}} = \varphi_{\mathrm{cool}} = \varphi_{\mathrm{heat}} = 4.55$ eV. Because experimental and numerical results concerning the cathode temperature agreed better taking $\varphi_{\mathrm{emission}} = 3.5$ eV, this value for the work function for electron emission was also used. According to experimental findings, the inner quartz wall temperature and the temperature of the electrode roots are set to 1,000 and 1,700 K, respectively.

First Impression

To give a first impression of the conditions in this high-pressure discharge lamp, Figs. 4.87 and 4.88 show the calculated temperature distribution within the plasma and the electrodes[84] for a lamp with 3 mm electrode gap, an operating pressure of 4 MPa, and a dc lamp current of 4 A obtained by using the self-consistent electrode–plasma model described in Sects. 4.4.1 and 4.4.2. Clearly visible are the extremely hot plasma spots very close to the electrode tips. These hot plasma spots have temperature maxima up to 11,000 K depending on lamp current, pressure, or mode of arc attachment to the electrodes (cf. Figs. 4.101 and 4.104). The electrodes have tip temperatures of about 3,600 K (cathode) and 2,600 K (anode) for the lamp configuration used in Figs. 4.87 and 4.88. This leads to extremely large temperature gradients between the hot plasma spots and the electrode tips, driving an intense thermal heat flux from the plasma to the electrodes. This power loss of the plasma together with the power loss due to the enhanced radiation by the hot plasma spots

---

[82] Lamps with two different electrodes are normally operated on dc currents, i.e., the two electrodes are optimized for anodic and cathodic operation, respectively. As usual in dc high-pressure lamps, the thicker electrode acts as the anode, the thinner one as the cathode.

[83] Strictly speaking, the lamp was filled with 20, 40, or 60 mg mercury, approximately resulting in the pressures given above. If not the entire dose of liquid mercury is evaporated during operation, the resulting pressure is lower. Moreover, different lamp currents and the resulting different temperature profiles change the mercury pressure.

[84] The temperature distribution of plasma and electrodes on the lamp axis are given in Figs. 4.101 and 4.104 for different operating conditions.

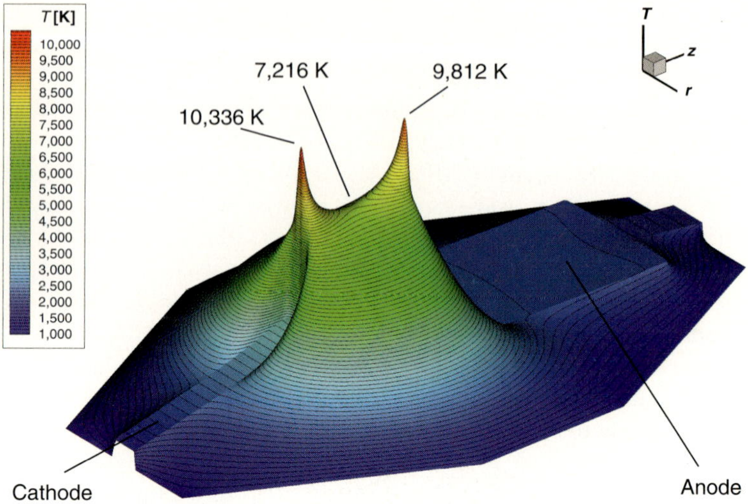

Fig. 4.87. Calculated temperature distribution (plasma and electrodes) with hot plasma spots and plasma column (ApolloHg-lamp, Hg, $p = 4$ MPa, $I = 4$ A (dc), $d_{\text{electrode}} = 3$ mm, $\varphi_{\text{emission}} = \varphi_{\text{cool/heat}} = 4.55$ eV, [62])

has to be balanced by an enlarged electric field near the electrodes. This enlarged electrical power density leads to the electrode fall voltages. Figure 4.89 shows the measured radiation intensity of the plasma and the electrode tips to give an impression of the conditions in the real lamp.

Calculated and Measured Lamp Voltage

Figure 4.90 shows the calculated electric potential on the lamp axis for ApolloHg-lamps with three different electrode gaps from 2 to 4 mm. The electrode fall voltages can be extracted easily by extrapolation of the linear part of the electric potential distribution towards anode and cathode, respectively. The influence of the electrode gap on the electric potential is clearly visible by comparing the different gradients of the linear part of the electric potential, i.e., the electric field in the so-called plasma column.[85] This electric field decreases for increasing electrode gap from 19.1 V mm^{-1} (2 mm electrode gap) to 15.8 V mm^{-1} (4 mm electrode gap). This is due to the constriction of the plasma column for decreasing electrode gaps as shown in Fig. 4.98. The increasing plasma temperature in the center of the plasma column for decreasing electrode gaps (cf. Figs. 4.95 and 4.98) and the thus increasing electrical conductivity is not able to balance the increase of the electrical current density $j = \sigma E$ associated with the constriction of the plasma column.

---

[85] Which is not really a plasma column in the case of small electrode gaps below approx. 5 mm.

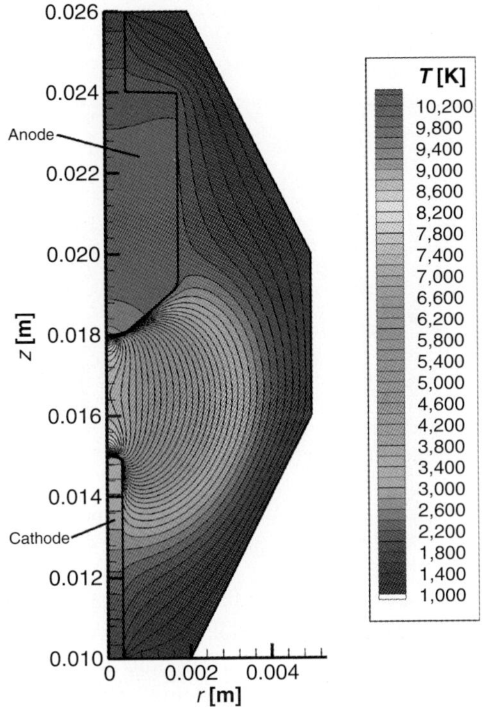

Fig. 4.88. Calculated temperature distribution (plasma and electrodes) with hot plasma spots and plasma column (ApolloHg-lamp, Hg, $p = 4$ MPa, $I = 4$ A (dc), $d_{\text{electrode}} = 3$ mm, $\varphi_{\text{emission}} = \varphi_{\text{cool/heat}} = 4.55$ eV, [62])

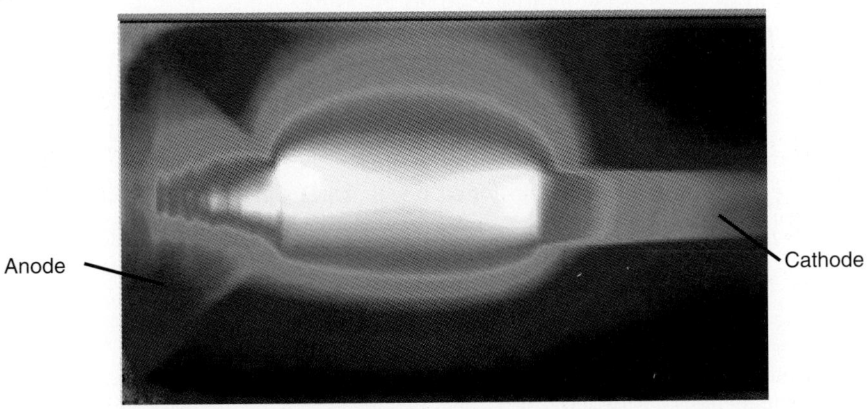

Fig. 4.89. Measured radiation intensity (ApolloHg-lamp, plasma and electrodes, [9, 128])

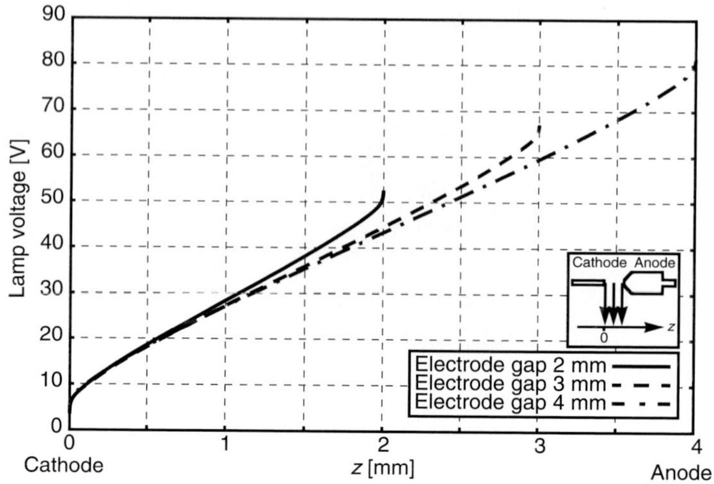

Fig. 4.90. Calculated electric potential on lamp axis (ApolloHg-lamp, Hg, $p = 4$ MPa, $I = 4$ A (dc), $d_{electrode} = 2$, 3, or 4 mm, $\varphi_{emission} = \varphi_{cool/heat} = 4.55$ eV, [62])

The corresponding anode and cathode fall voltages also depend on the electrode gap: The cathode fall voltage, i.e., the additional power consumption in the near cathode plasma due to the heat flux from the cathodic hot plasma spot to the cathode plus the power necessary to generate the ion current towards the cathode together with the power loss due to the enhanced radiation by the cathodic hot plasma spot, increases with increasing electrode gap from 9.2 V (2 mm electrode gap) to 11.7 V (4 mm electrode gap). The anode fall voltage also increases with increasing electrode gap from 4.6 V (2 mm electrode gap) to 6.4 V (4 mm electrode gap). These results concerning the dependence of the electrode fall voltages and the electric field in the plasma column on the electrode gap demonstrate clearly that it is necessary to consider the whole system consisting of anode, cathode, and the complete plasma especially in the case of short arc discharge lamps with electrode gaps of 5 mm and less.

The lamp voltage is straightforward to measure. Thus, it is interesting to compare the measured lamp voltages for different lamp parameters (lamp current, electrode gap, operating pressure) with calculated lamp voltages[86] using the self-consistent electrode–plasma model. Figure 4.91 shows the measured and the calculated lamp voltage for an ApolloHg-lamp with 3 mm electrode gap, an operating pressure of 4 MPa, and a dc lamp current between 3 and 5 A (left-hand side of Fig. 4.91, dashed and dash-dotted lines). Unfortunately, the measured and calculated results disagree: The measured lamp voltage rises with increasing lamp current, the calculated lamp voltage decreases with increasing lamp current, which looks like a principal problem of the numerical simulations at first glance. So we

[86] The calculated lamp voltages include the voltage drop over the (more or less) cylindrical plasma column, the constriction zones, and the near-electrode plasma, i.e., it can be compared with the measured lamp voltage (cf. e.g., Fig. 4.90).

have to think about an explanation for this apparent difference between measurements and calculations: The numerical simulations predict a decreasing lamp voltage for an increasing lamp current because the electrical conductivity of the plasma increases for increasing lamp current. This is due to the increasing electrical power input[87] (*UI*) into the plasma for increasing lamp current and the resulting increase in plasma temperature. This increase of the electrical conductivity with increasing lamp current overcompensates the required increase of the lamp voltage which would be necessary for an increasing lamp current if the electrical conductivity would be constant.[88] So, the question remains why the measured lamp voltage does not reveal this convincing behavior.

The answer to this problem is a principal difference between numerical simulations and experimental setup: The operating pressure is constant for the numerical simulations, whereas the mass of mercury within the discharge is constant for the experimental setup (unsaturated mercury vapor, cf. Sect. 1.3.6,

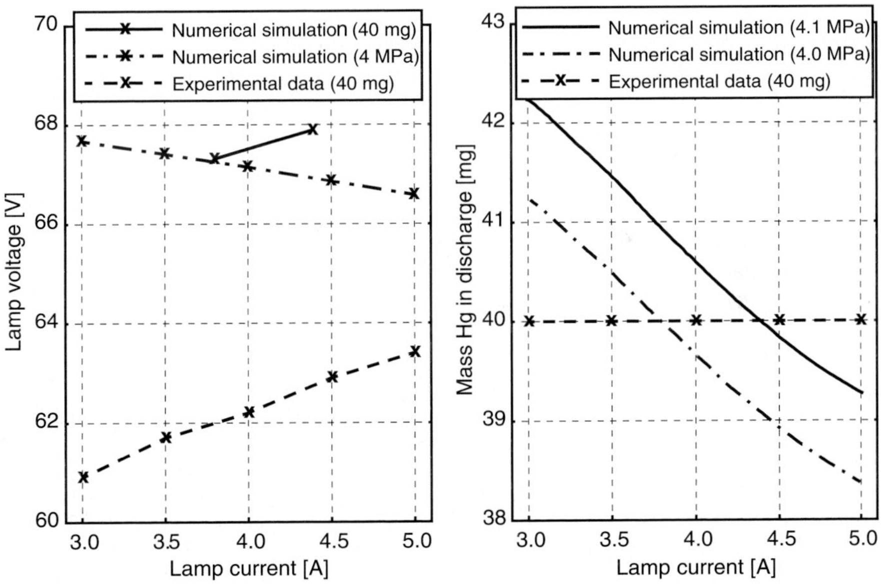

Fig. 4.91. Calculated and measured lamp voltage (left-hand side) and calculated mercury mass in the discharge (right-hand side) vs. lamp current (ApolloHg-lamp, Hg, $p = 4.0$ or 4.1 MPa, $I = 3$–5 A (dc), $d_{\text{electrode}} = 3$ mm, $\varphi_{\text{emission}} = \varphi_{\text{cool/heat}} = 4.55$ eV, [62])

---

[87] Despite the decreasing lamp voltage for increasing lamp current, the electrical power consumption (*UI*) rises with increasing lamp current.

[88] A higher lamp current corresponds to a higher electric current density $j = \sigma E$ in the plasma. If the electrical conductivity σ remains constant, the electric field must increase to ensure a higher electric current density.

especially Fig. 1.25). In the experimental setup (and the numerical simulations), the current increase causes an increase of the plasma temperature within the discharge volume. Because the total mass of mercury within the discharge is constant in the experimental setup, the operating pressure rises for increasing temperature, i.e., the pressure increases for increasing lamp current. An increase in the operating pressure, however, results in an increase of the lamp voltage (cf. Fig. 4.92, right-hand side). Thus, the measured increase of the lamp voltage for increasing lamp current is a superposition of an increase of the lamp voltage due to an increasing operating pressure and a decrease of the lamp voltage due to the increasing electrical conductivity as predicted by the numerical simulations (for constant operating pressure).

This looks like an explanation for the disagreement between numerical results and experimental measurements. We can analyze this difference in more detail: The operating pressure is fixed for the numerical simulations and we can calculate the corresponding mass of mercury within the discharge necessary to sustain this pressure. The results of such a calculation are shown on the right-hand side of Fig. 4.91 (solid and dash-dotted lines) vs. lamp current I. The lower curve corresponds to a numerical simulation assuming 4 MPa, the upper curve corresponds to 4.1 MPa. The dependence of the mercury mass in the discharge on lamp current I for the experimental setup is straightforward: It is constant and given by the lamp fill of 40 mg Hg (unsaturated mercury vapor).

So, what is the meaning of the points of intersection of the curves from the numerical simulations and the experimental setup? The intersection of the curve from the numerical simulation for 4.0 MPa with the 40 mg-line at $I = 3.79$ A means that the operating pressure in the experimental setup is 4.0 MPa at $I = 3.79$ A. Increasing the lamp current in the experimental setup from 3.79 A to 4.39 A (intersection 4.1 MPa-curve with 40 mg-line) increases the pressure from 4.0 to 4.1 MPa in the experimental setup. Thus, we have analyzed the pressure increase for increasing lamp current in the experimental setup.

The last step in our analysis is to transfer the results obtained from the mass–current graph (right-hand side of Fig. 4.91) to the voltage–current graph (left-hand side of Fig. 4.91): The points of intersection in the mass–current graph correspond to numerical results for a discharge with 40 mg mercury. So we can take the lamp voltage from the 4.0 MPa curve at $I = 3.79$ A (=67.31 V) and the lamp voltage from the 4.1 MPa curve at $I = 4.39$ A (=67.89 V) and transfer these values to the voltage–current graph. The resulting curve is shown on the left-hand side of Fig. 4.91 (solid line). It is labeled "Numerical simulation (40 mg)." Now we can compare the gradient of this curve with the experimental results: They agree very well.

The absolute values differ by approximately 10%, which is a good result bearing in mind all the (unknown) parameters which influence the total lamp voltage. There might be, for example, an uncertainty in the mass of mercury filled in the lamp, which would have a noticeable effect as it can be seen on the right-hand side of Fig. 4.92. Comparing this, for example, with the measured total lamp voltage for identical lamps shown in Fig. 3.49 and considering the results shown in Fig. 4.92, the agreement is very good.

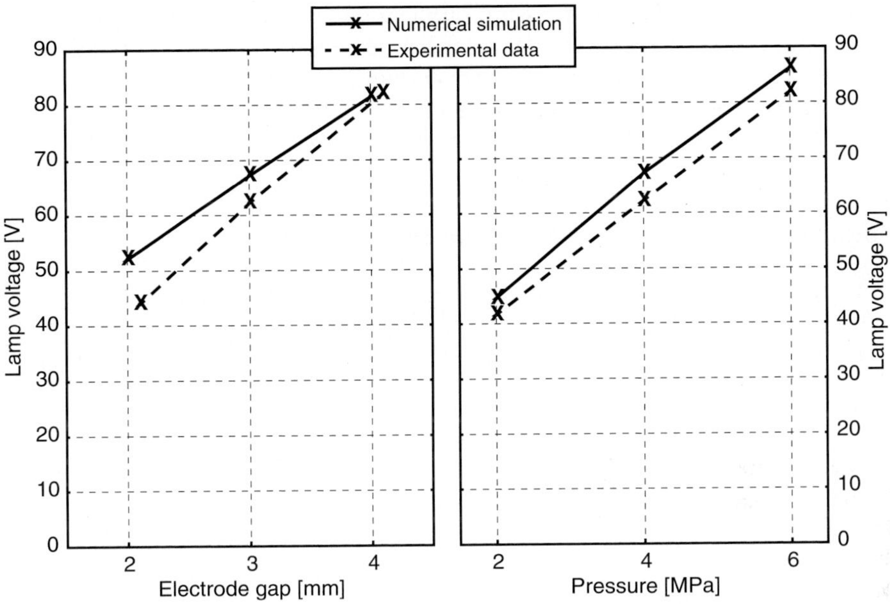

Fig. 4.92. Calculated and measured lamp voltage vs electrode gap (left-hand side) and vs. pressure (right-hand side) (ApolloHg-lamp, Hg, $p = 2$, 4, or 6 MPa, $I = 4$ A (dc), $d_{\text{electrode}} =$ 2, 3, or 4 mm, $\varphi_{\text{emission}} = \varphi_{\text{cool/heat}} = 4.55$ eV, [62])

Figure 4.92 shows the measured and the calculated lamp voltage for an ApolloHg-lamp with 2, 3, or 4 mm electrode gap, an operating pressure of 2, 4, or 6 MPa, and a dc lamp current of 4 A. The left-hand side of Fig. 4.92 depicts the dependence of the total lamp voltage on the electrode gap (pressure: 4 MPa), the right-hand side the dependence of the total lamp voltage on operating pressure (electrode gap: 3 mm).

The difference in the measured and calculated gradients of the total lamp voltage vs. electrode gap (left-hand side of Fig. 4.92) can be explained in the same way as the different dependence of the total lamp current on the lamp current: The mass of mercury filled in the lamp for the experimental investigations is constant, but the operating pressure is not equal in the lamps with different electrode gaps. The larger the electrode gap, the larger the average temperature of the plasma, and thus the larger the operating pressure. This increase of operating pressure with increasing electrode gap in the experimental setup results in a steeper gradient of the lamp voltage vs. electrode gap.

There is no difference in the measured and calculated gradients of the total lamp voltage vs. operating pressure (right-hand side of Fig. 4.92), which confirms the considerations discussed earlier. These results again prove the ability of the self-consistent electrode–plasma model to analyze the whole lamp including electrodes, plasma, and the interactions between them. The results concerning the dependence of the operating pressure on lamp current and electrode gap in HID lamps is a central finding. A similar result has already been discussed in Sect. 4.4.5 (end of section "Electrode Fall Voltages during External Laser Heating"), further investigations will be given in Sect. 4.4.8.

Calculated and Measured Electrode Temperature

A comparison of calculated and measured electrode temperatures is given in Figs. 4.93 (anode) and 4.94 (cathode). The measured and the calculated anode temperature agree very well. The difference of about 100 K agrees well with uncertainties in the measurements of the anode temperature. Besides the absolute values, the gradient of the anode temperature vs. lamp current agrees very well comparing experimental results and numerical simulations, i.e., the dependence of the anode temperature on the lamp current is well understood and modeled correctly by the numerical simulations.

The measured (dashed line with × in Fig. 4.94) and the calculated cathode temperatures ($\varphi_{\text{emission}} = 4.55$ eV, solid line in Fig. 4.94) do not match perfectly: There is a difference of about 300 K at the cathode tip ($z = 0.05$ and 1.0 mm) and 100 K some millimeters below the cathode tip. There might be many reason for this difference: The cathode tip might have a different geometry compared to the one assumed for the numerical simulations (compare with the results shown in Sect. 4.4.4), the crystal structure or microstructures of the cathode tip might enhance

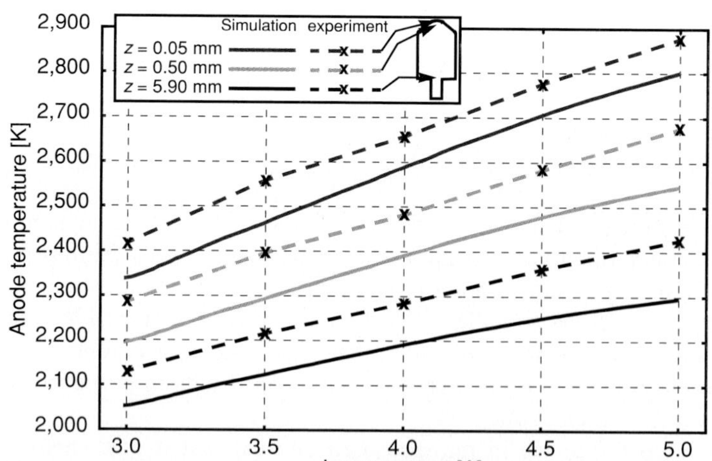

Fig. 4.93. Calculated and measured anode temperature at certain distances to anode tip vs. lamp current during a quasistationary increase of the lamp current from 3 to 5 A (ApolloHg-lamp, Hg, $p = 4$ MPa, $d_{\text{electrode}} = 3$ mm, $\varphi_{\text{emission}} = \varphi_{\text{cool/heat}} = 4.55$ eV, [62])

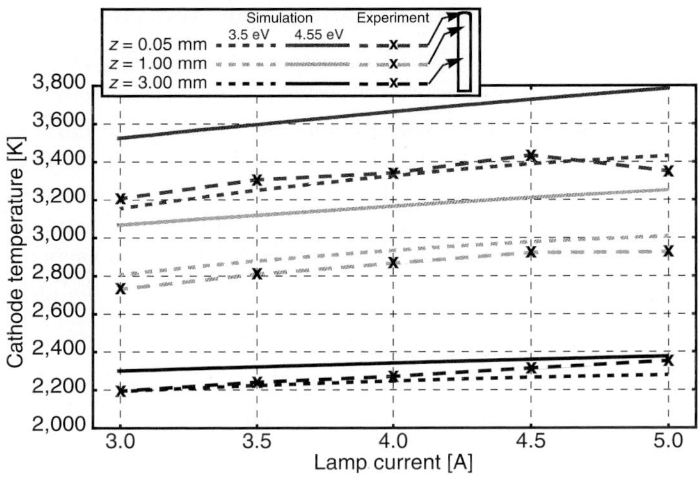

Fig. 4.94. Calculated and measured cathode temperature at certain distances to cathode tip vs. lamp current during a quasistationary increase of the lamp current from 3 to 5 A (ApolloHg-lamp, Hg, $p = 4$ MPa, $d_{\text{electrode}} = 3$ mm, $\varphi_{\text{emission}} = 3.5/4.55$ eV, $\varphi_{\text{cool/heat}} = 4.55$ eV, [62])

electron emission, or the diameter of the cathode in the experimental setup might be slightly larger than assumed for the numerical simulations. The different geometry is quite probable due to the high cathode tip temperature (about 3,400 K) measured in the experimental setup close to the melting point of tungsten (3,680 K), and thus the geometry might change during the experimental investigations (cf. Sect. 3.1.13).

Another explanation might be the work function for electron emission $\varphi_{\text{emission}}$, which is difficult to determine in operating HID lamps (cf. Sects. 3.1.11 and 4.4.6). Setting $\varphi_{\text{emission}} = 3.5$ eV (dashed line in Fig. 4.94), the numerical and the measured values agree very well. This is not only true for the absolute values of the cathode temperature, but also for the gradient of the cathode temperature vs. lamp current, i.e., the dependence of the cathode temperature on the lamp current. This demonstrates a dilemma all numerical simulations concerned with cathode behavior are in: The work function for electron emission plays a central role for the numerical simulations, but it is very difficult to measure in an operating lamp (cf. Sects. 3.1.11, 3.1.12, 4.4.3, and 4.4.6). We cannot deliver an explanation for the differences between measured and calculated cathode temperature (e.g., electrode tip geometry or work function), but we can deliver studies of the influence of certain parameters on the cathode (electrode) temperature, which is not straightforward possible in the experimental setup. The differences between anode and cathode behavior seem to be well understood and modeled correctly by the numerical simulations (see also Sects. 4.4.4–4.4.6).

Influence of Electrode Gap on Arc Attachment and Plasma Column

The influence of different electrode shapes on electrode and plasma properties has already been discussed in Sect. 4.4.4. Now we can use the numerical simulations obtained for the ApolloHg-lamp with different electrode gaps to further investigate the coupling of near-electrode plasma, constriction zones, hot plasma spots, and (cylindrical) plasma column. As we have already seen in Sect. 4.4.6 concerned with the D2 lamp with an electrode gap of 4 mm, that the transition from near-cathode plasma over the cathodic hot plasma spot to the anodic hot plasma spot and the near-anode plasma goes without clear-cut dividing lines. There is no distinct cylindrical plasma column, but only a transition from one hot plasma spot to the other (especially Figs. 4.78, 4.79, 4.84, and 4.85).

The thickness of this "transition zone" (instead of the cylindrical plasma column in HID lamps with large electrode gaps above 5–10 mm) is determined by the electrode gap, which has been varied in the ApolloHg-lamp between 2 and 4 mm. Figure 4.95 depicts the calculated plasma and electrode temperatures for 2, 3, and 4 mm electrode gap. It is obvious that the electrode gap has a wide influence on the plasma, i.e., the hot plasma spots and the plasma column. The smaller the electrode gap, the smaller the "transition zone" or plasma column. Only for electrode gaps above 5–10 mm, the temperature in the plasma column will be independent of the electrode properties. But Fig. 4.95 shows only one aspect of the influence of the electrode gap on plasma and electrode properties. Further insight into the influence of the electrode gap can be delivered by analyzing radial temperature profiles on the electrode tips and in the plasma column.

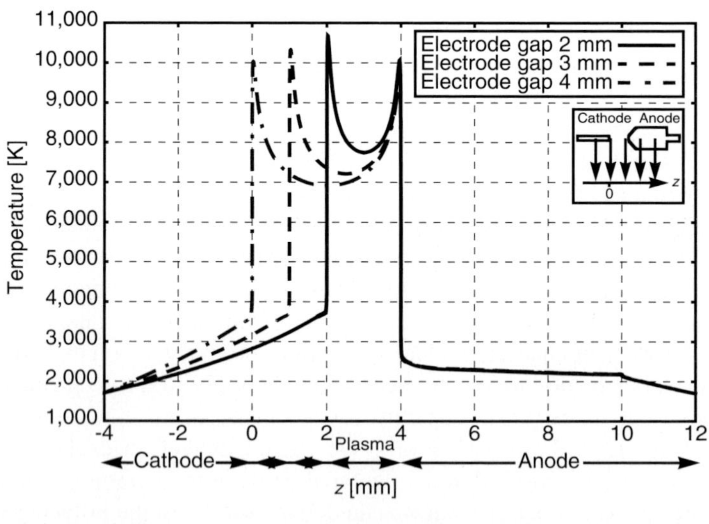

Fig. 4.95. Calculated plasma and electrode temperature on lamp axis (ApolloHg-lamp, Hg, $p = 4$ MPa, $I = 4$ A (dc), $d_{\text{electrode}} = 2$, 3, 4 mm, $\varphi_{\text{emission}} = \varphi_{\text{cool/heat}} = 4.55$ eV, spot mode, [62])

The temperature distribution on anode and cathode tip depending on the electrode gap is shown in Figs. 4.96 (anode) and 4.97 (cathode), respectively. The anode and cathode tip temperature rises by 150 K (anode) and 200 K (cathode), respectively, if the electrode gap is reduced from 4 to 2 mm. Thus, it can be seen that even the plasma far away[89] from the electrodes has an influence on the arc attachment to the electrode tips. The transitions from spot to diffuse arc attachments in Sect. 4.4.8 also indicate a strong coupling of near-electrode plasma and plasma between the hot plasma spots: If the electrode gap is only 2 mm, anode and cathode might change simultaneously from spot to diffuse arc attachment. This means that anode and cathode are coupled through the plasma between the hot plasma spots and the arc attachment mode seems to be sensible to minor changes even in the plasma 1 or 2 mm away from the electrode tips. This is important for electrode models which do not take the plasma with hot plasma spots and constriction zones into account: The predicted arc attachment modes might be wrong.

The change of the plasma temperature and the change in the temperature profile in the plasma between the electrodes depending on the electrode gap is given in Fig. 4.98. The plasma temperature increases by 850 K on the lamp axis and decreases by up to 1,200 K in the outer plasma region if the electrode gap is reduced from 4 to 2 mm. This means that the arc is much more constricted between the hot plasma spots ("transition zone") in the case of small electrode gaps. The temperature profile gets more parabolic for larger electrode gaps.

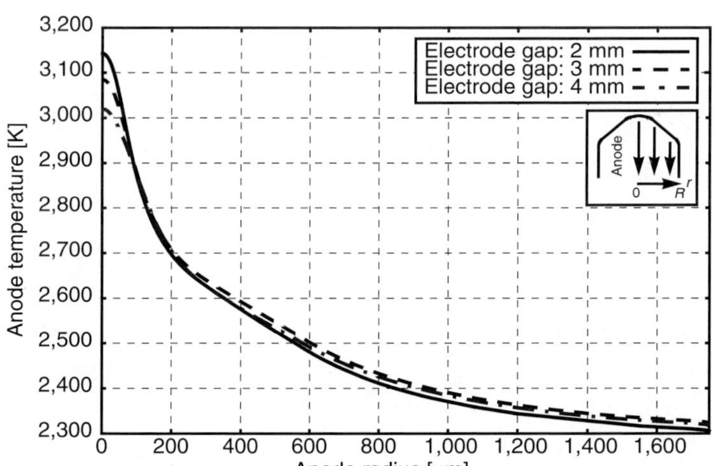

Fig. 4.96. Calculated anode temperature distribution vs. anode radius (ApolloHg-lamp, Hg, $p = 4$ MPa, $I = 4$ A (dc), $d_{\text{electrode}} = 2, 3, 4$ mm, $\varphi_{\text{emission}} = \varphi_{\text{cool/heat}} = 4.55$ eV, spot mode, [62])

[89] Distances larger than the typical distance between electrode tip and hot plasma spots of about 30–250 μm.

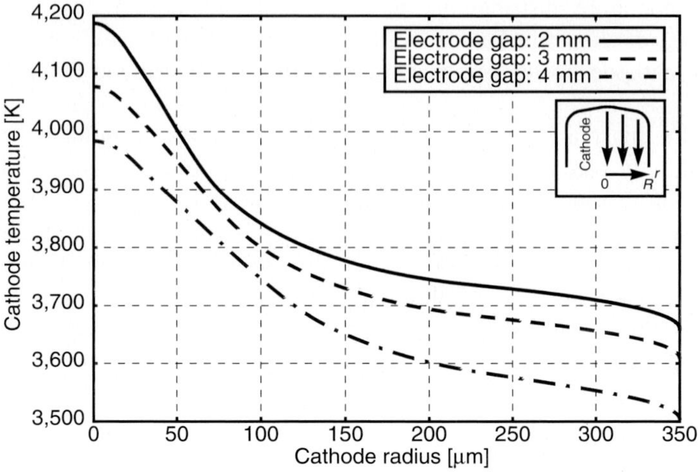

Fig. 4.97. Calculated cathode temperature distribution vs. cathode radius (ApolloHg-lamp, Hg, $p = 4$ MPa, $I = 4$ A (dc), $d_{\text{electrode}} = 2, 3, 4$ mm, $\varphi_{\text{emission}} = \varphi_{\text{cool/heat}} = 4.55$ eV, spot mode, [62])

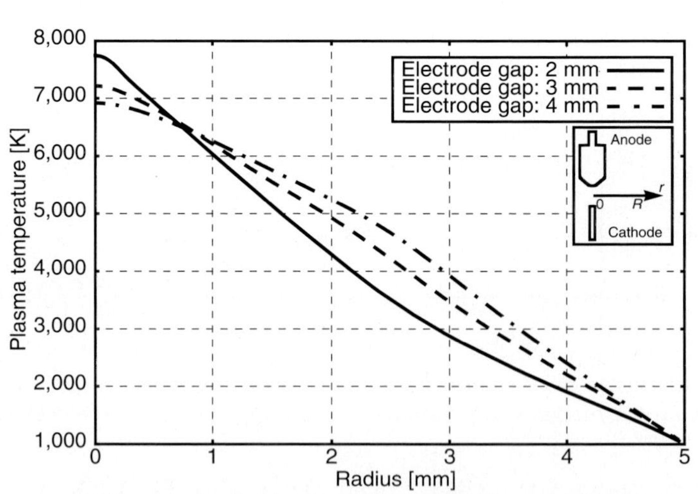

Fig. 4.98. Calculated temperature vs. radius in plasma column midway between electrodes (ApolloHg-lamp, Hg, $p = 4$ MPa, $I = 4$ A (dc), $d_{\text{electrode}} = 2, 3, 4$ mm, $\varphi_{\text{emission}} = \varphi_{\text{cool/heat}} = 4.55$ eV, spot mode, [62])

Summary

The comparison of numerical results and experimental measurements together with the detailed analysis of the conditions in the lamp, which is possible thanks to the numerical results, gives new insight into the interaction between electrodes and plasma. Many effects which are couple in an experimental setup (like lamp current and operating pressure), can be analyzed independently using the numerical simulations. The self-consistent electrode–plasma model is thus, for example, able to calculate the increase of the operating pressure for increasing lamp current and it can predict the influence of the electrode gap on temperature profiles on anode, cathode, and in the plasma between the hot plasma spots.

The results concerning the dependence of the electrode fall voltages and the electric field in the plasma column on the electrode gap again demonstrate that it is necessary to consider the whole system consisting of anode, cathode, and the complete plasma especially in the case of short arc discharge lamps with electrode gaps of 5 mm and less. The results concerning the dependence of the operating pressure on lamp current and electrode gap in HID lamps is a central finding. A similar result has already been discussed in Sect. 4.4.5 (end of section "Electrode Fall Voltages during External Laser Heating"), further investigations will be given in Sect. 4.4.8. This dependence is important for the interpretation of many experimental results to avoid wrong conclusions.

Besides the absolute values, numerical simulations and measured values agree very well concerning the dependence of the lamp properties on electrode gap, operating pressure, and lamp current. A clear advantage of the self-consistent electrode–plasma model is the consideration of the whole lamp, including anode, cathode, the complete discharge plasma, and interactions among all these components. Details of lamp properties can thus be investigated which are inaccessible for experimental methods. Further investigations concerned with the ApolloHg-lamp are described in Sect. 4.4.8.

4.4.8 Spot–Diffuse Transition and Time-Dependent Behavior

This section is concerned with the transition from the spot to the diffuse arc attachment to the electrode tips and the resulting changes in the plasma and electrode temperatures. Moreover, the time-dependent answer of the lamp properties following a current jump from 2.5 to 6 A will be analyzed and compared to experimental results. The lamp under consideration is the ApolloHg-lamp described in Sect. 4.4.7.

Spot–Diffuse Transition

Increasing the lamp current slowly (quasistationary) from 3 to 5 A, a transition from a constricted arc attachment (spot mode) to a broader arc attachment occurs (diffuse mode). This is possible at the cathode tip as well as at the anode tip. As an example, such a transition from spot to diffuse arc attachment on the anode tip is shown in

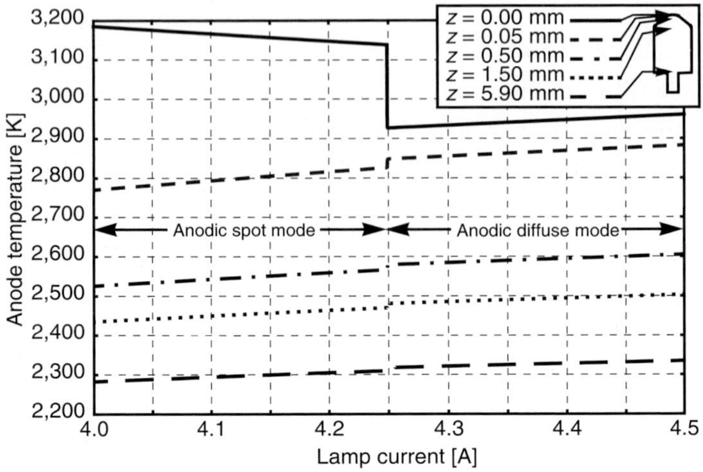

Fig. 4.99. Calculated anode temperature at certain distances to anode tip vs. lamp current during a quasistationary increase of the lamp current from 3 to 5 A (ApolloHg-lamp, Hg, $p = 6$ MPa, $d_{\mathrm{electrode}} = 3$ mm, $\varphi_{\mathrm{emission}} = \varphi_{\mathrm{cool/heat}} = 4.55$ eV)

Figs. 4.99 and 4.100. At $I = 4.25$ A, the arc attachment switches from spot to diffuse arc attachment on the anode tip, but decreasing the lamp current below 4.25 A does not restore the spot mode. Thus, we have a hysteresis as described in [109]: Below 4.25 A both arc attachment modes coexist, above 4.25 A only the diffuse arc attachment mode is possible on the anode tip (ApolloHg, Hg, 6 MPa, $d_{\mathrm{electrode}} = 3$ mm).

Figure 4.99 depicts the anode temperature vs. lamp current at different distances to the anode tip. At $I = 4.25$ A, the temperature difference between the temperature at the anode tip and the temperature 0.05 mm below the anode tip decreases suddenly from about 300 to less than 100 K. A more detailed analysis of the data shown in Fig. 4.99 reveals that the difference between the temperature at the anode tip and the temperature 0.05 mm below the anode tip already decreases continuously for lamp currents below 4.25 A: In contrast to all other temperature values, the anode tip temperature ($z = 0$ mm) decreases for increasing lamp current (below 4.25 A). After the transition from the spot to the diffuse arc attachment, the anode tip temperature ($z = 0$ mm) increases with increasing lamp current (as the temperatures below the anode tip). It seems that the falling of the temperature difference between the temperature at the anode tip and the temperature 0.05 mm below the anode tip below a certain limiting temperature induces the transition from spot to diffuse arc attachment.

The anode tip temperature suddenly decreases as soon as the current is increased beyond 4.25 A, but the temperature below the anode tip increases suddenly during the transition from spot to diffuse arc attachment. This increase of the "body temperature" is consistent with experimental findings (e.g., [109]). At first glance, it appears unusual that the anode "body temperature" increases because this increase is coupled to an increase of the power consumption of the anode (cf. Sect.

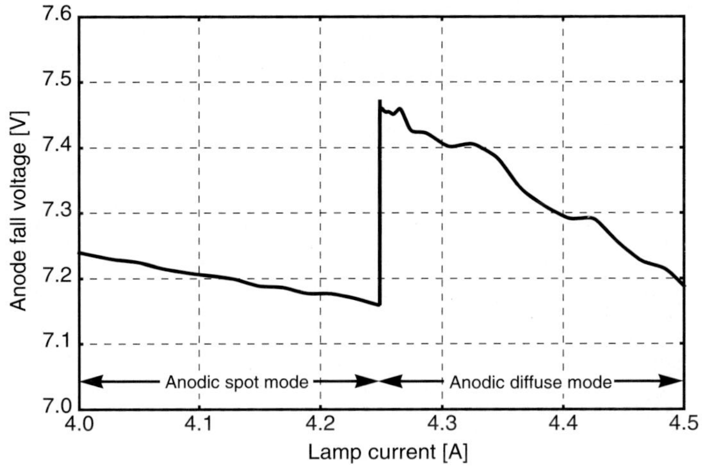

Fig. 4.100. Calculated anode fall voltage vs. lamp current during a quasistationary increase of the lamp current from 3 to 5 A (ApolloHg-lamp, Hg, $p = 6$ MPa, $d_{\text{electrode}} = 3$ mm, $\varphi_{\text{emission}} = \varphi_{\text{cool/heat}} = 4.55$ eV)

4.4.4). This can also be seen in Fig. 4.100, showing the sudden increase of the anode fall voltage at the transition from spot to diffuse arc attachment. We would normally suspect the anode to switch to the mode where power consumption of the anode is the lowest. Nevertheless, the gradient of the temperature increase vs. lamp current is lower in the case of the diffuse arc attachment ($I > 4.25$ A) compared to the gradient in the case of a spot-like arc attachment ($I < 4.25$ A). Thus, the "medium-term" power balance of the anode is better in the case of the diffuse arc attachment.

The gradient of the anode fall voltage vs. lamp current (Fig. 4.100) is lower in the case of the spot-like arc attachment ($I < 4.25$ A) compared to the gradient in the case of the diffuse arc attachment ($I > 4.25$ A). Thus, the "medium-term" power balance of the anode is also favorable in the case of the diffuse arc attachment. Nevertheless, not the power balance of single components of a HID lamp are important, but the total power balance of the lamp, i.e., the total lamp voltage. The total lamp voltage decreases by 0.1 V at 4.25 A, i.e., the total power consumption of the lamp decreases.[90]

Temperature Profiles during Different Arc Attachment Modes

The smaller the electrode gap, the smaller the plasma region between the hot plasma spots. In the case of an electrode gap of only 2 mm, the electrodes are even able to "feel" the influence of the opposite electrode: Both electrodes are even able

[90] In the numerical simulation, constant pressure is assumed. An analysis of the situation in the case of constant mass of mercury within the discharge volume is given in the section "Time-Dependent Behavior."

to switch simultaneously from spot to diffuse mode, so that there is no possibility to tell which electrode started the transition. This demonstrates again the importance of the plasma, the constriction zones, and the hot plasma spots for electrode models. The arc attachment is not only determined by electrode properties, but the plasma plays also an important role. The self-consistent electrode–plasma model used in Sects. 4.4.3–4.4.8 is able to consider this interaction because anode, cathode, plasma, and interactions among all these components are considered.

At $I = 4$ A the spot and diffuse arc attachment on cathode and anode can coexist (ApolloHg, Hg, 4 MPa, $d_{\text{electrode}} = 2$ mm). The plasma and electrode temperature distribution on the lamp axis for both types of arc attachments is shown in Fig. 4.101. It is clearly visible that the lamp has to be regarded as an integrated whole instead of treating electrodes and plasma independently. The plasma temperature in the case of diffuse arc attachments is considerably lower compared to the case of the spot modes. The plasma temperature profile, however, is broader in the case of diffuse arc attachments compared to the case of the spot modes. The resulting average plasma temperature is about 3% higher in the case of the diffuse arc attachments, i.e., in an experimental setup, the pressure in a lamp with diffuse arc attachments would be higher compared to a lamp with spot-like arc attachments.[91] This would result in a higher lamp voltage for the lamp with the diffuse arc attachments. Again, two effects are superimposed in the real lamp: The transition from spot to diffuse arc attachments leads (at constant pressure) to a decrease of the total lamp voltage. The associated increase in operating pressure, however, will result in a net increase of the total lamp voltage, which is also observed in experiments considering the spot–diffuse transition [109].

The temperature profiles on anode and cathode tip for spot and diffuse arc attachment are shown in Figs. 4.102 and 4.103, respectively. The electrode temperatures at the electrode tip might differ by more than 300 K comparing spot and diffuse arc attachment, but outside the arc attachment area, this difference is reduced to about 100 K. Since temperature measurements within the arc attachment area are nearly impossible, these differences are difficult to detect in an experimental setup. The temperature at the electrode tip is higher in the case of a spot-like arc attachment, but the temperature below the electrode is higher in the case of a diffuse arc attachment. This is also consistent with experimental observations[92] presented in [109].

---

[91] Such a small difference in the operating pressure cannot be measured, but it will influence the lamp voltage noticeably.

[92] The situation in the experimental setup described in [109] is quite different to the situation described here: instead of mercury argon is used and the operating pressure is 0.26 MPa in [109] compared to 4 MPa for the numerical simulations presented in Figs. 4.101–4.103. Nevertheless, the measured temperature differences at the electrode tip (about 400 K in [109]) and below the electrode tip (about 100–200 K in [109]) are quite close to the findings of the numerical simulations presented here.

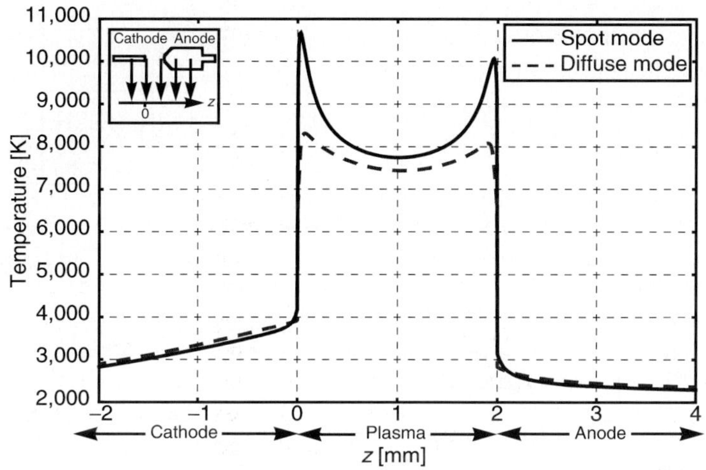

Fig. 4.101. Calculated electrode and plasma temperature on lamp axis for spot and diffuse arc attachments on anode and cathode (ApolloHg-lamp, Hg, $p = 4$ MPa, $I = 4$ A (dc), $d_{\text{electrode}} = 2$ mm, $\varphi_{\text{emission}} = \varphi_{\text{cool/heat}} = 4.55$ eV)

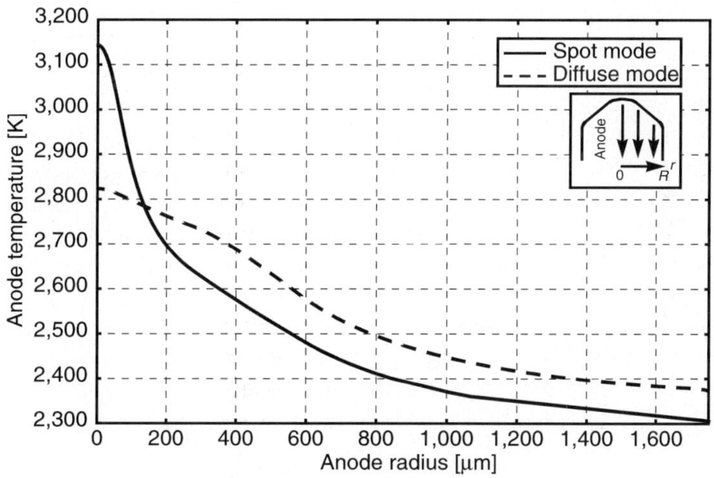

Fig. 4.102. Calculated anode temperature distribution vs. anode radius for spot and diffuse arc attachment on anode (ApolloHg-lamp, Hg, $p = 4$ MPa, $I = 4$ A (dc), $d_{\text{electrode}} = 2$ mm, $\varphi_{\text{emission}} = \varphi_{\text{cool/heat}} = 4.55$ eV)

Time-Dependent Behavior

To study the time-dependent behavior of the ApolloHg-lamp, a lamp current jump from 2.5 to 6.0 A has been used. The calculated electrode and plasma temperature distribution in a 2 MPa mercury discharge with an electrode gap of 3 mm at certain points in time after the lamp current jump is shown in Fig. 4.104. Compared to the

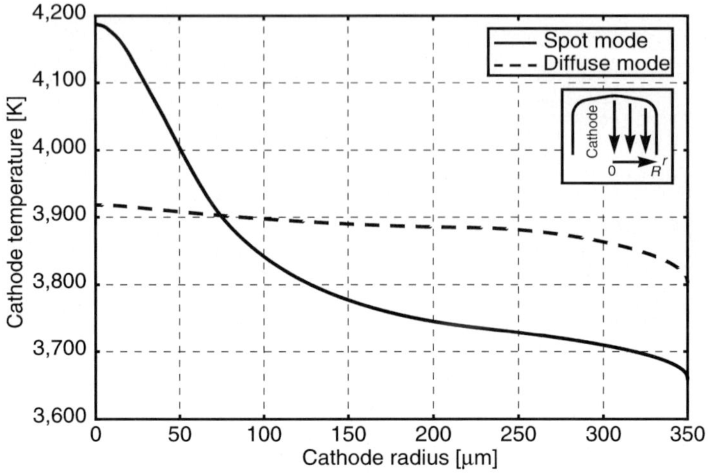

Fig. 4.103. Calculated cathode temperature distribution vs. cathode radius for spot and diffuse arc attachment on cathode (ApolloHg-lamp, Hg, $p = 4$ MPa, $I = 4$ A (dc), $d_{\text{electrode}} = 2$ mm, $\varphi_{\text{emission}} = \varphi_{\text{cool/heat}} = 4.55$ eV)

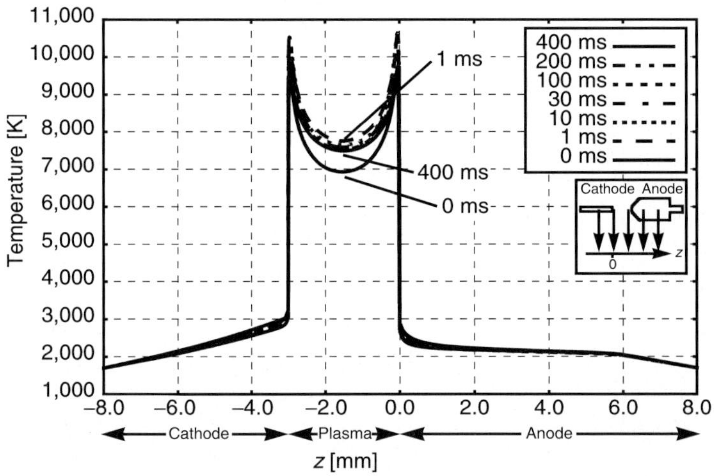

Fig. 4.104. Calculated electrode and plasma temperature on lamp axis at certain points in time after a lamp current jump from 2.5 to 6 A at $t = 0$ (ApolloHg-lamp, Hg, $p = 2$ MPa, $d_{\text{electrode}} = 3$ mm, $\varphi_{\text{emission}} = 3.5$ eV, $\varphi_{\text{cool/heat}} = 4.55$ eV)

electrodes, the plasma reacts quite fast. One ms after the current jump, the plasma temperature in the arc column has increased by about 1,000 K, then the plasma temperature relaxes to the new temperature distribution for $I = 6$ A with temperatures which are about 500 K higher than for $I = 2.5$ A.

A more detailed presentation of the anode temperature evolution is shown in Fig. 4.105: The calculated anode temperature distribution (right-hand side) is

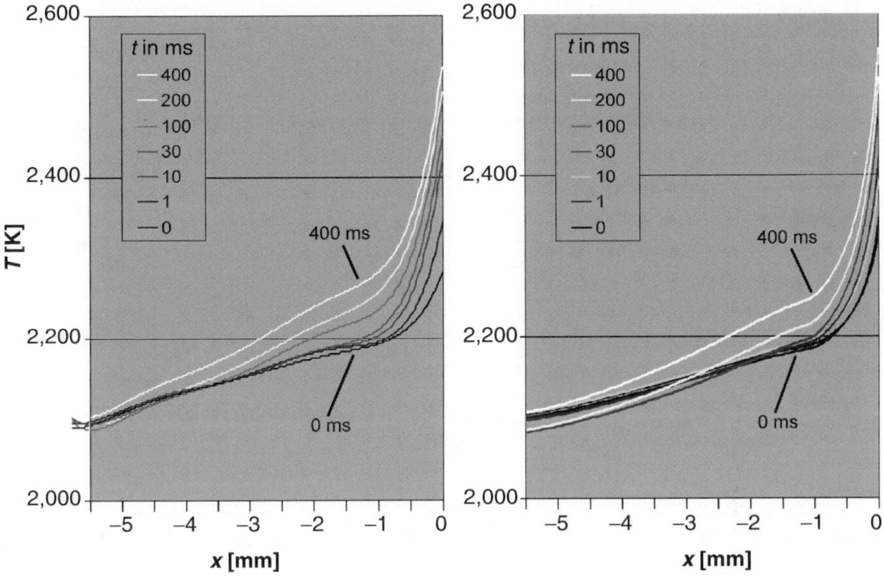

Fig. 4.105. Calculated (right-hand side) and measured (left-hand side) anode temperature distribution at certain points in time after a lamp current jump from 2.5 to 6 A at $t = 0$ (ApolloHg-lamp, Hg, $p = 2$ MPa, $d_{\text{electrode}} = 3$ mm, $\varphi_{\text{emission}} = 3.5$ eV, $\varphi_{\text{cool/heat}} = 4.55$ eV, [128])

compared to the measured anode temperature distribution (left-hand side). Due to the large mass of the anode and the corresponding thermal inertia, even after 400 ms no steady state is reached. The comparison between measured and calculated anode temperatures shows a very good agreement, especially if we take the large influence of arc attachment and electrode geometry into account (cf. Sect. 4.4.4 and as aforementioned).

The corresponding measured and calculated cathode temperature after the current jump from 2.5 to 6 A at $t = 0$ is shown in Figs. 4.106 (linear time scale) and 4.107 (logarithmic time scale). The agreement between experimental values and theoretical results for $z = 0.5$, 1.0, and 3.0 mm is quite good: Both the absolute change in cathode temperature and the time constants are reproduced very well. Comparing the results at the very cathode tip (measurements: "0 local" and "0 global," numerical simulations: "$z = 0.05$ mm" and "$z = 0.00$ mm"), the investigations described in Sect. 4.4.4 and in this section prove very helpful: The measurements labeled "0 local" were made within the arc attachment area at the cathode tip, whereas the measurements labeled "0 global" were also made at the cathode tip but not within the arc attachment area. The results obtained in Sect. 4.4.4 considering different shapes of the electrode tip demonstrate that especially the absolute temperature as well as the temperature distribution at the electrode tip is very sensitive to the local conditions at the electrode tip. Thus, we cannot expect an exact agreement of the temperature evolution at the electrode tip. If we

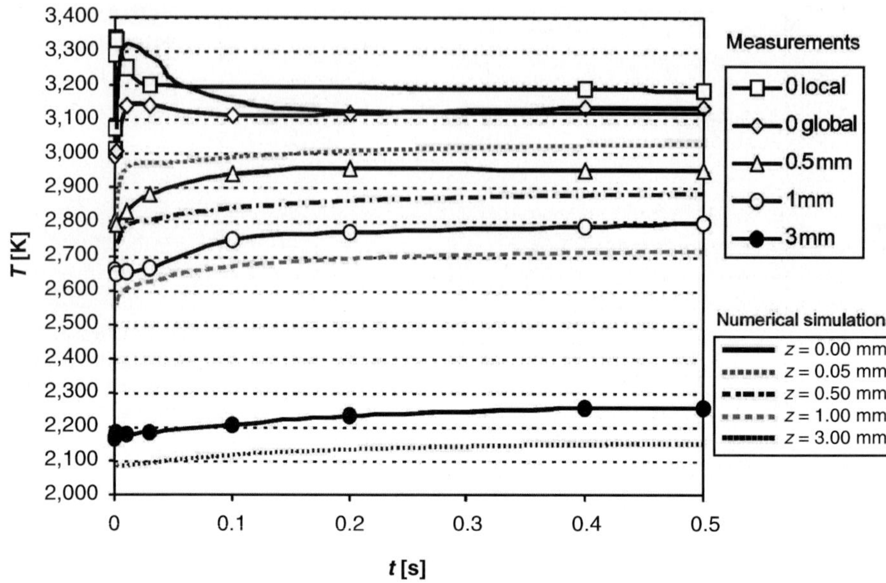

Fig. 4.106. Calculated and measured cathode temperature at certain distances to cathode tip vs. time after a lamp current jump from 2.5 to 6 A at $t = 0$ (ApolloHg-lamp, Hg, $p = 2$ MPa, $d_{electrode} = 3$ mm, $\varphi_{emission} = 3.5$ eV, $\varphi_{cool/heat} = 4.55$ eV, [128])

would have such an exact agreement, this would mean that we have "guessed" the real cathode tip geometry or, to put it another way, the real cathode tip geometry happened to coincide with the geometry assumed for the numerical simulation.[93]

Nevertheless, we can learn a lot from the comparison of measured and calculated cathode temperature at the cathode tip: The cathode temperature (measured and calculated) is quite close to the melting point of tungsten (3,680 K), i.e., we can expect that the shape of the cathode tip changes during operation. The measured temperature labeled "0 global" lies between the calculated cathode temperature for $z = 0.05$ and 0 mm, i.e., outside the arc attachment area a satisfactory agreement can be reached. The measured temperature within the arc attachment area (labeled "0 local"), however, has a much smaller time constant compared to the time constant obtained by the numerical simulations. This indicates that the structure on the cathode tip where the arc attached to has a small volume, i.e., there might be some kind of micro tip on the cathode which heats up very fast. Another possibility might be a short-time constriction of the arc attachment, maybe induced by small structures on the cathode tip.

---

[93] This is not only a result of the investigations presented in Sect. 4.4.4: Preliminary numerical examination with the self-consistent electrode–plasma model for the ApolloHg-lamp using slightly different cathode tip geometries revealed the same results.

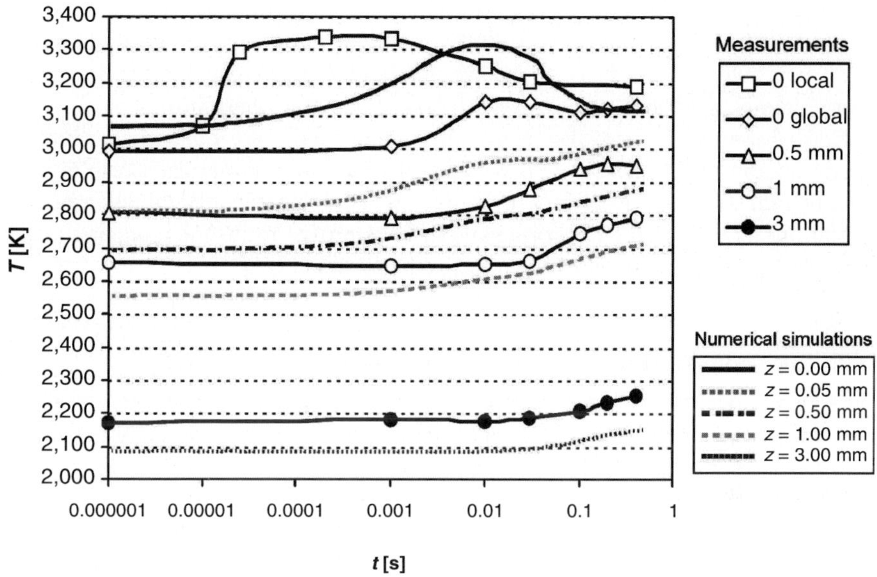

Fig. 4.107. Calculated and measured temperature distribution at certain points in time after a lamp current jump from 2.5 to 6 A at t = 0 (ApolloHg-lamp, Hg, $p = 2$ MPa, $d_{\text{electrode}} = 3$ mm, $\varphi_{\text{emission}} = 3.5$ eV, $\varphi_{\text{cool/heat}} = 4.55$ eV, [128])

Summary

The self-consistent electrode–plasma model is able to predict transitions from spot to diffuse arc attachments on anode and cathode tip. As in experimental findings, spot and diffuse arc attachment might coexist in a certain lamp current range, making it possible to compare the conditions associated with both types of arc attachment. Again, the self-consistent electrode–plasma model is able to point out the importance of the coupling of electrode and plasma: Different modes of arc attachment have a clear influence on the arc column and the arc column couples even anode and cathode behavior. For electrode gaps of 2 mm, the spot–diffuse transition on anode and cathode tip can happen simultaneously. Moreover, the numerical simulation can predict small changes in the operating pressure of a HID lamp induced by a transition of the arc attachment to the electrodes: This change in the operating pressure cannot be measured, but influences the total lamp voltage noticeably. Thus, the numerical investigations help to understand and analyze experimental findings in which different effects are superimposed inherently.

The investigation of the time-dependent behavior of an HID lamp after a current jump from 2.5 to 6 A gives further inside into the lamp properties. Experimental findings and numerical simulations agree very well. The influence of the electrode tip geometry on the electrode properties especially at the electrode tip as discussed in Sect. 4.4.4 proves to be very helpful for the interpretation of numerical and measured temperatures at the cathode tip. Small differences in the electrode tip geometry or in the operating conditions may have considerable influence on the lamp properties. Thus, numerical simulation will never be able to exactly predict all lamp properties in all details because many details of the real lamp are not accessible (see also Sect. 4.4.6). Nevertheless (or better: therefore), we can learn a lot from the numerical results, especially when details are concerned which are experimentally inaccessible. The comparison of numerical results with experimental findings is always a source of new insights: Both the numerical simulations as well as the experimental findings profit from such comparisons and the understanding of HID lamps is improved considerably.

4.4.9 Summary: Self–Consistent Electrode–Plasma Model

The self-consistent electrode–plasma model used within Sect. 4.4 consists of a set of only four equations: Two equations concerning power conservation of plasma and electrodes to determine the plasma temperature distribution $T_P(\vec{r}, t)$ (see (4.18)) and the electrode temperature distribution $T_E(\vec{r}, t)$ (see (4.19)), one equation for the electric potential $V(\vec{r}, t)$ (current conservation, (4.22)), and one additional equation accounting for deviations from LTE-conditions due to strong diffusion of electrons and ions in close proximity to the electrodes, resulting in a non-LTE electrical conductivity σ (see (4.42a) or (4.42b)). This last equation closes the gap between LTE plasma (plasma column, hot plasma spots) and non-LTE near-electrode plasma. The uniform treatment of the high-pressure plasma without using the "low-pressure" approach of different plasma layers in front of the electrodes (each with different sets of equations and different boundary conditions) contains all relevant physical processes. The self-regulating mechanisms are not "concealed" by many assumptions but can do their "job" to yield realistic results for the complete lamp system of electrodes and plasma. The given boundary conditions are reduced to a minimum, like temperature at the inside wall of the quartz vessel (not very important for the results) or temperature at the electrode roots. Geometric effects induced, for example, by different electrode tip geometries are allowed for because the plasma itself establishes its way from cathode tip over constriction zone and hot plasma spot to the arc column and further on to the anode with the related constriction zone and hot plasma spot.

Advantages of the numerical simulations performed with the self-consistent electrode–plasma model are the accuracy in space and time without any limitations as in experimental data and the considerable accessible information and data available from one single numerical simulation like total lamp voltage, cathode and anode fall voltages, electrode and plasma temperature distributions, heat fluxes to electrodes and quartz wall, etc. All these numerical results may be time dependent in the case of time dependent lamp currents (ac operation), current jumps, time-dependent external laser heating, etc. Furthermore, the numerical simulations offer the possibility to vary certain parameters like electrode geometry, work function, plasma composition, or external laser heating and separate the resulting effects, which are inherently superimposed (and not separable) in real lamps. In this way, the results of the numerical simulations improve our understanding of the principles of HID lamps and can finally lead to improved or new types of HID lamps. The results presented within Sect. 4.4 encourage further studies especially concerning the interaction between plasma and electrodes in order to further improve the understanding of HID lamps.

Numerous comparisons of numerical data with experimental results concerning plasma and electrode properties are presented in Sects. 4.4.3–4.4.8 for very different HID lamps. The agreement between theoretical and experimental results is very good and gives new insight into the conditions in HID lamps.

A comparison of results obtained with the self-consistent electrode–plasma model with the results obtained using three other cathode models (based on layer or sheath models) presented in Sect. 4.4.3 shows differences and common characteristics of the models under consideration. The most important result of this comparison is that the power balance of the cathode combined with the equation describing the electron emission current as a function of cathode temperature and local electric field is the most important part in modeling the interaction between plasma and cathode. But as soon as one is interested in details like, for example, the electric current density distribution on the cathode tip (arc attachment), it is important to include the constriction zone and the plasma column, where the current density distribution transforms from a rather broad profile in the plasma column to the small, constricted current attachment at the cathode tip. Moreover, it is necessary to consider the cathode at least in two dimensions if details concerning mode of arc attachment or temperature at the electrode tip are concerned. Models neglecting the thermal heat flux from the hot plasma spots to the relatively cool cathode will overestimate the ion current, which is the only heating mechanism used in these models for the cathode.

Numerical results for different electrode shapes are compared and discussed in Sect. 4.4.4 (electrode and plasma hot-spot temperature, electrode fall voltages, arc attachment on electrode tip) with a special focus on the interaction of plasma and electrodes. The most important result is that the shape of the electrode tip influences the lamp characteristics essentially. Electrode tip and electrode body temperature as well as plasma hot-spot temperature, electrode fall voltages, and type of arc attachment to the electrode tips are affected by the electrode tip geometry, both in anode and cathode phase. This is important for the comparison of theoretical and experimental lamp data. Either one knows the "real" electrode tip geometry from the experimental setup (which is often not the case) or one has to know the impact of the shape of the electrode tip on the lamp characteristics. For the second alternative, the results presented in Sect. 4.4.4 are very helpful. The use of one-dimensional electrode or plasma models or the use of models neglecting the hot plasma spots or the constriction zone might result in major errors concerning electrode and plasma hot-spot temperatures as well as electrode fall voltages. Furthermore, the analysis of different electrode tip geometries is important for the optimization of future electrodes. Choosing the best electrode tip geometry is a compromise between hot electrode tip temperatures and low electrode fall voltages.

The investigation of the influence of an external laser heating of the electrodes in a 0.6 MPa high-pressure mercury discharge lamp is presented in Sect. 4.4.5. The comparison of numerical results obtained with the self-consistent electrode–plasma model with different experimental data shows a very good agreement. Both the experimental and numerical results show a strong reaction of anode temperature and a weak reaction of cathode temperature during the external laser heating of the electrode. In contrast, there is a strong influence of the external laser heating on the cathode fall voltage and a barely observable influence on the anode fall voltage. Moreover, the numerical simulations and the experimental findings agree with respect to the response time of anode and cathode to the external laser heating.

The numerical simulations are able to explain all these effects in great detail. The strong influence of the external laser heating on the cathode fall voltage is due to the self-regulating mechanism of the cathode: The power brought into the electrode by the external laser heating partly “replaces” the heating by ion current. This leads to a reduced cathode fall voltage (less power needed for the generation of ions), but results also in an only moderately increased electrode temperature (more power input due to the external laser heating but less power input due to ion current). The anode reacts to the external laser heating with an increased electrode temperature, but the anode fall voltage is hardly influenced. The observed small increase of the anode fall voltage can be explained with the help of the numerical simulations and is due to a small increase of the operating pressure (<1%).

Furthermore, the numerical simulations explain the different response times during anode and cathode phase at current reversal or laser on/off: The cathode reacts faster because its self-regulating mechanism is determined by the fast adjustment of the ratio of electron to ion current, whereas the longer response time during the anode phase is determined by the thermal inertia of the anode. This leads, for example, to the tendency of the ac electrode temperature to the dc cathode temperature as described in Sect. 4.4.5. Moreover, from the comparison of experimental results and numerical simulations we can deduce that the electron emission current, which is essential at the cathode or during the cathode phase of the electrode, has no impact during the anode phase.

Together with the analysis of the impact of different electrode tip geometries on electrode and plasma temperature and electrode fall voltages (Sect. 4.4.4), the differences between anode and cathode phase described within Sect. 4.4.5 can be used for the optimization of electrode design. In the case of a dc lamp current, anode and cathode can be optimized independently, whereas the electrodes have to be optimized for anode and cathode phase simultaneously in the case of an ac lamp current. In the latter case, the different response times during anode and cathode phase might be of particular importance.

Investigations considering the commercially available D2 automotive headlight lamp are the focus of Sect. 4.4.6. From the complex plasma composition of this metal halide lamp complex plasma properties result, which are thoroughly discussed within Sect. 4.4.6. Moreover, the electrode properties and especially the work function for electron emission are discussed in Sect. 4.4.6. Beyond the presentation of numerical results concerning power balance or temperature distributions, a comparison with different experimental data is presented. The time-dependent numerical results for the D2 automotive headlight lamp agree very well with the measured total lamp voltage, the measured plasma temperature profiles, and the measured electrode temperature profile.

Investigations on high-pressure mercury discharge lamps with different operating pressures (2,4, and 6 MPa), different electrode gaps (2, 3, and 4 mm), and different lamp currents from 2 to 6 A are presented in Sects. 4.4.7 and 4.4.8. The comparison of numerical results and experimental measurements shows, for example, that there is an increase of the operating pressure for increasing lamp currents. This is a central finding which is important for the interpretation of experimental data and for the comparison of numerical and experimental results. Furthermore, it is shown in Sect. 4.4.7 that the electrode gap has an influence on temperature profiles on anode, cathode, and in the plasma between the hot plasma spots. Besides the absolute values, numerical simulations and measured values agree very well concerning the dependence of the lamp properties on electrode gap, operating pressure, and lamp current.

Finally, the self-consistent electrode–plasma model is able to predict transitions from spot to diffuse arc attachments on anode and cathode tip. As in experimental findings, spot and diffuse arc attachment might coexist in a certain lamp current range, making it possible to compare the conditions associated with both types of arc attachment, which is done in Sect. 4.4.8. Again, the self-consistent electrode–plasma model is able to point out the importance of the coupling of electrode and plasma: Different modes of arc attachment have a clear influence on the arc column and the arc column couples even anode and cathode behavior. For electrode gaps of 2 mm, the spot–diffuse transition on anode and cathode tip can happen simultaneously. Moreover, the numerical simulations can predict small changes in the operating pressure of a HID lamp induced by a transition of the arc attachment to the electrodes: This change in the operating pressure cannot be measured, but influences the total lamp voltage noticeable. Thus, the numerical investigations help to understand and analyze experimental findings, in which different effects are superimposed inherently.

The advantage of the self-consistent electrode–plasma model is the consideration of the whole lamp, including anode, cathode, the complete discharge plasma, and interactions among all these components. The coupling of plasma and electrodes, the consideration of different electrode shapes, electrode gaps, operating pressures, or lamp currents even in the time-dependent case are thus possible. Details of lamp properties which are inaccessible to experimental methods can be investigated. The comparison of numerical results and experimental measurements together with the detailed analysis of the conditions in the lamp, which is possible thanks to the numerical results, gives new insight into HID lamps. Both the numerical simulations as well as the experimental findings benefit from such comparisons and the understanding of HID lamps is improved considerably. Many effects which are couple in an experimental setup (like lamp current and operating pressure) can be analyzed independently using the numerical simulations obtained with the self-consistent electrode–plasma model.

Summary

This book deals with high-pressure or high-intensity discharge (HID) lamps. An introduction to artificial light sources with a special focus on HID lamps is given as well as a discussion of the underlying physical laws for the description of high-pressure discharge lamps (Chaps. 1 and 2). Based on these fundamentals, experimental and theoretical methods used for the investigation of HID lamps are thoroughly discussed (Chaps. 3 and 4). The focus of interest for the understanding of high-intensity discharge lamps is directed toward plasma and electrodes. Thus, Chaps. 2–4 are divided into a plasma and an electrode section. However, electrodes and plasma are coupled in real lamps. This is of particular importance for the numerical simulations, where the influence of the plasma on the electrode behavior and the effect electrodes have on the plasma properties must be considered.

In an experimental setup, this coupling between electrodes and plasma is inherently included. Experimental investigations are necessary to understand, improve, or develop new high-pressure discharge lamps. The overview of the manifold experimental methods to explore electrodes and plasma given within Chap. 3 improve the principal understanding of HID lamps and their components. The experimental determination of the electrode temperature using different methods is explained exemplarily in great detail in Chap. 3. This includes, for example, the mastery of interfering plasma radiation or the problem of uncertain emissivity properties of the electrodes. This is to visualize the challenges and difficulties of experimental methods and to give an impression of the possible accuracy of experimental investigations. Furthermore, experimental methods concerned with other electrode properties like work function for electron emission or the deformation of the electrodes during lamp operation are discussed in Chap. 3. Concerning the plasma, the determination of plasma temperature and investigations on the electrical properties of the plasma like electrode fall voltages and plasma potential are addressed.

Within the framework of numerical simulations, which are the subject of Chap. 4, electrode models and plasma models are of great importance. Thus, Chap. 4 starts with the detailed description of these models. However, the important interaction between plasma and electrodes is not included in pure electrode and pure plasma models. A huge step toward the understanding of HID lamps as a whole is thus the consideration of electrodes and plasma including the interactions between them, which is actually the main field of research of the author of this book. Hence, a self–consistent electrode-plasma model occupies a large part of Chap. 4.

The derivation of this electrode–plasma model and many numerical results obtained with this model are treated in Sect. 4.4: Different cathode models are compared in Sect. 4.4.3, the influence of different electrode shapes on anode and cathode properties as well as the effect of the electrode geometry on the plasma are discussed in Sect. 4.4.4. This investigation of the consequences of different electrode shapes proves to be very important for the interpretation of numerical and experimental results, and for the comparison of measured and simulated electrode and plasma properties. Section 4.4.5 compares experimental and numerical findings in the case of external laser heating of electrodes, showing the differences of anode and cathode characteristics with respect to electrode temperature, electrode fall voltage, power balance, and response time. The numerical simulations are of particular importance for the analysis of the experimental data and reveal new insight into electrode behavior. Investigations considering the commercially available D2 automotive headlight lamp are the focus of Sect. 4.4.6. Here, the complex plasma composition of a metal halide lamp and the profound characteristics of electrodes are discussed. For example, the influence of the plasma composition on the electron emission of the cathode is analyzed. Furthermore, comparisons between experimental and numerical findings concerning electrode and plasma temperature are given. Section 4.4.7 is concerned with the variation of electrode gap, operating pressure, and lamp current and the resulting changes in plasma and electrode properties. Again, the strong coupling between electrodes and plasma is demonstrated by the detailed numerical results. A comparison of experimental and numerical data concerning the total lamp voltage shows that, for example, an increasing lamp current leads to an (not measurable) increase of the operating pressure, leading to an increase of the total lamp voltage. Thus, some effects may be superimposed in an experimental setup and can only be analyzed in detail by numerical simulations. This is, for example, important for the interpretation of experimental data. Finally, spot-diffuse transitions of the arc attachment to the electrodes and the time-dependent response of lamp properties to a current jump are discussed in Sect. 4.4.8 with a special focus on electrode–plasma interactions.

A fundamental result of the investigations described in this book and especially in Sect. 4.4 is the coupling between plasma and electrodes: The shape of the electrodes does not only change electrode properties, but also influences considerably the near-electrode plasma. Moreover, different arc attachments to electrodes with the same geometry have a distinct influence on the plasma column even some millimeters away from the electrode tips. The plasma between the electrodes can even couple anode and cathode, for example with respect to the mode of arc attachment to the electrodes, and the plasma composition may effect the electron emission current of the cathode. Variations of electrode geometry, electrode gap, operating pressure, or lamp current always influence plasma, electrodes, and the interactions between them. Another result is the importance of the power balance of the electrodes for the interaction between electrodes and plasma. Further investigations are concerned with the time-dependent behavior of plasma and electrodes, the difference between anode and cathode characteristics, and the comparison of experimental and numerical findings. A more detailed summary of all these results can be found in Sect. 4.4.9 and, of course, in Sects. 4.4.1 to 4.4.8.

Bibliography

1. J. Almanstoetter et al., J. Phys. D.: Appl. Phys. **35**, 1751–1756 (2002)
2. A. Anders, *A Formulary for Plasma Physics* (Akademie, Berlin, 1990) ISBN 3-05-501263-1
3. N.W. Ashcroft, N. Mermin, *Solid State Physics* (Saunders, London, 1976), ISBN 0-03-049346-3
4. G. Babucke, G. Hartel, H.-G. Kloss, J. Phys. D: Appl. Phys. **24**, 1316–1321 (1991)
5. W.L. Bade, J.M. Yos, *A Theoretical and Experimental Study of Thermionic Arc Cathodes* (Research and Advanced Development Division, AVCO Corporation, Wilmington, Massachusetts, USA, 1962), Technical Report RAD-TR-62-23
6. M. Baeva et al., *Spectroscopic Studies on High-Pressure Hg–Xe Discharge*. Proceedings ICPIG XXVI International Conference on Phenomena in Ionized Gases, Greifswald, 15–20 July 2003, vol 2, p. 63–64
7. H. Bartels, Z. Phys. **127**, 243 (Part I) (1950) and **128**, 546 (Part II) (1950)
8. C. Bauer, *Photographs of High Pressure Mercury Lamps and Transmission Versus Wavelength of Dichroic Mirrors*, personal communication (2003)
9. C. Bauer, *Experimentelle Untersuchungen des Elektrodenverhaltens von Hochdruck-Plasmalampen bei zeitabhängigem Betrieb*, PhD thesis, Universität Karlsruhe (TH), Tenea Verlag Berlin, ISBN 3-86504-043-8 (2004)
10. A. Bauer, P. Schulz, Z. Phys. **139**, 197–211 (1954)
11. C. Bauer, M. Neiger, H. Mönch, T. Krücken, *Experimental Investigations of Locally and Temporally Resolved Electrode Temperature Distributions of HID-Lamps Using a 3-Wavelength Pyrometric Method*. Proceedings of the 9th International Symposium on the Science and Technology of Light Sources (LS-9), Cornell University Press, Ithaca, NY, 2001, pp. 197–198, ISBN 0-9713422-0-2
12. M.L. Beks et al., *2D-Model for HID Lamps Using PLASIMO*, Light Sources 2004. Proceedings of the 10th International Symposium on the Science and Technology of Light Sources (LS-10), Toulouse, 18–22 July 2004 (Institute of Physics Publishing (IoP), Bristol, 2004), pp. 295–296, ISBN 0-7503-1007-3
13. M.S. Benilov, J. Phys. D: Appl. Phys. **30**, 1115–1119 (1997)
14. M.S. Benilov, Phys. Rev. E **58**(5), 6480–6494 (1998)
15. M.S. Benilov, Phys. Rev. E **57**(2), 2230–2241 (1998)
16. M.S. Benilov, J. Phys. D: Appl. Phys. **32**, 257–262 (1999)
17. M.S. Benilov, S. Coulombe, Phys. Plasmas **8**(9), 4227–4233 (2001)

18. M.S. Benilov, M.D. Cunha, J. Phys. D: Appl. Phys. **35**, 1736–1750 (2002)
19. M.S. Benilov, M.D. Cunha, Phys. Rev. E **68**, 056407 (2003)
20. M.S. Benilov, A. Marotta, J. Phys. D: Appl. Phys. **28**, 1869–1882 (1995)
21. M.S. Benilov, G.G. Bochkarev, B.V. Rogov, IEEE Trans. Plasma Sci. **23**(4), 742–749 (1995)
22. M.S. Benilov, M.D. Cunha, G.V. Naidis, *Modelling Interaction of Metal Halide Plasmas with a Thermionic Cathode*, Light Sources 2004. Proceedings of the 10th International Symposium on the Science and Technology of Light Sources (LS-10), Toulouse, 18–22 July 2004 (Institute of Physics Publishing (IoP), Bristol, 2004), pp. 537–538, ISBN 0-7503-1007-3
23. D. Berson, F. Dunn, M. Takao, Science **295** (2002)
24. W. Bleakney, Phys. Rev. **36**, 1303–1308 (1930)
25. W. van Bommel, *The Third Photoreceptor in the Human Eye and its Meaning for Lighting*. Progress in Automobile Lighting (PAL) Symposium, Darmstadt, 2003
26. M. Born, J. Phys. D: Appl. Phys. **32**, 876–885 (1999)
27. M. Born et al., *Mercury Free HID Lamps*. Proceedings ICPIG XXVI International Conference on Phenomena in Ionized Gases, Greifswald, 15–20 July 2003, vol. 2, pp. 121–122
28. M. Born et al., *Mercury-Free Automotive Discharge Lamps*, Light Sources 2004. Proceedings of the 10th International Symposium on the Science and Technology of Light Sources (LS-10), Toulouse, 18–22 July 2004 (Institute of Physics Publishing (IoP), Bristol, 2004), pp. 179–180, ISBN 0-7503-1007-3
29. P. Börner et al., *Adaptation of a Nuclear Fusion Radiation Transport Monte Carlo Code for HID Lamp Modelling*, Light Sources 2004. Proceedings of the 10th International Symposium on the Science and Technology of Light Sources (LS-10), Toulouse, 18–22 July 2004 (Institute of Physics Publishing (IoP), Bristol, 2004), pp. 407–408, ISBN 0-7503-1007-3
30. R. Bötticher, W. Bötticher, J. Phys. D: Appl. Phys. **34**, 1110–1115 (2001)
31. R. Bötticher, W. Bötticher, J. Phys. D: Appl. Phys. **33**, 367–374 (2000)
32. R. Bötticher, W. Graser, A. Kloss, J. Phys. D: Appl. Phys. **37**, 55–63 (2004)
33. M. Boulos, P. Fauchais, E. Pfender, *Thermal Plasmas. Fundamentals and Applications*, vol. 1 (Plenum Press, New York, 1994), ISBN 0306446073
34. G. Brainard et al., J. Neurosci. **21**(16), 6405–6412 (2001)
35. S.C. Brown, *Basic Data of Plasma Physics* (MIT and Wiley, New York, 1959)
36. M. Carpaij, M. Neiger, Tungsten transport processes in UHP lamps. Forschungsbericht LTI 2004. http://www.lti.uni-karlsruhe.de/veroeffentlichungen.php, 2005
37. K. Charrada, G. Zissis, J. Phys. D: Appl. Phys. **33**, 968–976 (2000)
38. K. Charrada, G. Zissis, M. Aubes, J. Phys. D: Appl. Phys. **29**, 2432–2438 (1996)
39. K. Charrada, G. Zissis, M. Stambouli, J. Phys. D: Appl. Phys. **29**, 753–760 (1996)
40. G. Crabb, M. Burtwell, *The Application of Mesopic Vision Models to Street Lighting*, Light Sources 2004. Proceedings of the 10th International Symposium on the Science and Technology of Light Sources (LS-10), Toulouse, 18–22 July 2004 (Institute of Physics Publishing (IoP), Bristol, 2004), pp. 513–514, ISBN 0-7503-1007-3
41. G. Crabb, M. Burtwell, R. Beaumont, *Reflectance Measurements on Three Pavement Surfaces Using CMH and HPS Lamps*, Light Sources 2004. Proceedings of the 10th International Symposium on the Science and Technology of Light Sources (LS-10), Toulouse, 18–22 July 2004 (Institute of Physics Publishing (IoP), Bristol, 2004), pp. 515–516, ISBN 0-7503-1007-3
42. J.J. Curry et al., Plasma Sources Sci. Technol. **13**, 403–408 (2004)

43. L. Dabringhausen et al., J. Phys. D: Appl. Phys. **35**, 1621–1630 (2002)
44. G. Derra, E. Fischer, H. Mönch, *UHP-Lampen: Lichtquellen extrem hoher Leuchtdichte für das Projektionsfernsehen*, Physikalische Blätter, 54. Jahrgang, Heft 9 (Wiley-VCH, Weinheim 1998)
45. G. Derra et al., *New UHP Lamp Technologies for Video Projection*. Proceedings of the 9th International Symposium on the Science and Technology of Light Sources (LS-9) (Cornell University Press, Ithaca, NY, 2001), pp. 31–41, ISBN 0-9713422-0-2
46. G. Derra et al., J. Phys. D: Appl. Phys. **38**, 2995–3010 (2005)
47. W.P. Dyke, W.W. Dolan, *Field Emission*, Advances in Electronics and Electron Physics VIII (Academic, New York, 1956), pp. 89–185
48. W. Elenbaas, *Light Sources* (Philips Technical Library, Macmillan, 1972), ISBN 0-8448-0057-0
49. K.G. Emeleus, R.W. Lunt, C.A. Meek, Proc. R. Soc. (London) **A156**, 394–411 (1936)
50. A. Erraki et al., *Modeling of Radiative Transfer in Mercury-Argon HID DC Lamp*, Light Sources 2004. Proceedings of the 10th International Symposium on the Science and Technology of Light Sources (LS-10), Toulouse, 18–22 July 2004 (Institute of Physics Publishing (IoP), Bristol, 2004), pp. 383–384, ISBN 0-7503-1007-3
51. E. Fischer, Philips J. Res. **42** (1), 58–85 (1987); E. Fischer, *The Electrode Regions of High-Pressure Gas Discharges*. Proceedings XVIII International Conference on Phenomena in Ionized Gases, Part II (Invited Lectures), Budapest, 1985
52. E. Fischer, *Ultra High Performance Discharge Lamps for Projection TV Systems.* Proceedings of the 8th International Symposium on the Science & Technology of Light Sources (LS-8), 1998, pp. 36–42, ISBN 3-00-003105-7
53. P. Flesch, *Selbstkonsistente Behandlung von Elektroden und Plasma in Hochdruckgasentladungslampen*, PhD thesis, Universität Karlsruhe (TH), Logos Verlag Berlin, 2000, ISBN 3-89722-525-5
54. P. Flesch, *Market Survey*, November–December 2004
55. P. Flesch, M. Neiger, *Time-Dependent Modelling of High Pressure Discharge Lamps Including Electrodes*. Proceedings of the 9th International Symposium on the Science and Technology of Light Sources (LS-9) (Cornell University Press, Ithaca, NY, 2001), pp. 307–308, ISBN 0-9713422-0-2, Landmark Lecture
56. P. Flesch, M. Neiger, J. Phys. D: Appl. Phys. **35**, 1681–1694 (2002)
57. P. Flesch, M. Neiger, J. Phys. D: Appl. Phys. **36**, 849–860 (2003)
58. P. Flesch, M. Neiger, *Time Dependent Simulation of Plasma and Electrodes in HID-Lamps with Different Electrode Shapes*. Proceedings ICPIG XXVI International Conference on Phenomena in Ionized Gases, Greifswald, 15–20 July 2003, vol 2, pp. 61–62
59. P. Flesch, M. Neiger, J. Phys. D: Appl. Phys. **37**, 2848–2862 (2004)
60. P. Flesch, M. Neiger, *Investigating the Impact of External Laser Heating of Electrodes in HID Lamps*, Light Sources 2004. Proceedings of the 10th International Symposium on the Science and Technology of Light Sources (LS-10), Toulouse, 18–22 July 2004 (Institute of Physics Publishing (IoP), Bristol, 2004), pp. 363–364, ISBN 0-7503-1007-3
61. P. Flesch, M. Neiger, *Modelling of D2 Automotive HID Lamps Including Plasma and Electrodes*, Light Sources 2004. Proceedings of the 10th International Symposium on the Science and Technology of Light Sources (LS-10), Toulouse, 18–22 July 2004 (Institute of Physics Publishing (IoP), Bristol, 2004), pp. 361–362, ISBN 0-7503-1007-3
62. P. Flesch, M. Neiger, J. Phys. D: Appl. Phys. **38**, 3792–3803 (2005)
63. P. Flesch, M. Neiger, J. Phys. D: Appl. Phys. **38**, 3098–3111 (2005)

64. P. Flesch, M. Neiger, IEEE Trans. Plasma Sci. **33**(2), 508–509 (2005)
65. P. Flesch, R. Bötticher, H. Schmitz, J. Wendelstorf, *Vergleich verschiedener Kathodenmodelle*, report in the context of the network *Grundlegende Charakterisierung von Elektroden für zeitveränderliche Energieeinkopplung in umweltfreundliche Hochdruck-Plasmalampen*, FKZ 13N7107/0, 2001, 32 pp.
66. V. S. Fomenko, *Handbook of Thermionic Properties*, Samsonov Ed. (Plenum, New York 1966)
67. T. Fröhlich, M. Hamm, *Impact on Future Vehicle Electrical Systems Caused by Adaptive Headlamp Technologies*, VDI Berichte Nr. 1789, 2003, ISBN 3-18-091789-X
68. M. Galvez, *3-Dimensional Radiation Transport for LTE Plasmas*, Light Sources 2004. Proceedings of the 10th International Symposium on the Science and Technology of Light Sources (LS-10), Toulouse, 18–22 July 2004 (Institute of Physics Publishing (IoP), Bristol, 2004), pp. 505–506, ISBN 0-7503-1007-3
69. M. Galvez, *3-Dimensional LTE Modelling of HID Lamps with Electrode–Plasma Interaction*, Light Sources 2004. Proceedings of the 10th International Symposium on the Science and Technology of Light Sources (LS-10), Toulouse, 18–22 July 2004 (Institute of Physics Publishing (IoP), Bristol, 2004), pp. 459–460, ISBN 0-7503-1007-3
70. M. Gendre, *Two Centuries of Electric Light Source Innovations*, from [71], 2003
71. M. Gendre, Homepage with electric light source directory. http://www.lampreview.net/
72. C. Gerthsen, H. Kneser, H. Vogel, *Physik*, Springer Berlin Heidelberg New York, 16. Auflage, 1989, ISBN 3-540-51196-2
73. H. Giese, *Theoretische Untersuchungen zur Konvektion in Quecksilber-Hochdruckgasentladungslampen*, PhD thesis, Universität Karlsruhe (TH), Verlag der Augustinus Buchhandlung, Aachen, 1997, ISBN 3-86073-573-X
74. H. Giese, *3-D HID Lamp Model Including Radiative Transfer*, Light Sources 2004. Proceedings of the 10th International Symposium on the Science and Technology of Light Sources (LS-10), Toulouse, 18–22 July 2004 (Institute of Physics Publishing (IoP), Bristol, 2004), pp. 245–246, ISBN 0-7503-1007-3
75. H. Giese, T. Kruecken, U. Niemann, F. Noertemann, *Understanding HID Lamp Properties*. Proceedings Spie: The International Society for Optical Engineering, Vol. 4775, Modeling and Characterization of Light Sources, 2002, pp. 1–21
76. G. Glickman et al., J. Biol. Rhythms **18**(1), 71–79 (2003)
77. V. Godyak, J. Shaffer, *Endura: A New High Output Electrodeless Fluorescent Light Source*. Proceedings of the 8th International Symposium on the Science & Technology of Light Sources (LS-8), 1998, pp. 14–23, ISBN 3-00-003105-7
78. E. B. Goldstein, *Wahrnehmungspsychologie* (Spektrum Akademischer, 2002), ISBN 3-8274-1083-5, E.B. Goldstein, *Sensation and Perception* (Wadsworth 2001), ISBN 0534539645
79. S. Gotoh, T. Kasuya, M. Wada, *Work Function Measurement of an Electrode in a Low Pressure Hg Discharge by Photoelectric Method*, Light Sources 2004. Proceedings of the 10th International Symposium on the Science and Technology of Light Sources (LS-10), Toulouse, 18–22 July 2004 (Institute of Physics Publishing (IoP), Bristol, 2004), pp. 529–530, ISBN 0-7503-1007-3
80. J. de Groot, J. van Vliet, *The High-Pressure Sodium Lamp* (Philips Technical Library, Kluwer, Deventer, 1986), ISBN 90-201-1902-8
81. K. Günther, T. Hartmann, H. Sarroukh, *Hg Free Ceramic Automotive Headlight Lamps*, Light Sources 2004. Proceedings of the 10th International Symposium on the Science and Technology of Light Sources (LS-10), Toulouse, 18–22 July 2004 (Institute of Physics Publishing (IoP), Bristol, 2004), pp. 219–220, ISBN 0-7503-1007-3

82. H. Haken, H.C. Wolf, *Atomic and Quantum Physics* (Springer, Berlin Heidelberg New York, 1984), ISBN 3-540-13137-X
83. G. Hartel, H. Schöpp, J. Phys. D: Appl. Phys. **29**, 2881–2884 (1996)
84. G. Hartel, H. Schöpp, H. Hess, L. Hitzschke, J. Appl. Phys. **85**(10), 7076–7088 (1999)
85. T. Hartmann, *Die Leistungsbilanz von Elektroden in effizienten Hochdruck-Entladungslampen*, PhD thesis, Technische Universität Berlin, Tenea Verlag Berlin, 2002, ISBN 3-936582-01-7
86. T. Hartmann et al., J. Phys. D: Appl. Phys. **35**, 1657–1667 (2002)
87. M. Haverlag, *Modelling of High Intensity Discharge Lamps*. Proceedings ICPIG XXVI International Conference on Phenomena in Ionized Gases, Greifswald, 15–20 July 2003, vol. 2, pp. 119–120
88. U. Hechtfischer et al., *Physics and Chemistry of Ultra-High-Pressure Mercury Discharge Lamps*, 55th APS Gaseous Electronic Conference (GEC02), Minneapolis, Minnesota, October 15–18 October 2002, APS Bulletin, Volume 47, (7), 2002
89. R. Hilbig et al., *Molecular Discharge as Light Sources*, Light Sources 2004. Proceedings of the 10th International Symposium on the Science and Technology of Light Sources (LS-10), Toulouse, 18–22 July 2004 (Institute of Physics Publishing (IoP), Bristol, 2004), pp. 75–84, ISBN 0-7503-1007-3
90. J.O. Hirschfelder, C.F. Curtiss, R.B. Bird, *Molecular Theory of Gases and Liquids* (Wiley, New York, 1964)
91. S. Hollo, *Personal Communication*, 12 January 2004 and 2 February 2004
92. J.D. Hooker, Lamp Development Manager of Sylvania Lighting International, and *Museum of Historic Lamp Technology*. http://www.lamptech.co.uk/
93. R. Hoppstock, *Xenon-Hochdruckkurzbogenlampen mit Metallhalogenid-Dotierung und Wolfram-Halogen-Kreisprozeß*, PhD thesis, Universität Karlsruhe (TH), 1987
94. J. Hust, A. Lankford, *Update of Thermal Conductivity and Electrical Resistivity of Electrolytic Iron, Tungsten, and Stainless Steel*, National Bureau of Standards Special Publication 260–90, Boulder, CO 80303, September 1984
95. R. Iffländer, *Gas Discharge Lamps for Laser Excitation*, self-publisher: http://www.ifflaender.net, Schramberg, 2003
96. C. de Jager, Handbuch der Physik (Encyclopedia of Physics) (Springer, Berlin Heidelberg New York, 1959), Vol. 52
97. S. Jüngst, D. Lang, M. Galvez, *Improved Arc Tubes for Ceramic Metal Halide Lamps*, Light Sources 2004. Proceedings of the 10th International Symposium on the Science and Technology of Light Sources (LS-10), Toulouse, 18–22 July 2004 (Institute of Physics Publishing (IoP), Bristol, 2004), pp. 115–124, ISBN 0-7503-1007-3
98. T. Kappen, *Ceramic Metal Halide Lamps; A World of Lighting*, Light Sources 2004. Proceedings of the 10th International Symposium on the Science and Technology of Light Sources (LS-10), Toulouse, 18–22 July 2004 (Institute of Physics Publishing (IoP), Bristol, 2004), pp. 43–52, ISBN 0-7503-1007-3
99. M. Kettlitz, R. Großjohann, J. Phys. D: Appl. Phys. **35**, 1702–1706 (2002)
100. M. Kettlitz, H. Pursch, *Determination of Plasma Parameters Near the Electrodes of Short Arc Lamps*. Proceedings of the 9th International Symposium on the Science and Technology of Light Sources (LS-9) (Cornell University Press, Ithaca, NY, 2001), pp. 193–194, ISBN 0-9713422-0-2
101. E. Kindel et al., J. Phys. D: Appl. Phys. **31**, 1352–1361 (1998)
102. A. Kloss et al., J. Appl. Phys. **88**(3), 1271–1275 (2000)
103. R. Knerr, *Knaurs Lexikon der Physik* (Droemersche Verlagsanstalt München, 1988), ISBN 3-426-26357-1

104. S. Kokoschka, *Skript zur Vorlesung Grundlagen der Lichttechnik*, Karlsruhe, Juli 2003
105. A. Koerber, U. Weichmann, *Continuum Radiation in High-Pressure Mercury Lamps*, Light Sources 2004. Proceedings of the 10th International Symposium on the Science and Technology of Light Sources (LS-10), Toulouse, 18–22 July 2004 (Institute of Physics Publishing (IoP), Bristol, 2004), pp. 199–200, ISBN 0-7503-1007-3
106. E. Krishnakumar, S.K. Srivastava, J. Phys. B: At. Mol. Opt. Phys. **21**, 1055–1082 (1988)
107. H. Kuchling, *Taschenbuch der Physik* (Fachbuchverlag, Leipzig 1991), ISBN3-343-00759-5
108. Lenkungskreis Optische Technologien für das 21. Jahrhundert, *Deutsche Agenda Optische Technologien für das 21. Jahrhundert*, WAZ-Druck, Duisburg, Mai 2000, ISBN 3-00-006083-9
109. S. Lichtenberg, J. Phys. D: Appl. Phys. **35**, 1648–1656 (2002)
110. P.H. Lindsay, D.A. Norman, *Einführung in die Psychologie* (Springer, Berlin Heidelberg New York, 1981), ISBN 3-540-09874-7
111. G.G. Lister, J.E. Lawler, W.P. Lapatovich, V.A. Godyak, Rev. Mod. Phys. **76**, 541–598 (2004)
112. J.J. Lowke, R. Morrow, J. Haidar, J. Phys. D: Appl. Phys. **30**, 2033–2042 (1997)
113. J.J. Lowke, J.C. Quartel, Aust. J. Phys. **50**(3), 539–552 (1997)
114. J. Luhmann, *Charakterisierung thermisch emittierender Kathoden für Hochdruckgasentladungslampen durch elektrische Messungen*, PhD thesis, Ruhr-Universität Bochum, Tenea Verlag Berlin, 2001, ISBN 3-932274-70-9
115. J. Luhmann et al., J. Phys. D: Appl. Phys. **35**, 1631–1638 (2002)
116. W. Martin, W. Wiese, *Atomic Spectroscopy, A Compendium of Basic Ideas, Notation, Data, and Formulas*, NIST. http://physics.nist.gov/Pubs/AtSpec/AtSpec.PDF, 2003
117. A. Martin et al., *Physical Basic Properties of a High Pressure Na Lamp at Several Voltages and Frequencies*, Light Sources 2004. Proceedings of the 10th International Symposium on the Science and Technology of Light Sources (LS-10), Toulouse, 18–22 July 2004 (Institute of Physics Publishing (IoP), Bristol, 2004), pp. 479–480, ISBN 0-7503-1007-3
118. C. Meyer, H. Nienhuis, *Discharge Lamps* (Philips Technical Library, Kluwer Deventer, 1988), ISBN 90-201-2147-2
119. H. Mönch et al., *Controlled Electrodes in UHP Lamps*, Light Sources 2004. Proceedings of the 10th International Symposium on the Science and Technology of Light Sources (LS-10), Toulouse, 18–22 July 2004 (Institute of Physics Publishing (IoP), Bristol, 2004), pp. 171–172, ISBN 0-7503-1007-3
120. R. Morrow, J.J. Lowke, J. Phys. D: Appl. Phys. **26**, 634–642 (1993)
121. D. Nandelstädt, *Charakterisierung thermisch emittierender Kathoden für Hochdruckgasentladungslampen durch optische Untersuchungen*, PhD thesis, Ruhr-Universität Bochum, Tenea Berlin, 2002, ISBN 3-932274-94-6
122. D. Nandelstädt, *Personal Communication*
123. D. Nandelstädt et al., J. Phys. D: Appl. Phys. **35**, 1639–1647 (2002)
124. NASA, *Earth's City Lights*. http://visibleearth.nasa.gov/
125. National Electrical Manufacturers Association (NEMA), USA, *Fluorescent Lamps and the Environment*. http://www.nema.org/lamprecycle/nemafluorfinal.pdf, 1999
126. National Electrical Manufacturers Association (NEMA), USA, *NEMA Lamp Manufacturers Reduce Use of Mercury by 67 Percent*. http://www.nema.org/index_nema.cfm/656/#may02, Press Release May 2002
127. C.R. Nave, *HyperPhysics*. Department of Physics and Astronomy, Georgia State University, USA. http://hyperphysics.phy-astr.gsu.edu/hbase/hframe.html

128. M. Neiger, P. Flesch, C. Bauer, *Experimentelle und theoretische Untersuchungen des Elektrodenverhaltens*, final report of a BMBF project in the context of the network *Grundlegende Charakterisierung von Elektroden für zeitveränderliche Energieeinkopplung in umweltfreundliche Hochdruck-Plasmalampen*, FKZ 13N7765, February 2004, 76 pp.
129. M. Neiger, P. Flesch, H. Strobel, C. Bauer, *Zeitaufgelöste Simulationen und pyrometrische Messungen des Elektrodenverhaltens*, final report of a BMBF project in the context of the network *Grundlegende Untersuchungen von quecksilberfreien Höchstdruck-Kompaktlampen hoher Leuchtdichte und Lichtausbeute in Quarztechnologie*, FKZ 13N8073, 2005, 76 pp.
130. W. Neumann, *The Mechanism of the Thermoemitting Arc Cathode* (Akademie, Berlin, 1987)
131. W.B. Nottingham, Phys. Rev. **55**, 203–219 (1939)
132. M. Oettler, *Experimentelle Untersuchungen an Kurzbogenlampen bei pulsüberlagertem Betrieb*, PhD thesis, Universität Karlsruhe (TH), Logos Verlag Berlin, 1997, ISBN 3-931216-95-0
133. Osram, *Lumilux Skywhite Product Information*, http://www.osram.de, April 2004
134. A.C. Palacin, *Numerische Simulation von Edelgas-Hochdruck-Kurzbogenlampen*, PhD thesis, Universität Karlsruhe (TH), 1988
135. J. Paulini, T. Klein, G. Simon, J. Phys. D: Appl. Phys. **26**, 1310–1315 (1993)
136. P. Pekarski et al., *UHP Lamps for Projection Systems.* Proceedings ICPIG XXVI International Conference on Phenomena in Ionized Gases, Greifswald, 15–20 July 2003, vol. 2, pp. 125–126
137. M.H. Pirenne, *Vision and the eye* Chapman and Hall, (London, 1967)
138. H.-P. Popp, Hochdruckgasentladungsstrahler, in *Optische Strahlungsquellen*, ed. by Hans Albrecht et al. (Lexika ed. by, Grafenau, 1977), ISBN 3-88146-112-4
139. B. Preston, *New Concepts for Roadway Lighting*, Light Sources 2004. Proceedings of the 10th International Symposium on the Science and Technology of Light Sources (LS-10), Toulouse, 18–22 July 2004 (Institute of Physics Publishing (IoP), Bristol, 2004), p. 53, ISBN 0-7503-1007-3
140. P. Reiser, *Image from Wikipedia* http://en.wikipedia.org/wiki/CIE_1931_color_space, image http://commons.wikimedia.org/wiki/Image:CIExy1931.png
141. W. Rieder, *Plasma und Lichtbogen* (Friedr. Vieweg & Sohn, Braunschweig, 1967)
142. A. Ritz, H. Mönch, *UHP Lamps with Dichroic Coating*, Light Sources 2004. Proceedings of the 10th International Symposium on the Science and Technology of Light Sources (LS-10), Toulouse, 18–22 July 2004 (Institute of Physics Publishing (IoP), Bristol, 2004), pp. 301–302, ISBN 0-7503-1007-3
143. N. Robinson, *Solar Radiation* (Elsevier Publishing Company Amsterdam, London, 1966), citing [96]
144. L. Sansonnens, J. Haidar, J.J. Lowke, J. Phys. D: Appl. Phys. **33**, 148–157 (2000)
145. W. Schlager, *Optogalvanische In-Situ Messung der Elektrodenaustrittsarbeit von Hochdrucklampen*, PhD thesis, Universität Karlsruhe (TH), 1994
146. W. Schlager, M. Neiger, J. Phys. D: Appl. Phys. **33**, 3083–3093 (2000)
147. W. Schlager, E. Fischer, R. Snijkers, *3-wavelength pyrometry to determine the temperature of electrodes in HID lamps*, internal report, Philips, 1999
148. H. Schmitz, *Die Randschicht eines Saha-Plasmas – Theorie und Anwendung*, PhD thesis, Ruhr-Universität Bochum, 2000
149. H. Schmitz, K.-U. Riemann, *Theoretical Modelling of the Cathodic Region of High Intensity Discharge Lamps.* Proceedings of the 8th International Symposium on the

Science & Technology of Light Sources (LS-8), 1998, pp. 390–391, ISBN 3-00-003105-7
150. H. Schmitz, K.-U. Riemann, J. Phys. D: Appl. Phys. **35**, 1727–1735 (2002)
151. W. Schottky, Ann. Phys. **44**, 1011–1032 (1914)
152. A. Schreiber et al., *Radiation Resistance of Quartz Glass for VUV Discharge Lamps*, Light Sources 2004. Proceedings of the 10th International Symposium on the Science and Technology of Light Sources (LS-10), Toulouse, 18–22 July 2004 (Institute of Physics Publishing (IoP), Bristol, 2004), pp. 105–114, ISBN 0-7503-1007-3
153. M. Sieg, *Personal Communication*, 8 July 2003, 14 July 2003, and 23 January 2004
154. M. Sieg, M. Kettlitz, H. Hess, *Experimental Data, Personal Communication*, Institute of Low-Temperature Plasma Physics (INP) at Greifswald, Germany, 2002
155. M. Sieg et al., *On the Electrode Sheath Voltage in High-Pressure Argon, Xenon and Mercury Discharges.* Proceedings ICPIG XXVI International Conference on Phenomena in Ionized Gases, Greifswald, 15–20 July 2003, vol. 4, pp. 104–105
156. M. Sieg et al., *Laser Exposure of High-Pressure Arc Electrodes.* Proceedings ICPIG XXVI International Conference on Phenomena in Ionized Gases, Greifswald, 15–20 July 2003, vol. 4, pp. 103–104
157. H.P. Stormberg, R. Schäfer, J. Appl. Phys. **54**, 4338 (1983)
158. H. Strobel, M. Neiger, *Time-Dependent Simulation of Plasma and Electrodes in Mercury-Free High-Intensity Discharge Lamps*, Light Sources 2004. Proceedings of the 10th International Symposium on the Science and Technology of Light Sources (LS-10), Toulouse, 18–22 July 2004 (Institute of Physics Publishing (IoP), Bristol, 2004), pp. 151–152, ISBN 0-7503-1007-3
159. T. Thrum, *Program for calculating plasma radiation*, Lichttechnisches Institut, Universität Karlsruhe (TH), 1996
160. T. Thrum, *Numerische Modellierung gepulst betriebener Hochdruckentladungslampen*, PhD thesis, Universität Karlsruhe (TH), Logos Berlin, 1997, ISBN 3-89722-036-9
161. T. Thrum, *Personal Communication*
162. P. Tielemans, F. Oostvogels, Philips J. Res. **38**(4/5), 214–223 (1983)
163. L. Troudi, R. Ben Ahmed, S. El Aissi, K. Charrada, G. Zissis, M. Sassi, J. Phys. D: Appl. Phys. **37**, 610–622 (2004)
164. D. Uhrlandt et al., *Low-Pressure Mercury-Free Plasma Light Sources: Theoretical and Experimental Perspectives*, Light Sources 2004. Proceedings of the 10th International Symposium on the Science and Technology of Light Sources (LS-10), Toulouse, 18–22 July 2004 (Institute of Physics Publishing (IoP), Bristol, 2004), pp. 15–24, ISBN 0-7503-1007-3
165. J.F. Waymouth, J. Light Visual Environ. **6**(2), 53–64 (1982)
166. G. Wei, *Transparent Ceramic Lamp Envelope Materials*, Light Sources 2004. Proceedings of the 10th International Symposium on the Science and Technology of Light Sources (LS-10), Toulouse, 18–22 July 2004 (Institute of Physics Publishing (IoP), Bristol, 2004), pp. 169–170, ISBN 0-7503-1007-3
167. J. Wendelstorf, *Two-temperature, Two-Dimensional Modeling of Cathode–Plasma Interaction in Electric Arcs.* Proceedings ICPIG XXIV International Conference on Phenomena in Ionized Gases, Warsaw, 11–16 July 1999, vol. 2, pp. 227–228, ISBN 83-902319-5-6
168. J. Wendelstorf, *Ab initio modelling of thermal plasma gas discharges (electric arcs)*, PhD thesis, University of Braunschweig, http://www.wendelstorf.de, 2000

169. H. Wiesmann, *Zweidimensionale Simulation von Hochdruckgasentladungslampen unter Berücksichtigung der Elektrodenbereiche mit der Finite-Elemente-Methode*, PhD thesis, Universität Karlsruhe (TH), Shaker Verlag Aachen, 1997, ISBN 3-8265-2632-5
170. Wikimedia Foundation, Inc, *Wikipedia, The free encyclopedia.* http://en.wikipedia.org/wiki/Main_Page
171. R. Wilhelm, *Numerische Modellierung der Transporteigenschaften dotierter Hochdrucklampen-Plasmen*, Master's thesis, Lichttechnisches Institut, Universität Karlsruhe (TH), 1996
172. R. Wilhelm, *Berechnung des Strahlungsdiffusionsanteils der Wärmeleitfähigkeit von Hochdrucklampen-Plasmen*, Internal Report, Lichttechnisches Institut, Universität Karlsruhe (TH), 1996
173. H.F. Winters et al., Phys. Rev. B **41**, 6240–6256 (1990)
174. S. Yih, C. Wang, *Tungsten – Sources, Metallurgy, Properties, and Applications* (Plenum, New York, 1979), pp. 247–280
175. X. Zhu et al., X-*Ray Absorption Measurements to Determine the Gas Temperature in HID lamps*, Light Sources 2004. Proceedings of the 10th International Symposium on the Science and Technology of Light Sources (LS-10), Toulouse, 18–22 July 2004 (Institute of Physics Publishing (IoP), Bristol, 2004), pp. 251–252, ISBN 0-7503-1007-3

Index

Printing: Krips bv, Meppel
Binding: Stürtz, Würzburg